T0275553

Techniques of Functional Analysis for Differential and Integral Equations

Mathematics in Science and Engineering

Techniques of Functional Analysis for Differential and Integral Equations

Paul Sacks
Department of Mathematics, Iowa State University,
Ames, IA, United States

Series Editor
Goong Chen

Academic Press is an imprint of Elsevier
125 London Wall, London EC2Y 5AS, United Kingdom
525 B Street, Suite 1800, San Diego, CA 92101-4495, United States
50 Hampshire Street, 5th Floor, Cambridge, MA 02139, United States
The Boulevard, Langford Lane, Kidlington, Oxford OX5 1GB, United Kingdom

Library of Congress Cataloging-in-Publication Data
A catalog record for this book is available from the Library of Congress

British Library Cataloguing-in-Publication Data
A catalogue record for this book is available from the British Library

ISBN: 978-0-12-811426-1

For information on all Academic Press publications
visit our website at https://www.elsevier.com/books-and-journals

Publisher: Candice Janco
Acquisition Editor: Graham Nisbet
Editorial Project Manager: Susan Ikeda
Production Project Manager: Paul Prasad Chandramohan
Cover Designer: Alan Studholme

Typeset by SPi Global, India

Contents

Preface

The purpose of this book is to provide a textbook option for a course in modern methods of applied mathematics suitable for first year graduate students in Mathematics and Applied Mathematics, as well as for students in other science and engineering disciplines with good undergraduate-level mathematics preparation. While the term "applied mathematics" has a very broad meaning, the scope of this textbook is much more limited, namely to present techniques of mathematical analysis which have been found to be particularly useful in understanding certain kinds of mathematical problems which commonly occur in the scientific and technological disciplines, especially physics and engineering. These methods, which are often regarded as belonging to the realm of functional analysis, have been motivated most specifically in connection with the study of ordinary differential equations, partial differential equations, and integral equations. The mathematical modeling of physical phenomena typically involves one or more of these types of equations, and insight into the physical phenomenon itself may result from a deep understanding of the underlying mathematical properties which the models possess. All concepts and techniques discussed in this book are ultimately of interest because of their relevance for the study of these three general types of problems. Of course there is a great deal of beautiful mathematics, which has grown out of these ideas, and so intrinsic mathematical motivation cannot be denied or ignored.

The background the student will obtain by studying this material is sufficient preparation for more advanced graduate-level courses in differential equations and numerical analysis, as well as more specialized topics such as fluid dynamics. The presentation avoids overly technical material from measure and function theory, while at the same time maintaining a high standard of mathematical rigor. This is accomplished by including careful presentation of the statements of such results whose proofs are beyond the scope of the text, and precise references to other resources where complete proofs can be found. As the book is meant as textbook for a course, which is typically the basis for a written qualifying examination for PhD students, a great many examples and exercises are given.

The topics presented in this book have served as the basis for a two-semester course for first year graduate students at Iowa State University. A normal division is that Chapters 1–8 are covered in the first semester, and Chapters 9–16 in the second. A partial exception is Chapter 14, all of whose topics are optional in relation to later material, and so may be easily omitted.

I would like to thank past and present friends and colleagues for many helpful discussions and suggestions about the teaching of Applied Mathematics at this level and points of detail in earlier drafts of this book: Tuncay Aktosun, Jim Evans, Scott Hansen, Fritz Keinert, Howard Levine, Gary Lieberman, Hailiang Liu, Tasos Matzavinos, Hien Nguyen, John Schotland, Mike Smiley, Pablo Stinga and Jue Yan. All mistakes and shortcomings are of course my own doing.

Paul Sacks
October 28, 2016

Chapter 1

Some Basic Discussion of Differential and Integral Equations

In this chapter we will discuss "standard problems" in the theory of ordinary differential equations (ODEs), integral equations, and partial differential equations (PDEs). The techniques developed in this book are all meant to have some relevance for one or more of these kinds of problems, so it seems best to start with some awareness of exactly what the problems are. In each case there are some relatively elementary methods, which the reader may well have seen before, or which rely only on simple calculus considerations, which we will review. At the same time we establish terminology and notations, and begin to get some sense of the ways in which problems are classified.

1.1 ORDINARY DIFFERENTIAL EQUATIONS

An nth order ordinary differential equation for an unknown function $u = u(t)$ on an interval $(a,b) \subset \mathbb{R}$ is any equation of the form

$$F(t,u,u',u'',\dots,u^{(n)}) = 0 \tag{1.1.1}$$

where we use the usual notations $u', u'', \dots$ for derivatives of order $1, 2, \dots$ and also $u^{(n)}$ for derivative of order n. Unless otherwise stated, we will assume that the ODE can be solved for the highest derivative, that is, written in the form

$$u^{(n)} = f(t,u,u',\dots,u^{(n-1)}) \tag{1.1.2}$$

For the purpose of this discussion, a solution of either equation will mean a real valued function on (a,b) possessing continuous derivatives up through order n, and for which the equation is satisfied at every point of (a,b). While it is easy to write down ODEs in the form (1.1.1) without any solutions (e.g., $(u')^2 + u^2 + 1 = 0$), we will see that ODEs of the type (1.1.2) essentially always have solutions, subject to some very minimal assumptions on f.

Techniques of Functional Analysis for Differential and Integral Equations
http://dx.doi.org/10.1016/B978-0-12-811426-1.00001-5

The ODE is *linear* if it can be written as

$$\sum_{j=0}^{n} a_j(t)u^{(j)}(t) = g(t) \tag{1.1.3}$$

for some coefficients $a_0, \ldots, a_n, g$, and *homogeneous* if also $g(t) \equiv 0$. It is common to use operator notation for derivatives, especially in the linear case. Set

$$D = \frac{d}{dt} \tag{1.1.4}$$

so that $u' = Du$, $u'' = D(Du) = D^2u$, etc., in which case Eq. (1.1.3) may be given as

$$Lu := \sum_{j=0}^{n} a_j(t)D^j u = g(t) \tag{1.1.5}$$

By standard calculus properties L is a *linear operator*, meaning that

$$L(c_1u_1 + c_2u_2) = c_1Lu_1 + c_2Lu_2 \tag{1.1.6}$$

for any scalars c_1, c_2 and any n times differentiable functions u_1, u_2.

An ODE normally has infinitely many solutions—the collection of all solutions is called the *general solution* of the given ODE.

Example 1.1. By elementary calculus considerations, the simple ODE $u' = 0$ has general solution $u(t) = c$, where c is an arbitrary constant. Likewise $u' = u$ has the general solution $u(t) = ce^t$ and $u'' = 2$ has the general solution $u(t) = t^2 + c_1t + c_2$, where c_1, c_2 are arbitrary constants. □

1.1.1 Initial Value Problems

The general solution of an nth order ODE typically contains exactly n arbitrary constants, whose values may be then chosen so that the solution satisfies n additional, or side, conditions. The most common kind of side conditions of interest for an ODE are *initial conditions*,

$$u^{(j)}(t_0) = \gamma_j \quad j = 0, 1, \ldots, n-1 \tag{1.1.7}$$

where t_0 is a given point in (a, b) and $\gamma_0, \ldots, \gamma_{n-1}$ are given constants. Thus we are prescribing the value of the solution and its derivatives up through order $n-1$ at the point t_0. The problem of solving Eq. (1.1.2) together with the initial conditions (1.1.7) is called an *initial value problem* (IVP). It is a very important fact that under fairly unrestrictive hypotheses a unique solution exists. In stating conditions on f, we regard it as a function $f = f(t, y_1, \ldots, y_n)$ defined on some domain in $\mathbb{R}^{n+1}$.

Theorem 1.1. *Assume that*

$$f, \frac{\partial f}{\partial y_1}, \dots, \frac{\partial f}{\partial y_n} \tag{1.1.8}$$

are defined and continuous in a neighborhood of the point $(t_0, \gamma_0, \dots, \gamma_{n-1}) \in \mathbb{R}^{n+1}$. *Then there exists* $\epsilon > 0$ *such that the IVP* (1.1.2), (1.1.7) *has a unique solution on the interval* $(t_0 - \epsilon, t_0 + \epsilon)$.

A proof of this theorem may be found in standard ODE textbooks (see, e.g., [4] or [7]). A slightly weaker version of this theorem will be proved in Section 3.5. As will be discussed there, thc condition of continuity of the partial derivatives of f with respect to each of the variables y_i can actually be replaced by the weaker assumption that f is Lipschitz continuous with respect to each of these variables. If we assume only that f is continuous in a neighborhood of the point $(t_0, \gamma_0, \dots, \gamma_{n-1})$ then it can be proved that at least one solution exists, but it may not be unique, see Exercise 1.3. Similar results are valid for systems of ODEs.

It should also be emphasized that the theorem asserts a *local* existence property, that is, only in some sufficiently small interval centered at t_0. It has to be this way, first of all, since the assumptions on f are made only in the vicinity of $(t_0, \gamma_0, \dots, \gamma_{n-1})$. But even if the continuity properties of f were assumed to hold throughout $\mathbb{R}^{n+1}$, then as the following example shows, it would still only be possible to prove that a solution exists for points t close enough to t_0.

Example 1.2. Consider the first order IVP

$$u' = u^2 \quad u(0) = \gamma \tag{1.1.9}$$

for which the assumptions of Theorem 1.1 hold for any γ. It may be checked that the solution of this problem is

$$u(t) = \frac{\gamma}{1 - \gamma t} \tag{1.1.10}$$

which is only a valid solution for $t < \frac{1}{\gamma}$, which can be arbitrarily small. □

With more restrictions on f it may be possible to show that the solution exists on *any* interval containing t_0, in which case we would say that the solution exists *globally*. This is the case, for example, for the linear ODE (1.1.3).

Whenever the conditions of Theorem 1.1 hold, the set of all possible solutions may be regarded as being parametrized by the n constants $\gamma_0, \dots, \gamma_{n-1}$, so that as mentioned above, the general solution will contain exactly n arbitrary parameters. In the special case of the linear equation (1.1.3) it can be shown that the general solution may be given as

$$u(t) = \sum_{j=1}^{n} c_j u_j(t) + u_p(t) \tag{1.1.11}$$

where u_p is any particular solution of Eq. (1.1.3), and $u_1, \ldots, u_n$ are any n linearly independent solutions of the corresponding homogeneous equation $Lu = 0$. Any such set of functions $u_1, \ldots, u_n$ is also called a *fundamental set* for $Lu = 0$.

Example 1.3. If $Lu = u'' + u$ then by direct substitution we see that $u_1(t) = \sin t$, $u_2(t) = \cos t$ are solutions, and they are clearly linearly independent. Thus $\{\sin t, \cos t\}$ is a fundamental set for $Lu = 0$ and $u(t) = c_1 \sin t + c_2 \cos t$ is the general solution of $Lu = 0$. For the inhomogeneous ODE $u'' + u = e^t$ one may check that $u_p(t) = \frac{1}{2}e^t$ is a particular solution, so the general solution is $u(t) = c_1 \sin t + c_2 \cos t + \frac{1}{2}e^t$. □

1.1.2 Boundary Value Problems

For an ODE of degree $n \geq 2$ it may be of interest to impose side conditions at more than one point, typically the endpoints of the interval of interest. We will then refer to the side conditions as *boundary conditions* and the problem of solving the ODE subject to the given boundary conditions as a *boundary value problem* (BVP). Since the general solution still contains n parameters, we still expect to be able to impose a total of n side conditions. However we can see from simple examples that the situation with regard to existence and uniqueness in such BVPs is much less clear than for IVPs.

Example 1.4. Consider the BVP

$$u'' + u = 0 \quad 0 < t < \pi \qquad u(0) = 0 \quad u(\pi) = 1 \tag{1.1.12}$$

Starting from the general solution $u(t) = c_1 \sin t + c_2 \cos t$, the two boundary conditions lead to $u(0) = c_2 = 0$ and $u(\pi) = c_2 = 1$. Since these are inconsistent, the BVP has no solution. □

Example 1.5. For the BVP

$$u'' + u = 0 \quad 0 < t < \pi \qquad u(0) = 0 \quad u(\pi) = 0 \tag{1.1.13}$$

we have solutions $u(t) = C \sin t$ for any constant C, that is, the BVP has infinitely many solutions. □

The topic of BVPs will be studied in much more detail in Chapter 13.

1.1.3 Some Exactly Solvable Cases

Let us finally review explicit solution methods for some commonly occurring types of ODEs.

- For the first order linear ODE

$$u' + p(t)u = q(t) \tag{1.1.14}$$

define the so-called *integrating factor* $\rho(t) = e^{P(t)}$ where P is any function satisfying $P' = p$. Multiplying the equation through by ρ we then get the equivalent equation

$$(\rho u)' = \rho q \tag{1.1.15}$$

so if we pick Q such that $Q' = \rho q$, the general solution may be given as

$$u(t) = \frac{Q(t) + C}{\rho(t)} \tag{1.1.16}$$

- For the linear homogeneous constant coefficient ODE

$$Lu = \sum_{j=0}^{n} a_j u^{(j)} = 0 \tag{1.1.17}$$

if we look for solutions in the form $u(t) = e^{\lambda t}$ then by direct substitution we find that u is a solution provided λ is a root of the corresponding *characteristic polynomial*

$$P(\lambda) = \sum_{j=0}^{n} a_j \lambda^j \tag{1.1.18}$$

We therefore obtain as many linearly independent solutions as there are distinct roots of P. If this number is less than n, then we may seek further solutions of the form $te^{\lambda t}, t^2 e^{\lambda t}, \dots$, until a total of n linearly independent solutions have been found. In the case of complex roots, equivalent expressions in terms of trigonometric functions are often used in place of complex exponentials.

- Finally, closely related to the previous case, is the so-called Cauchy-Euler type equation

$$Lu = \sum_{j=0}^{n} (t - t_0)^j a_j u^{(j)} = 0 \tag{1.1.19}$$

for some constants $a_0, \dots, a_n$. In this case we look for solutions in the form $u(t) = (t - t_0)^\lambda$ with λ to be found. Substituting into Eq. (1.1.19) we will find again an nth order polynomial whose roots determine the possible values of λ. The interested reader may refer to any standard undergraduate level ODE book for the additional considerations which arise in the case of complex or repeated roots.

1.2 INTEGRAL EQUATIONS

In this section we discuss the basic set-up for the study of linear integral equations. See, for example, [16, 22] as general references in the classical theory of integral equations. Let $\Omega \subset \mathbb{R}^N$ be an open set, K a given function on $\Omega \times \Omega$ and set

$$Tu(x) = \int_\Omega K(x,y)u(y)\,dy \tag{1.2.20}$$

Here the function K is called the *kernel* of the *integral operator* T, which is linear since Eq. (1.1.6) obviously holds.

A class of associated integral equations is then

$$\lambda u(x) - \int_\Omega K(x,y)u(y)\,dy = f(x) \quad x \in \Omega \tag{1.2.21}$$

for some scalar λ and given function f in some appropriate class. If $\lambda = 0$ then Eq. (1.2.21) is said to be a *first kind* integral equation, otherwise it is *second kind*. Let us consider some simple examples which may be studied by elementary means.

Example 1.6. Let $\Omega = (0,1) \subset \mathbb{R}$ and $K(x,y) \equiv 1$. The corresponding first kind integral equation is therefore

$$-\int_0^1 u(y)\,dy = f(x) \quad 0 < x < 1 \tag{1.2.22}$$

For simplicity here we will assume that f is a continuous function. The left-hand side is independent of x, thus a solution can exist only if $f(x)$ is a constant function. When f is constant, on the other hand, infinitely many solutions will exist, since we just need to find any u with the given definite integral.

For the corresponding second kind equation,

$$\lambda u(x) - \int_0^1 u(y)\,dy = f(x) \tag{1.2.23}$$

a solution, if one exists, must have the specific form $u(x) = (f(x) + C)/\lambda$ for some constant C. Substituting into the equation then gives, after obvious algebra, that

$$C(\lambda - 1) = \int_0^1 f(y)\,dy \tag{1.2.24}$$

Thus, for any continuous function f and $\lambda \neq 0, 1$, there exists a unique solution of the integral equation, namely

$$u(x) = \frac{f(x)}{\lambda} + \frac{\int_0^1 f(y)\,dy}{\lambda(\lambda - 1)} \tag{1.2.25}$$

In the remaining case that $\lambda = 1$, it is immediate from Eq. (1.2.24) that a solution can exist only if $\int_0^1 f(y)\,dy = 0$, in which case $u(x) = f(x) + C$ is a solution for any choice of C. □

This very simple example already exhibits features which turn out to be common to a much larger class of integral equations of this general type. These are

- The first kind integral equation will require much more restrictive conditions on f in order for a solution to exist.
- For most $\lambda \neq 0$ the second kind integral equation has a unique solution for any f.
- There may exist a few exceptional values of λ for which either existence or uniqueness fails in the corresponding second kind equation.

All of these points will be elaborated and made precise in Chapter 12.

Example 1.7. Let $\Omega = (0, 1)$ and

$$Tu(x) = \int_0^x u(y)\,dy \tag{1.2.26}$$

corresponding to the kernel

$$K(x, y) = \begin{cases} 1 & y < x \\ 0 & x \leq y \end{cases} \tag{1.2.27}$$

The corresponding integral equation may then be written as

$$\lambda u(x) - \int_0^x u(y)\,dy = f(x) \tag{1.2.28}$$

This is the prototype of an integral operator of so-called *Volterra type*, see Definition 1.1 below.

In the first kind case, $\lambda = 0$, we see that $f(0) = 0$ is a necessary condition for solvability, in which case the solution is $u(x) = -f'(x)$, provided that f is differentiable in some suitable sense. For $\lambda \neq 0$ we note that differentiation of Eq. (1.2.28) with respect to x gives

$$u' - \frac{1}{\lambda}u = \frac{f'(x)}{\lambda} \tag{1.2.29}$$

This is an ODE of the type (1.1.14), and so may be solved by the method given there. The result, after some obvious algebraic manipulation, is

$$u(x) = \frac{e^{x/\lambda}}{\lambda}f(0) + \frac{1}{\lambda}\int_0^x e^{(x-y)/\lambda}f'(y)\,dy \tag{1.2.30}$$

Note, however, that by an integration by parts, this formula is seen to be equivalent to

$$u(x) = \frac{f(x)}{\lambda} + \frac{1}{\lambda^2}\int_0^x e^{(x-y)/\lambda}f(y)\,dy \tag{1.2.31}$$

Observe that Eq. (1.2.30) seems to require differentiability of f even though Eq. (1.2.31) does not, thus Eq. (1.2.31) would be the preferred solution formula. It may be verified directly by substitution that Eq. (1.2.31) is a valid solution of Eq. (1.2.28) for all $\lambda \neq 0$, assuming only that f is continuous on $[0, 1]$. □

Concerning the two simple integral equations just discussed, there are again some features which will turn out to be generally true.

- For the first kind equation, there are fewer restrictions on f needed for solvability in the Volterra case (1.2.28) than in the non-Volterra case (1.2.23).
- There are no exceptional values $\lambda \neq 0$ in the Volterra case, that is, a unique solution exists for every $\lambda \neq 0$ and every continuous f.

Finally let us mention some of the more important ways in which integral operators, or the corresponding integral equations, are classified:

Definition 1.1. The kernel $K(x, y)$ is called

- *symmetric* if $K(x, y) = \overline{K(y, x)}$
- *Volterra type* if $N = 1$ and $K(x, y) = 0$ for $x > y$ or $x < y$
- *convolution type* if $K(x, y) = F(x - y)$ for some function F
- *Hilbert-Schmidt type* if $\int_{\Omega\times\Omega} |K(x, y)|^2\, dxdy < \infty$
- *singular* if $K(x, y)$ is unbounded on $\Omega \times \Omega$

Important examples of integral operators, some of which will receive much more attention later in the book, are the Fourier transform

$$Tu(x) = \frac{1}{(2\pi)^{N/2}} \int_{\mathbb{R}^N} e^{-ix\cdot y} u(y)\, dy \tag{1.2.32}$$

the Laplace transform

$$Tu(x) = \int_0^\infty e^{-xy} u(y)\, dy \tag{1.2.33}$$

the Hilbert transform

$$Tu(x) = \frac{1}{\pi} \int_{-\infty}^{\infty} \frac{u(y)}{x - y}\, dy \tag{1.2.34}$$

and the Abel operator

$$Tu(x) = \int_0^x \frac{u(y)}{\sqrt{x - y}}\, dy \tag{1.2.35}$$

1.3 PARTIAL DIFFERENTIAL EQUATIONS

An mth order partial differential equation (PDE) for an unknown function $u = u(x)$ on a domain $\Omega \subset \mathbb{R}^N$ is any equation of the form

$$F(x, \{D^\alpha u\}_{|\alpha| \le m}) = 0 \tag{1.3.36}$$

Here we are using the so-called *multiindex* notation for partial derivatives which works as follows. A multiindex is vector of nonnegative integers

$$\alpha = (\alpha_1, \alpha_2, \dots, \alpha_N) \qquad \alpha_i \in \{0, 1, \dots\} \tag{1.3.37}$$

In terms of α we define

$$|\alpha| = \sum_{i=1}^{N} \alpha_i \tag{1.3.38}$$

the order of α, and

$$D^\alpha u = \frac{\partial^{|\alpha|} u}{\partial_{x_1}^{\alpha_1} \partial_{x_2}^{\alpha_2} \dots \partial_{x_N}^{\alpha_N}} \tag{1.3.39}$$

the corresponding α derivative of u. For later use it is also convenient to define the factorial of a multiindex

$$\alpha! = \alpha_1! \alpha_2! \dots \alpha_N! \tag{1.3.40}$$

The PDE (1.3.36) is linear if it can be written as

$$Lu(x) = \sum_{|\alpha| \le m} a_\alpha(x) D^\alpha u(x) = g(x) \tag{1.3.41}$$

for some coefficient functions a_α.

1.3.1 First Order PDEs and the Method of Characteristics

Let us start with the simplest possible example.

Example 1.8. When $N = 2$ and $m = 1$ consider

$$\frac{\partial u}{\partial x_1} = 0 \tag{1.3.42}$$

By elementary calculus considerations it is clear that u is a solution if and only if u is independent of x_1, that is,

$$u(x_1, x_2) = f(x_2) \tag{1.3.43}$$

for some function f. This is then the general solution of the given PDE, which we note contains an arbitrary function f. □

Example 1.9. Next consider, again for $N = 2$, $m = 1$, the PDE

$$a \frac{\partial u}{\partial x_1} + b \frac{\partial u}{\partial x_2} = 0 \tag{1.3.44}$$

where a, b are fixed constants, at least one of which is not zero. The equation amounts precisely to the condition that u has directional derivative 0 in the direction $\boldsymbol{\theta} = \langle a, b \rangle$, so u is constant along any line parallel to $\boldsymbol{\theta}$. This in turn leads to the conclusion that $u(x_1, x_2) = f(ax_2 - bx_1)$ for some arbitrary function f, which at least for the moment would seem to need to be differentiable. □

The collection of lines parallel to $\boldsymbol{\theta}$, that is, lines $ax_2 - bx_1 = C$ obviously play a special role in the previous example, they are the so-called *characteristics*, or *characteristic curves* associated to this particular PDE. The general concept of characteristic curve will now be described for the case of a first order linear PDE in two independent variables (with a temporary change of notation)

$$a(x,y)u_x + b(x,y)u_y = c(x,y) \tag{1.3.45}$$

Consider the associated ODE system

$$\frac{dx}{dt} = a(x,y) \qquad \frac{dy}{dt} = b(x,y) \tag{1.3.46}$$

and suppose we have some solution pair $x = x(t), y = y(t)$ which we regard as a parametrically given curve in the (x, y) plane. Any such curve is then defined to be a characteristic curve for Eq. (1.3.45). The key observation now is that if $u(x, y)$ is a differentiable solution of Eq. (1.3.45) then

$$\begin{aligned}\frac{d}{dt}u(x(t),y(t)) &= a(x(t),y(t))u_x(x(t),y(t)) + b(x(t),y(t))u_y(x(t),y(t))\\ &= c(x(t),y(t))\end{aligned} \tag{1.3.47}$$

so that u satisfies a certain first order ODE along any characteristic curve. For example, if $c(x,y) \equiv 0$ then, as in the previous example, any solution of the PDE is constant along any characteristic curve.

We now use this property to construct solutions of Eq. (1.3.45). Let $\Gamma \subset \mathbb{R}^2$ be some curve, which we assume can be parametrized as

$$x = f(s), \quad y = g(s), \qquad s_0 < s < s_1 \tag{1.3.48}$$

The *Cauchy problem* for Eq. (1.3.45) consists in finding a solution of Eq. (1.3.45) with values prescribed on Γ, that is,

$$u(f(s), g(s)) = h(s) \quad s_0 < s < s_1 \tag{1.3.49}$$

for some given function h. Assuming for the moment that such a solution u exists, let $x(t,s), y(t,s)$ be the characteristic curve passing through $(f(s), g(s)) \in \Gamma$ when $t = 0$, that is,

$$\begin{cases} \frac{\partial x}{\partial t} = a(x,y) & x(0,s) = f(s) \\ \frac{\partial y}{\partial t} = b(x,y) & y(0,s) = g(s) \end{cases} \tag{1.3.50}$$

We must then have

$$\frac{\partial}{\partial t}u(x(t,s),y(t,s)) = c(x(t,s),y(t,s)) \qquad u(x(0,s),y(0,s)) = h(s) \tag{1.3.51}$$

This is a first order IVP in t, depending on s as a parameter, which is guaranteed to have a solution at least for $|t| < \epsilon$ for some $\epsilon > 0$, provided that c is continuously differentiable. The three relations $x = x(t,s), y = y(t,s), z = u(x(t,s), y(t,s))$ generally amounts to the parametric description of a surface in $\mathbb{R}^3$ containing Γ. If we can eliminate the parameters s,t to obtain the surface in nonparametric form $z = u(x,y)$ then u is the sought after solution of the Cauchy problem.

Example 1.10. Let Γ denote the x axis and let us solve

$$xu_x + u_y = 1 \tag{1.3.52}$$

with $u = h$ on Γ. Introducing $f(s) = s, g(s) = 0$ as the parametrization of Γ, we must then solve

$$\begin{cases} \frac{\partial x}{\partial t} = x & x(0,s) = s \\ \frac{\partial y}{\partial t} = 1 & y(0,s) = 0 \\ \frac{\partial}{\partial t} u(x(t,s), y(t,s)) = 1 & u(0,s) = h(s) \end{cases} \tag{1.3.53}$$

We then easily obtain

$$x(t,s) = se^t \quad y(t,s) = t \quad u(x(t,s), y(t,s)) = t + h(s) \tag{1.3.54}$$

and eliminating t,s yields the solution formula

$$u(x,y) = y + h(xe^{-y}) \tag{1.3.55}$$

The characteristics in this case are the curves $x = se^t, y = t$ for fixed s, or $x = se^y$ in nonparametric form. Note here that the solution is defined throughout the x,y plane even though nothing in the preceding discussion guarantees that. Since h has not been otherwise prescribed we may also regard Eq. (1.3.55) as the general solution of Eq. (1.3.52), again containing one arbitrary function. □

The attentive reader may already realize that this procedure cannot work in all cases, as is made clear by the following consideration: if $c \equiv 0$ and Γ is itself a characteristic curve, then the solution on Γ would have to simultaneously be equal to the given function h and to be constant, so that no solution can exist except possibly in the case that h is a constant function. From another, more general, point of view we must eliminate the parameters s,t by inverting the relations $x = x(s,t), y = y(s,t)$ to obtain s,t in terms of x,y, at least near Γ. According to the inverse function theorem this should require that the Jacobian matrix

$$\begin{bmatrix} \frac{\partial x}{\partial t} & \frac{\partial y}{\partial t} \\ \frac{\partial x}{\partial s} & \frac{\partial y}{\partial s} \end{bmatrix}\Bigg|_{t=0} = \begin{bmatrix} a(f(s), g(s)) & b(f(s), g(s)) \\ f'(s) & g'(s) \end{bmatrix} \tag{1.3.56}$$

be nonsingular for all s. Equivalently the direction $\langle f', g' \rangle$ should not be parallel to $\langle a, b \rangle$, and since $\langle a, b \rangle$ must be tangent to the characteristic curve, this amounts to the requirement that Γ itself should have a noncharacteristic tangent direction at every point. We say that Γ is noncharacteristic for the PDE (1.3.45) when this condition holds.

The following precise theorem can be established, see, for example, Chapter 1 of [19], or Chapter 3 of [11]. The proof amounts to showing that the method of constructing a solution just described can be made rigorous under the stated assumptions.

Theorem 1.2. *Let $\Gamma \subset \mathbb{R}^2$ be a continuously differentiable curve, which is noncharacteristic for Eq.* (1.3.45)*, h a continuously differentiable function on Γ, and let a, b, c be continuously differentiable functions in a neighborhood of Γ. Then there exists a unique continuously differentiable function $u(x, y)$ defined in a neighborhood of Γ which is a solution of Eq.* (1.3.45).

The method of characteristics is capable of a considerable amount of generalization, in particular to first order PDEs in any number of independent variables, and to fully nonlinear first PDEs, see the references just mentioned previously.

1.3.2 Second Order Problems in $\mathbb{R}^2$

In order to better understand what can be known about solutions of a potentially complicated looking PDE, one natural approach is to try to obtain another equation which is equivalent to the original one, but is somehow simpler in structure. We illustrate with the following special type of second order PDE in two independent variables x, y:

$$Au_{xx} + Bu_{xy} + Cu_{yy} = 0 \tag{1.3.57}$$

where A, B, C are real constants, not all zero. Consider introducing new coordinates ξ, η by means of a linear change of variable

$$\xi = \alpha x + \beta y \quad \eta = \gamma x + \delta y \tag{1.3.58}$$

with $\alpha\delta - \beta\gamma \neq 0$, so that the transformation is invertible. Our goal is to make a good choice of $\alpha, \beta, \gamma, \delta$ so as to achieve a simpler looking, but equivalent PDE to study.

Given any PDE and any change of coordinates, we obtain the expression for the PDE in the new coordinate system by straightforward application of the chain rule. In the case at hand we have

$$\frac{\partial u}{\partial x} = \frac{\partial u}{\partial \xi}\frac{\partial \xi}{\partial x} + \frac{\partial u}{\partial \eta}\frac{\partial \eta}{\partial x} = \alpha\frac{\partial u}{\partial \xi} + \gamma\frac{\partial u}{\partial \eta} \tag{1.3.59}$$

$$\begin{aligned} \frac{\partial^2 u}{\partial x^2} &= \left(\alpha\frac{\partial}{\partial \xi} + \gamma\frac{\partial}{\partial \eta}\right)\left(\alpha\frac{\partial u}{\partial \xi} + \gamma\frac{\partial u}{\partial \eta}\right) \\ &= \alpha^2\frac{\partial^2 u}{\partial \xi^2} + 2\alpha\gamma\frac{\partial^2 u}{\partial \xi \partial \eta} + \gamma^2\frac{\partial^2 u}{\partial \eta^2} \end{aligned} \tag{1.3.60}$$

with similar expressions for u_{xy} and u_{yy}. Substituting into Eq. (1.3.57) the resulting PDE is

$$au_{\xi\xi} + bu_{\xi\eta} + cu_{\eta\eta} = 0 \tag{1.3.61}$$

where

$$a = \alpha^2 A + \alpha\beta B + \beta^2 C \tag{1.3.62}$$
$$b = 2\alpha\gamma A + (\alpha\delta + \beta\gamma)B + 2\beta\delta C \tag{1.3.63}$$
$$c = \gamma^2 A + \gamma\delta B + \delta^2 C \tag{1.3.64}$$

We now seek to make special choices of $\alpha, \beta, \gamma, \delta$ to achieve as simple a form as possible for the transformed PDE (1.3.61).

Suppose first that $B^2 - 4AC > 0$, so that there exist two real and distinct roots r_1, r_2 of $Ar^2 + Br + C = 0$. If $\alpha, \beta, \gamma, \delta$ are chosen so that

$$\frac{\alpha}{\beta} = r_1 \qquad \frac{\gamma}{\delta} = r_2 \tag{1.3.65}$$

then $a = c = 0$, and $\alpha\delta - \beta\gamma \neq 0$, so that the transformed PDE is simply, after division by a constant, $u_{\xi\eta} = 0$. The general solution of this second order PDE is easily obtained: u_ξ must be a function of ξ alone, so integrating with respect to ξ and observing that the "constant of integration" could be any function of η, we get

$$u(\xi, \eta) = F(\xi) + G(\eta) \tag{1.3.66}$$

for any differentiable functions F, G. Finally reverting to the original coordinate system, the result is

$$u(x, y) = F(\alpha x + \beta y) + G(\gamma x + \delta y) \tag{1.3.67}$$

an expression for the general solution containing two arbitrary functions.

The lines $\alpha x + \beta y = C$, $\gamma x + \delta y = C$ are called the characteristics for Eq. (1.3.57). Characteristics are an important concept for this and some more general second order PDEs, but they don't play as central a role as in the first order case.

Example 1.11. For the PDE

$$u_{xx} - u_{yy} = 0 \tag{1.3.68}$$

we have $B^2 - 4AC > 0$ and the roots r satisfy $r^2 - 1 = 0$. We may then choose, for example, $\alpha = \beta = \gamma = 1$, $\delta = -1$, to get the general solution

$$u(x, y) = F(x + y) + G(x - y) \tag{1.3.69}$$

□

Next assume that $B^2 - 4AC = 0$. If either of A or C is 0, then so is B, in which case the PDE already has the form $u_{\xi\xi} = 0$ or $u_{\eta\eta} = 0$, say the first of these without loss of generality. Otherwise, choose

$$\alpha = -\frac{B}{2A} \quad \beta = 1 \quad \gamma = 1 \quad \delta = 0 \tag{1.3.70}$$

to obtain $a = b = 0$, $c = A$, so that the transformed PDE in all cases may be taken to be $u_{\xi\xi} = 0$.

Finally, if $B^2 - 4AC < 0$ then $A \neq 0$ must hold, and we may choose

$$\alpha = \frac{2A}{\sqrt{4AC - B^2}} \quad \beta = \frac{-B}{\sqrt{4AC - B^2}} \quad \gamma = 0 \quad \delta = 1 \tag{1.3.71}$$

in which case the transformed equation is

$$u_{\xi\xi} + u_{\eta\eta} = 0 \tag{1.3.72}$$

We have therefore established that any PDE of the type (1.3.57) can be transformed, by means of a linear change of variables, to one of the three simple types,

$$u_{\xi\eta} = 0 \qquad u_{\xi\xi} = 0 \qquad u_{\xi\xi} + u_{\eta\eta} = 0 \tag{1.3.73}$$

each of which then leads to a prototype for a certain larger class of PDEs. If we allow lower order terms

$$Au_{xx} + Bu_{xy} + Cu_{yy} + Du_x + Eu_y + Fu = 0 \tag{1.3.74}$$

then after the transformation (1.3.58) it is clear that the lower order terms remain as lower order terms. Thus any PDE of the type (1.3.74) is, up to a change of coordinates, one of the three types (1.3.73), up to lower order terms, and only the value of the discriminant $B^2 - 4AC$ needs to be known to determine which of the three types is obtained.

The preceding discussion motivates the following classification: The PDE (1.3.74) is said to be:

- hyperbolic if $B^2 - 4AC > 0$
- parabolic if $B^2 - 4AC = 0$
- elliptic if $B^2 - 4AC < 0$

The terminology comes from an obvious analogy with conic sections, that is, the solution set of $Ax^2 + Bxy + Cy^2 + Dx + Ey + F = 0$ is respectively a hyperbola, parabola, or ellipse (or a degenerate case) according as $B^2 - 4AC$ is positive, zero, or negative.

We can also allow the coefficients $A, B, \ldots, F$ to be variable functions of x, y, and in this case the classification is done pointwise, so the type can change. An important example of this phenomenon is the so-called Tricomi equation (see, e.g., Chapter 12 of [14])

$$u_{xx} - xu_{yy} = 0 \tag{1.3.75}$$

which is hyperbolic for $x > 0$ and elliptic for $x < 0$. One might refer to the equation as being parabolic for $x = 0$ but generally speaking we do not do this, since it is not really meaningful to speak of a PDE being satisfied in a set without interior points.

The preceding discussion is special to the case of $N = 2$ independent variables—in the case of $N \geq 3$ there is no such complete classification. As we will see there are still PDEs referred to as being hyperbolic, parabolic, or elliptic, but there are others which are not of any of these types, although these tend to be of less physical importance.

1.3.3 Further Discussion of Model Problems

According to the previous discussion, we should focus our attention on a representative problem for each of the three types, since then we will also gain considerable information about other problems of the given type.

Wave Equation

For the hyperbolic case we consider the *wave equation*

$$u_{tt} - c^2 u_{xx} = 0 \tag{1.3.76}$$

where $c > 0$ is a constant. Here we have changed the name of the variable y to t, following the usual convention of regarding $u = u(x,t)$ as depending on a "space" variable x and "time" variable t. This PDE arises in the simplest model of wave propagation in one space dimension, where u represents, for example, the displacement of a vibrating medium from its equilibrium position, and c is the wave speed.

Following the procedure outlined at the beginning of this section, an appropriate change of coordinates is $\xi = x + ct, \eta = x - ct$, and we obtain the expression, also known as d'Alembert's formula, for the general solution,

$$u(x,t) = F(x+ct) + G(x-ct) \tag{1.3.77}$$

for arbitrary twice differentiable functions F, G. The general solution may be viewed as the superposition of two waves of fixed shape, moving to the right and to the left with speed c.

The IVP for the wave equation consists in solving Eq. (1.3.76) for $x \in \mathbb{R}$ and $t > 0$ subject to the side conditions

$$u(x,0) = f(x) \quad u_t(x,0) = g(x) \qquad x \in \mathbb{R} \tag{1.3.78}$$

where f, g represent the initial displacement and initial velocity of the vibrating medium. This problem may be completely and explicitly solved by means of d'Alembert's formula. Setting $t = 0$ and using the prescribed side conditions, we must have

$$F(x) + G(x) = f(x) \quad c(F'(x) - G'(x)) = g(x) \qquad x \in \mathbb{R} \tag{1.3.79}$$

Integrating the second relation gives $F(x) - G(x) = \frac{1}{c}\int_0^x g(s)\,ds + C$ for some constant C, and combining with the first relation yields

$$\begin{aligned} F(x) &= \frac{1}{2}\left(f(x) + \frac{1}{c}\int_0^x g(s)\,ds + C\right) \\ G(x) &= \frac{1}{2}\left(f(x) - \frac{1}{c}\int_0^x g(s)\,ds - C\right) \end{aligned} \tag{1.3.80}$$

Substituting into Eq. (1.3.77) and doing some obvious simplification we obtain

$$u(x,t) = \frac{1}{2}\left(f(x+ct) + f(x-ct)\right) + \frac{1}{2c}\int_{x-ct}^{x+ct} g(s)\,ds \tag{1.3.81}$$

We remark that a general solution formula like Eq. (1.3.77) can be given for any PDE which is exactly transformable to $u_{\xi\eta} = 0$, that is to say, any hyperbolic PDE of the form (1.3.57), but once lower order terms are allowed such a simple solution method is no longer available. For example, the so-called *Klein-Gordon equation* $u_{tt} - u_{xx} + u = 0$ may be transformed to $u_{\xi\eta} + 4u = 0$ which unfortunately cannot be solved in so transparent a form. Thus the d'Alembert solution method, while very useful when applicable, is limited in its scope.

Heat Equation

Another elementary method, which may be used in a wide variety of situations, is the separation of variables technique. We illustrate with the case of the *initial and boundary value problem*

$$u_t = u_{xx} \qquad 0 < x < 1 \quad t > 0 \tag{1.3.82}$$

$$u(0,t) = u(1,t) = 0 \qquad t > 0 \tag{1.3.83}$$

$$u(x,0) = f(x) \qquad 0 < x < 1 \tag{1.3.84}$$

Here Eq. (1.3.82) is the *heat equation*, a parabolic equation modeling, for example, the temperature in a one-dimensional medium $u = u(x,t)$ as a function of location x and time t, Eq. (1.3.83) are the boundary conditions, stating that the temperature is held at temperature zero at the two boundary points $x = 0$ and $x = 1$ for all t, and Eq. (1.3.84) represents the initial condition, that is, that the initial temperature distribution is given by the prescribed function $f(x)$.

We begin by ignoring the initial condition and otherwise looking for special solutions of the form $u(x,t) = \phi(t)\psi(x)$. Obviously $u = 0$ is such a solution, but cannot be of any help in eventually solving the full stated problem, so we insist that neither of ϕ or ψ is the zero function. Inserting into Eq. (1.3.82) we obtain immediately that

$$\phi'(t)\psi(x) = \phi(t)\psi''(x) \tag{1.3.85}$$

must hold, or equivalently

$$\frac{\phi'(t)}{\phi(t)} = \frac{\psi''(x)}{\psi(x)} \tag{1.3.86}$$

Since the left side depends on t alone and the right side on x alone, it must be that both sides are equal to a common constant which we denote by $-\lambda$ (without yet at this point ruling out the possibility that λ itself is negative or even complex). We have therefore obtained ODEs for ϕ and ψ

$$\phi'(t) + \lambda\phi(t) = 0 \qquad \psi''(x) + \lambda\psi(x) = 0 \tag{1.3.87}$$

linked via the separation constant λ. Next, from the boundary condition (1.3.83) we get $\phi(t)\psi(0) = \phi(t)\psi(1) = 0$, and since ϕ is nonzero we must have $\psi(0) = \psi(1) = 0$.

The ODE and side conditions for ψ, namely

$$\psi''(x) + \lambda\psi(x) = 0 \quad 0 < x < 1 \qquad \psi(0) = \psi(1) = 0 \tag{1.3.88}$$

is the simplest example of a so-called *Sturm-Liouville problem*, a topic which will be studied in detail in Chapter 13, but this particular case can be handled by elementary considerations. We emphasize that our goal is to find nonzero solutions of Eq. (1.3.88), along with the values of λ these correspond to, and as we will see, only certain values of λ will be possible.

Considering first the case that $\lambda > 0$, the general solution of the ODE is

$$\psi(x) = c_1 \sin\sqrt{\lambda}x + c_2 \cos\sqrt{\lambda}x \tag{1.3.89}$$

The first boundary condition $\psi(0) = 0$ implies that $c_2 = 0$ while the second gives $c_1 \sin\sqrt{\lambda} = 0$. We are not allowed to have $c_1 = 0$, since otherwise $\psi = 0$, so instead $\sin\sqrt{\lambda} = 0$ must hold, that is, $\sqrt{\lambda} = \pi, 2\pi, \ldots$. Thus we have found one collection of solutions of Eq. (1.3.88), which we denote $\psi_k(x) = \sin k\pi x$, $k = 1, 2, \ldots$. Since they were found under the assumption that $\lambda > 0$, we should next consider other possibilities, but it turns out that we have already found all possible solutions of Eq. (1.3.88). For example, if we suppose $\lambda < 0$ and $k = \sqrt{-\lambda}$ then to solve Eq. (1.3.88) we must have $\psi(x) = c_1e^{kx} + c_2e^{-kx}$. From the boundary conditions

$$c_1 + c_2 = 0 \quad c_1e^k + c_2e^{-k} = 0 \tag{1.3.90}$$

we see that the unique solution is $c_1 = c_2 = 0$ for any $k > 0$. Likewise we can check that $\psi = 0$ is the only possible solution for $k = 0$ and for nonreal k.

For each allowed value of λ we obviously have the corresponding function $\phi(t) = e^{-\lambda t}$, so that

$$u_k(x,t) = e^{-k^2\pi^2 t} \sin k\pi x \quad k = 1, 2, \ldots \tag{1.3.91}$$

represents, aside from multiplicative constants, all possible product solutions of Eqs. (1.3.82), (1.3.83).

To complete the solution of the initial and boundary value problem, we observe that any sum $\sum_{k=1}^{\infty} c_k u_k(x,t)$ is also a solution of Eqs. (1.3.82), (1.3.83) as long as $c_k \to 0$ sufficiently rapidly, and we try to choose the coefficients c_k to achieve the initial condition (1.3.84). This amounts to the requirement that

$$f(x) = \sum_{k=1}^{\infty} c_k \sin k\pi x \tag{1.3.92}$$

hold. For any f for which such a sine series representation is valid, we then have the solution of the given PDE problem

$$u(x,t) = \sum_{k=1}^{\infty} c_k e^{-k^2\pi^2 t} \sin k\pi x \tag{1.3.93}$$

The question then becomes to characterize this set of f's in some more straightforward way, and this is done, among many other things, within the theory of Fourier series, which will be discussed in Chapter 7. Roughly speaking the conclusion will be that essentially any reasonable function can be represented this way, but there are many aspects to this, including elaboration of the precise sense in which the series converges. One other fact concerning this series which we can easily anticipate at this point, is a formula for the coefficient c_k: If we assume that Eq. (1.3.92) holds, we can multiply both sides by $\sin m\pi x$ for some integer m and integrate with respect to x over $(0, 1)$, to obtain

$$\int_0^1 f(x) \sin m\pi x \, dx = c_m \int_0^1 \sin^2 m\pi x \, dx = \frac{c_m}{2} \tag{1.3.94}$$

since $\int_0^1 \sin k\pi x \sin m\pi x \, dx = 0$ for $k \neq m$. Thus, if f is representable by a sine series, there is only one possibility for the kth coefficient, namely

$$c_k = 2\int_0^1 f(x) \sin k\pi x \, dx \tag{1.3.95}$$

Laplace Equation

Finally we discuss a model problem of elliptic type,

$$u_{xx} + u_{yy} = 0 \quad x^2 + y^2 < 1 \tag{1.3.96}$$

$$u(x,y) = f(x,y) \quad x^2 + y^2 = 1 \tag{1.3.97}$$

where f is a given function. The PDE in Eq. (1.3.96) is known as *Laplace's equation*, and is commonly written as $\Delta u = 0$ where $\Delta = \frac{\partial^2}{\partial x^2} + \frac{\partial^2}{\partial y^2}$ is the Laplace operator, or Laplacian. A function satisfying Laplace's equation in some set is said to be a *harmonic* function on that set, thus we are solving the boundary value problem of finding a harmonic function in the unit disk $x^2 + y^2 < 1$ subject to a prescribed boundary condition on the boundary of the disk.

One should immediately recognize that it would be natural here to make use of polar coordinates (r, θ), where according to the usual calculus notations,

$$r = \sqrt{x^2 + y^2} \quad \tan\theta = \frac{y}{x} \quad x = r\cos\theta \quad y = r\sin\theta \tag{1.3.98}$$

and we regard $u = u(r, \theta)$ and $f = f(\theta)$.

To begin we need to find the expression for Laplace's equation in polar coordinates. Again this is a straightforward calculation with the chain rule, for example,

$$\frac{\partial u}{\partial x} = \frac{\partial u}{\partial r}\frac{\partial r}{\partial x} + \frac{\partial u}{\partial \theta}\frac{\partial \theta}{\partial x} \tag{1.3.99}$$

$$= \frac{x}{\sqrt{x^2+y^2}}\frac{\partial u}{\partial r} - \frac{y}{x^2+y^2}\frac{\partial u}{\partial \theta} \tag{1.3.100}$$

$$= \cos\theta\frac{\partial u}{\partial r} - \frac{\sin\theta}{r}\frac{\partial u}{\partial \theta} \tag{1.3.101}$$

and similar expressions for $\frac{\partial u}{\partial y}$ and the second derivatives. The end result is

$$u_{xx} + u_{yy} = u_{rr} + \frac{1}{r}u_r + \frac{1}{r^2}u_{\theta\theta} = 0 \tag{1.3.102}$$

We may now try separation of variables, looking for solutions in the special product form $u(r,\theta) = R(r)\Theta(\theta)$. Substituting into Eq. (1.3.102) and dividing by $R\Theta$ gives

$$r^2\frac{R''(r)}{R(r)} + r\frac{R'(r)}{R(r)} = -\frac{\Theta''(\theta)}{\Theta(\theta)} \tag{1.3.103}$$

so both sides must be equal to a common constant λ. Therefore R and Θ must be nonzero solutions of

$$\Theta'' + \lambda\Theta = 0 \qquad r^2R'' + rR' - \lambda R = 0 \tag{1.3.104}$$

Next it is necessary to recognize that there are two "hidden" side conditions which we must make use of. The first of these is that Θ must be 2π periodic, since otherwise it would not be possible to express the solution u in terms of the original variables x, y in an unambiguous way. We can make this explicit by requiring

$$\Theta(0) = \Theta(2\pi) \qquad \Theta'(0) = \Theta'(2\pi) \tag{1.3.105}$$

As in the case of Eq. (1.3.88) we can search for allowable values of λ by considering the various cases $\lambda > 0, \lambda < 0$, etc. The outcome is that nontrivial solutions exist precisely if $\lambda = k^2, k = 0, 1, 2, \ldots$, with corresponding solutions being, up to multiplicative constant,

$$\psi_k(x) = \begin{cases} 1 & k = 0 \\ \sin kx \text{ or } \cos kx & k = 1, 2, \ldots \end{cases} \tag{1.3.106}$$

If one is willing to use the complex valued solutions, we could replace $\sin kx, \cos kx$ by $e^{\pm ikx}$ for $k = 1, 2, \ldots$.

With λ determined we must next solve the corresponding R equation,

$$r^2R'' + rR' - k^2R = 0 \tag{1.3.107}$$

which is of the Cauchy-Euler type (1.1.19). The general solution is

$$R(r) = \begin{cases} c_1 + c_2 \log r & k = 0 \\ c_1 r^k + c_2 r^{-k} & k = 1, 2, \ldots \end{cases} \tag{1.3.108}$$

and here we encounter the second hidden condition: the solution R should not be singular at the origin, since otherwise the PDE would not be satisfied throughout the unit disk. Thus we should choose $c_2 = 0$ in each case, leaving $R(r) = r^k, k = 0, 1, \ldots$.

Summarizing, we have found all possible product solutions $R(r)\Theta(\theta)$ of Eq. (1.3.96), and they are

$$1, r^k \sin k\theta, r^k \cos k\theta \quad k = 1, 2, \ldots \tag{1.3.109}$$

up to constant multiples. Any sum of such terms is also a solution of Eq. (1.3.96), so we may seek a solution of Eqs. (1.3.96), (1.3.97) in the form

$$u(r,\theta) = a_0 + \sum_{k=1}^{\infty} a_k r^k \cos k\theta + b_k r^k \sin k\theta \tag{1.3.110}$$

The coefficients must then be determined from the requirement that

$$f(\theta) = a_0 + \sum_{k=1}^{\infty} a_k \cos k\theta + b_k \sin k\theta \tag{1.3.111}$$

This is another problem in the theory of Fourier series, which is very similar to that associated with Eq. (1.3.92), and again will be studied in detail in Chapter 7. Exact formulas for the coefficients in terms of f may be given, as in Eq. (1.3.95), see Exercise 1.19.

1.3.4 Standard Problems and Side Conditions

Let us now formulate a number of typical PDE problems which will recur throughout this book, and which are for the most part variants of the model problems discussed in the previous section. Let Ω be some domain in $\mathbb{R}^N$ and let $\partial\Omega$ denote the boundary of Ω. For any sufficiently differentiable function u, the Laplacian of u is defined to be

$$\Delta u = \sum_{k=1}^{N} \frac{\partial^2 u}{\partial x_k^2} \tag{1.3.112}$$

- The PDE

$$\Delta u = h \quad x \in \Omega \tag{1.3.113}$$

is *Poisson's equation*, or *Laplace's equation* in the special case that $h = 0$. It is regarded as being of elliptic type, by analogy with the $N = 2$ case discussed in the previous section, or on account of a more general definition of ellipticity which will be given in Chapter 8. The most common type of side conditions associated with this PDE are

  - Dirichlet, or first kind, boundary conditions

$$u(x) = g(x) \quad x \in \partial\Omega \tag{1.3.114}$$

  - Neumann, or second kind, boundary conditions

$$\frac{\partial u}{\partial n}(x) = g(x) \quad x \in \partial\Omega \tag{1.3.115}$$

  where $\frac{\partial u}{\partial n}(x) = (\nabla u \cdot n)(x)$ is the directional derivative in the direction of the outward normal direction $n(x)$ for $x \in \partial\Omega$.

  - Robin, or third kind, boundary conditions

$$\frac{\partial u}{\partial n}(x) + \sigma(x)u(x) = g(x) \quad x \in \partial\Omega \tag{1.3.116}$$

  for some given function σ.

- The PDE

$$\Delta u + \lambda u = h \quad x \in \Omega \tag{1.3.117}$$

where λ is some constant, is the *Helmholtz equation*, also of elliptic type. The three types of boundary condition mentioned for the Poisson equation may also be imposed in this case.

- The PDE

$$u_t = \Delta u \qquad x \in \Omega \quad t > 0 \tag{1.3.118}$$

is the *heat equation* and is of parabolic type. Here $u = u(x,t)$, where x is regarded as a spatial variable and t a time variable. By convention, the Laplacian acts only with respect to the N spatial variables $x_1, \dots, x_N$. Appropriate side conditions for determining a solution of the heat equation are an initial condition

$$u(x,0) = f(x) \quad x \in \Omega \tag{1.3.119}$$

and boundary conditions of the Dirichlet, Neumann, or Robin type mentioned above. The only needed modification is that the functions involved may be allowed to depend on t, for example, the Dirichlet boundary condition for the heat equation is

$$u(x,t) = g(x,t) \qquad x \in \partial\Omega \quad t > 0 \tag{1.3.120}$$

and similarly for the other two types.

- The PDE

$$u_{tt} = \Delta u \qquad x \in \Omega \quad t > 0 \tag{1.3.121}$$

is the *wave equation* and is of hyperbolic type. Since it is second order in t it is natural that there be two initial conditions, usually given as

$$u(x,0) = f(x) \quad u_t(x,0) = g(x) \qquad x \in \Omega \tag{1.3.122}$$

Suitable boundary conditions for the wave equation are precisely the same as for the heat equation.

- Finally, the PDE

$$iu_t = \Delta u \qquad x \in \mathbb{R}^N \quad t > 0 \tag{1.3.123}$$

is the Schrödinger equation. Even when $N = 1$ it does not fall under the classification scheme of Section 1.3.2 because of the complex coefficient $i = \sqrt{-1}$. It is nevertheless one of the fundamental PDEs of mathematical physics, and we will have some things to say about it in later chapters. The spatial domain here is taken to be all of $\mathbb{R}^N$ rather than a subset Ω because this is by far the most common situation and the only one which will arise in this book. Since there is no spatial boundary, the only needed side condition is an initial condition for u, $u(x,0) = f(x)$, as in the heat equation case.

1.4 WELL-POSED AND ILL-POSED PROBLEMS

All of the PDEs and associated side conditions discussed in the previous section turn out to be natural, in the sense that they lead to what are called *well-posed problems*, a somewhat imprecise concept which we explain next. Roughly speaking a problem is well-posed if

- A solution exists.
- The solution is unique.
- The solution depends continuously on the data.

Here by "data" we mean any of the ingredients of the problem which we might imagine being changed, to obtain another problem of the same general type. For example, in the Dirichlet problem for the Poisson equation

$$\Delta u = f \quad x \in \Omega \qquad u = 0 \quad x \in \partial\Omega \tag{1.4.124}$$

the term $f = f(x)$ would be regarded as the given data. The idea of continuous dependence is that if a "small" change is made in the data, then the resulting solution should also undergo only a small change. For such a notion to be made precise, it is necessary to have some specific idea in mind of how we would measure the magnitude of a change in f. As we shall see, there may be many natural ways to do so, and no precise statement about well-posedness can be given until such choices are made. In fact, even the existence and uniqueness requirements, which may seem more clear cut, may also turn out to require much clarification in terms of what the exact meaning of "solution" is.

A problem which is not well-posed is called *ill-posed*. A classical problem in which ill-posedness can be easily recognized is Hadamard's example:

$$u_{xx} + u_{yy} = 0 \qquad -\infty < x < \infty \quad y > 0 \tag{1.4.125}$$

$$u(x,0) = 0 \quad u_y(x,0) = g(x) \qquad -\infty < x < \infty \tag{1.4.126}$$

Note that it is *not* of one of the standard types mentioned above.

If $g(x) = \alpha \sin kx$ for some $\alpha, k > 0$ then a corresponding solution is

$$u(x,y) = \alpha \frac{\sin kx}{k} e^{ky} \tag{1.4.127}$$

This is known to be the unique solution, but notice that a change in α (i.e., of the data g) of size ϵ implies a corresponding change in the solution for, say, $y = 1$ of size ϵe^k. Since k can be arbitrarily large, it follows that the problem is ill-posed, that is, small changes in the data do not necessarily lead to small changes in the solution.

Note that in this example if we change the PDE from $u_{xx} + u_{yy} = 0$ to $u_{xx} - u_{yy} = 0$ then (aside from the name of a variable) we have precisely the problem (1.3.76), (1.3.78), which from the explicit solution (1.3.81) may be seen to be well-posed under any reasonable interpretation. This serves to emphasize that some care must be taken in recognizing what are the "correct" side conditions for a given PDE. Other interesting examples of ill-posed problems are given in Exercises 1.23 and 1.26, see also [26].

1.5 EXERCISES

1.1. Find a fundamental set and the general solution of $u''' + u'' + u' = 0$.

1.2. Let $L = aD^2 + bD + c$ $(a \neq 0)$ be a constant coefficient second order linear differential operator, and let $p(\lambda) = a\lambda^2 + b\lambda + c$ be the associated characteristic polynomial. If λ_1, λ_2 are the roots of p, show that we can express the operator L as $L = a(D - \lambda_1)(D - \lambda_2)$. Use this factorization to obtain the general solution of $Lu = 0$ in the case of repeated roots, $\lambda_1 = \lambda_2$.

1.3. Show that the solution of the IVP $y' = \sqrt[3]{y}$, $y(0) = 0$ is not unique. (Hint: $y(t) = 0$ is one solution, find another one.) Why doesn't this contradict the assertion in Theorem 1.1 about unique solvability of the IVP?

1.4. Solve the IVP for the Cauchy-Euler equation

$$(t+1)^2 u'' + 4(t+1)u' - 10u = 0 \qquad u(1) = 2 \quad u'(1) = -1$$

1.5. Consider the integral equation

$$\lambda u(x) - \int_0^1 K(x,y)u(y)\,dy = g(x)$$

for the kernel

$$K(x,y) = \frac{x^2}{1+y^3}$$

(a) For what values of $\lambda \in \mathbb{C}$ does there exist a unique solution for any function g which is continuous on $[0,1]$?

(b) Find the solution set of the equation for all $\lambda \in \mathbb{C}$ and continuous functions g. (Hint: For $\lambda \neq 0$ any solution must have the form $u(x) = \frac{g(x)}{\lambda} + Cx^2$ for some constant C.)

1.6. Find a kernel $K(x,y)$ such that $u(x) = \int_0^1 K(x,y)f(y)\,dy$ is the unique solution of

$$u'' + u = f(x) \qquad u(0) = u'(0) = 0$$

(Hint: Review the variation of parameters method in any undergraduate ODE textbook.)

1.7. If $f \in C([0,1])$,

$$K(x,y) = \begin{cases} y(x-1) & 0 < y < x < 1 \\ x(y-1) & 0 < x < y < 1 \end{cases}$$

and

$$u(x) = \int_0^1 K(x,y)f(y)\,dy$$

show that

$$u'' = f \quad 0 < x < 1 \qquad u(0) = u(1) = 0$$

1.8. For each of the integral operators in Eqs. (1.2.26) and (1.2.32)–(1.2.35), discuss the classification(s) of the corresponding kernel, according to Definition 1.1.

1.9. Find the general solution of $(1+x^2)u_x + u_y = 0$. Sketch some of the characteristic curves.

1.10. The general solution in Example 1.10 was found by solving the corresponding Cauchy problem with Γ being the x axis. But the general solution should not actually depend on any specific choice of Γ. Show that the same general solution is found if instead we take Γ to be the line $x = 1$.

1.11. Find the solution of

$$yu_x + xu_y = 1 \qquad u(0,y) = e^{-y^2}$$

Discuss why the solution you find is only valid for $|y| \geq |x|$.

1.12. The method of characteristics developed in Section 1.3.1 for the linear PDE (1.3.45) can be easily extended to the so-called *semilinear* equation

$$a(x,y)u_x + b(x,y)u_y = c(x,y,u) \tag{1.5.128}$$

We simply replace Eq. (1.3.47) by

$$\frac{d}{dt}u(x(t),y(t)) = c(x(t),y(t),u(x(t),y(t))) \tag{1.5.129}$$

which is still an ODE along a characteristic. With this in mind, solve

$$u_x + xu_y + u^2 = 0 \qquad u(0,y) = \frac{1}{y} \quad y > 0 \tag{1.5.130}$$

1.13. Find the general solution of $u_{xx} - 4u_{xy} + 3u_{yy} = 0$.

1.14. Find the regions of the xy plane where the PDE

$$yu_{xx} - 2u_{xy} + xu_{yy} - 3u_x + u = 0$$

is elliptic, parabolic, and hyperbolic.

1.15. Find a solution formula for the half line wave equation problem

$$u_{tt} - c^2u_{xx} = 0 \qquad x > 0 \quad t > 0 \tag{1.5.131}$$

$$u(0,t) = h(t) \quad t > 0 \tag{1.5.132}$$

$$u(x,0) = f(x) \quad x > 0 \tag{1.5.133}$$

$$u_t(x,0) = g(x) \quad x > 0 \tag{1.5.134}$$

Note where the solution coincides with Eq. (1.3.81) and explain why this should be expected.

1.16. Complete the details of verifying Eq. (1.3.102).

1.17. If u is a twice differentiable function on $\mathbb{R}^N$ depending only on $r = |x|$, show that

$$\Delta u = u_{rr} + \frac{N-1}{r}u_r$$

(Spherical coordinates in $\mathbb{R}^N$ are reviewed in Section A.4, but the details of the angular variables are not needed for this calculation. Start by showing that $\frac{\partial u}{\partial x_j} = u'(r)\frac{x_j}{r}$.)

1.18. Verify in detail that there are no nontrivial solutions of Eq. (1.3.88) for nonreal $\lambda \in \mathbb{C}$.

1.19. Assuming that Eq. (1.3.111) is valid, find the coefficients a_k, b_k in terms of f. (Hint: multiply the equation by $\sin m\theta$ or $\cos m\theta$ and integrate from 0 to 2π.)

1.20. In the two-dimensional case, solutions of Laplace's equation $\Delta u = 0$ may also be found by means of analytic function theory. Recall that if $z = x + iy$ then a function $f(z)$ is analytic in an open set Ω if $f'(z)$ exists at every point of Ω. If we think of $f = u + iv$ and u, v as functions of x, y

then $u = u(x, y)$, $v = v(x, y)$ must satisfy the Cauchy-Riemann equations $u_x = v_y, u_y = -v_x$. Show in this case that u, v are also solutions of Laplace's equation. Find u, v if $f(z) = z^3$ and $f(z) = e^z$.

1.21. Find all of the product solutions $u(x, t) = \phi(t)\psi(x)$ that you can which satisfy the damped wave equation

$$u_{tt} + \alpha u_t = u_{xx} \qquad 0 < x < \pi \quad t > 0$$

and the boundary conditions

$$u(0, t) = u_x(\pi, t) = 0 \qquad t > 0$$

Here $\alpha > 0$ is the damping constant. What is the significance of the condition $\alpha < 1$?

1.22. Show that any solution of the wave equation $u_{tt} - u_{xx} = 0$ has the "four point property"

$$u(x, t) + u(x + h - k, t + h + k) = u(x + h, t + h) + u(x - k, t + k)$$

for any h, k. (Suggestion: Use d'Alembert's formula.)

1.23. In the Dirichlet problem for the wave equation

$$\begin{aligned} u_{tt} - u_{xx} &= 0 \qquad 0 < x < 1 \quad 0 < t < 1 \\ u(0, t) = u(1, t) &= 0 \qquad 0 < t < 1 \\ u(x, 0) = 0 \quad u(x, 1) &= f(x) \qquad 0 < x < 1 \end{aligned}$$

show that neither existence nor uniqueness holds. (Hint: For the nonexistence part, use Exercise 1.22 to find an f for which no solution exists.)

1.24. Let Ω be the rectangle $[0, a] \times [0, b]$ in $\mathbb{R}^2$. Find all possible product solutions

$$u(x, y, t) = \phi(t)\psi(x)\zeta(y)$$

satisfying

$$\begin{aligned} u_t - \Delta u &= 0 \qquad (x, y) \in \Omega \quad t > 0 \\ u(x, y, t) &= 0 \qquad (x, y) \in \partial\Omega \quad t > 0 \end{aligned}$$

1.25. Find a solution of the Dirichlet problem for $u = u(x, y)$ in the unit disc $\Omega = \{(x, y) : x^2 + y^2 < 1\}$,

$$\Delta u = 1 \quad (x, y) \in \Omega \qquad u(x, y) = 0 \quad (x, y) \in \partial\Omega$$

(Suggestion: look for a solution in the form $u = u(r)$ and recall Eq. 1.3.102.)

1.26. The problem

$$u_t = u_{xx} \qquad 0 < x < 1 \quad t < T \tag{1.5.135}$$

$$u(0, t) = u(1, t) = 0 \qquad t > 0 \tag{1.5.136}$$

$$u(x, T) = f(x) \qquad 0 < x < 1 \tag{1.5.137}$$

is sometimes called a *final value problem* for the heat equation.

(a) Show that this problem is ill-posed.

(b) Show that this problem is equivalent to Eqs. (1.3.82)–(1.3.84) except with the heat equation (1.3.82) replaced by the *backward heat equation* $u_t = -u_{xx}$.

Chapter 2

Vector Spaces

We will be working frequently with function spaces which are themselves special cases of more abstract spaces. Most such spaces which are of interest to us have both *linear structure* and *metric structure*. This means that given any two elements of the space it is meaningful to speak of (i) a linear combination of the elements, and (ii) the distance between the two elements. These two kinds of concepts are abstracted in the definitions of vector space and metric space. In this chapter we focus on the first of these aspects.

2.1 AXIOMS OF A VECTOR SPACE

Definition 2.1. A vector space is a set $\mathbf{X}$ such that whenever $x, y \in \mathbf{X}$ and λ is a scalar we have $x+y \in \mathbf{X}$ and $\lambda x \in \mathbf{X}$, and for which the following axioms hold.

[V1] $x+y=y+x$ for all $x, y \in \mathbf{X}$
[V2] $(x+y)+z=x+(y+z)$ for all $x, y, z \in \mathbf{X}$
[V3] There exists an element $0 \in \mathbf{X}$ such that $x+0=x$ for all $x \in \mathbf{X}$
[V4] For every $x \in \mathbf{X}$ there exists an element $-x \in \mathbf{X}$ such that $x+(-x)=0$
[V5] $\lambda(x+y)=\lambda x+\lambda y$ for all $x, y \in \mathbf{X}$ and any scalar λ
[V6] $(\lambda+\mu)x=\lambda x+\mu x$ for any $x \in \mathbf{X}$ and any scalars λ, μ
[V7] $\lambda(\mu x)=(\lambda\mu)x$ for any $x \in \mathbf{X}$ and any scalars λ, μ
[V8] $1x=x$ for any $x \in \mathbf{X}$

Here the field of scalars may be either the real numbers $\mathbb{R}$ or the complex numbers $\mathbb{C}$, and we may refer to $\mathbf{X}$ as a real or complex vector space accordingly, if a distinction needs to be made.

By an obvious induction argument, if $x_1, \dots, x_m \in \mathbf{X}$ and $\lambda_1, \dots, \lambda_m$ are scalars, then the linear combination $\sum_{j=1}^m \lambda_j x_j$ must also be an element of $\mathbf{X}$.

Example 2.1. Ordinary N-dimensional Euclidean space

$$\mathbb{R}^N := \{x=(x_1, x_2, \dots, x_N) : x_j \in \mathbb{R}\}$$

is a real vector space with the usual operations of vector addition and scalar multiplication,

Techniques of Functional Analysis for Differential and Integral Equations
http://dx.doi.org/10.1016/B978-0-12-811426-1.00002-7

$$(x_1, x_2, \ldots, x_N) + (y_1, y_2, \ldots, y_N) = (x_1 + y_1, x_2 + y_2, \ldots, x_N + y_N)$$
$$\lambda(x_1, x_2, \ldots, x_N) = (\lambda x_1, \lambda x_2, \ldots, \lambda x_N) \quad \lambda \in \mathbb{R}$$

If we allow the components x_j as well as the scalars λ to be complex, we obtain instead the complex vector space $\mathbb{C}^N$. □

Example 2.2. If $E \subset \mathbb{R}^N$, let

$$C(E) = \{f : E \to \mathbb{R} : f \text{ is continuous at } x \text{ for every } x \in E\}$$

denote the set of real valued continuous functions on E. Clearly $C(E)$ is a real vector space with the ordinary operations of function addition and scalar multiplication

$$(f + g)(x) = f(x) + g(x) \qquad (\lambda f)(x) = \lambda f(x) \quad \lambda \in \mathbb{R}$$

If we allow the range space in the definition of $C(E)$ to be $\mathbb{C}$ then $C(E)$ becomes a complex vector space.

Spaces of differentiable functions likewise may be naturally regarded as vector spaces, for example,

$$C^m(E) = \{f : D^\alpha f \in C(E),\ |\alpha| \le m\}$$

and

$$C^\infty(E) = \{f : D^\alpha f \in C(E),\ \text{for all } \alpha\}$$

□

Example 2.3. If $0 < p < \infty$ and E is a measurable subset of $\mathbb{R}^N$, the space $L^p(E)$ is defined to be the set of measurable functions $f : E \to \mathbb{R}$ or $f : E \to \mathbb{C}$ such that

$$\int_E |f(x)|^p \, dx < \infty \tag{2.1.1}$$

Here the integral is defined in the Lebesgue sense. The reader unfamiliar with measure theory and Lebesgue integration should consult a standard textbook such as [30, 32], or see a brief summary in Section A.1.

We may now verify that $L^p(E)$ is vector space for any $0 < p < \infty$. To see this we use the known fact that if f, g are measurable then so are $f + g$ and λf for any scalar λ, and the numerical inequality $(a + b)^p \le C_p(a^p + b^p)$ holds for $a, b \ge 0$, where $C_p = \max(2^{p-1}, 1)$. It follows from these facts that $f + g \in L^p(E)$ whenever $f, g \in L^p(E)$ and checking the remaining axioms is routine.

The related space $L^\infty(E)$ is defined as the set of measurable functions f for which

$$\text{ess sup}_{x \in E} |f(x)| < \infty \tag{2.1.2}$$

Here $M = \text{ess sup}_{x \in E} |f(x)|$ (the *essential supremum of* $|f|$) if $|f(x)| \le M$ a.e. and there is no smaller constant with this property. We leave the verification of the vector space axioms as an exercise. □

Definition 2.2. If $\mathbf{X}$ is a vector space, a subset $M \subset \mathbf{X}$ is a *subspace* of $\mathbf{X}$ if

(i) $x + y \in M$ whenever $x, y \in M$
(ii) $\lambda x \in M$ whenever $x \in M$ and λ is a scalar

That is to say, a subspace is a subset of $\mathbf{X}$ which is closed under formation of linear combinations. Clearly a subspace of a vector space is itself a vector space.

Example 2.4. The subset $M = \{x \in \mathbb{R}^N : x_j = 0\}$ is a subspace of $\mathbb{R}^N$ for any fixed j. □

Example 2.5. If $E \subset \mathbb{R}^N$ then $C^\infty(E)$ is a subspace of $C^m(E)$ for any m, which in turn is a subspace of $C(E)$. □

Example 2.6. If $\mathbf{X}$ is any vector space and $S \subset \mathbf{X}$, then span(S), the span of S, is the set of all finite linear combinations of elements of S. That is to say, $x \in$ span(S) if there exist scalars $\lambda_1, \lambda_2, \dots \lambda_m$ and elements $x_1, \dots x_m \in S$ such that

$$x = \sum_{j=1}^{m} \lambda_j x_j \tag{2.1.3}$$

It is obviously a subspace of $\mathbf{X}$, also sometimes referred to as the subspace generated by S, or the linear envelope of S. □

Example 2.7. If we take $\mathbf{X} = C([a,b])$ and $f_j(x) = x^{j-1}$ for $j = 1, 2, \dots$ then the subspace generated by $\{f_j\}_{j=1}^{N+1}$ is $\mathcal{P}_N$, the vector space of polynomials of degree less than or equal to N. Likewise, the subspace generated by $\{f_j\}_{j=1}^{\infty}$ is $\mathcal{P}$, the vector space of all polynomials. □

2.2 LINEAR INDEPENDENCE AND BASES

Definition 2.3. We say that $S \subset \mathbf{X}$ is *linearly independent* if whenever $x_1, \dots, x_m \in S$, $\lambda_1, \dots, \lambda_m$ are scalars and $\sum_{j=1}^m \lambda_j x_j = 0$ then $\lambda_1 = \lambda_2 = \dots = \lambda_m = 0$. Otherwise S is linearly dependent.

Equivalently, S is linearly dependent if it is possible to express at least one of its elements as a linear combination of the remaining ones. In particular any set containing the zero element is linearly dependent.

Definition 2.4. We say that $S \subset \mathbf{X}$ is a *basis* of $\mathbf{X}$ if for any $x \in \mathbf{X}$ there exists unique scalars $\lambda_1, \lambda_2, \dots, \lambda_m$ and elements $x_1, \dots, x_m \in S$ such that $x = \sum_{j=1}^m \lambda_j x_j$.

The following characterization of a basis is then immediate:

Proposition 2.1. *$S \subset X$ is a basis of X if and only if S is linearly independent and* span$(S) = X$.

It is important to emphasize that in this definition of basis it is required that every $x \in \mathbf{X}$ be expressible as a *finite* linear combination of the basis elements.

This notion of basis will be inadequate for later purposes, and will be replaced by one which allows infinite sums, but this cannot be done until a meaning of convergence is available. The notion of basis in Definition 2.4 is called a *Hamel basis* if a distinction is necessary.

Definition 2.5. We say that dim (**X**), the dimension of **X**, is m if there exist m linearly independent vectors in **X** but any collection of $m+1$ elements of **X** is linearly dependent. If there exists m linearly independent vectors for any positive integer m, then we say dim $(\mathbf{X}) = \infty$.

Proposition 2.2. *The elements* $\{x_1, x_2, \dots, x_m\}$ *form a basis for* span $(\{x_1, x_2, \dots, x_m\})$ *if and only if they are linearly independent.*

Proposition 2.3. *The dimension of* **X** *is the number of vectors in any basis of* **X**.

The proof of both of these propositions is left for the exercises.

Example 2.8. $\mathbb{R}^N$ or $\mathbb{C}^N$ has dimension N. We will denote by e_j the standard unit vector with a one in the jth position and zero elsewhere. Then $\{e_1, e_2, \dots, e_N\}$ will be referred to as the standard basis for either $\mathbb{R}^N$ or $\mathbb{C}^N$. □

Example 2.9. In the vector space $C([a,b])$ the elements $f_j(t) = t^{j-1}$ are linearly independent (see Exercise 2.6), so that the dimension is ∞, as is the dimension of the subspace $\mathcal{P}$. Also evidently the subspace $\mathcal{P}_N$ is of dimension $N+1$. □

Example 2.10. The set of solutions of the ordinary differential equation $u'' + u = 0$ is precisely the set of linear combinations $u(t) = \lambda_1 \sin t + \lambda_2 \cos t$. Since $\sin t, \cos t$ are linearly independent functions, they form a basis for this two-dimensional space. □

The following is an interesting result, although not of great practical significance for us. Its proof, which is not at all obvious in the infinite dimensional case, relies on the Axiom of Choice and will not be given here.

Theorem 2.1. *Every vector space has a basis.*

2.3 LINEAR TRANSFORMATIONS OF A VECTOR SPACE

If **X** and **Y** are vector spaces, a mapping $T : \mathbf{X} \longmapsto \mathbf{Y}$ is called *linear* if

$$T(\lambda_1 x_1 + \lambda_2 x_2) = \lambda_1 T(x_1) + \lambda_2 T(x_2) \tag{2.3.4}$$

for all $x_1, x_2 \in \mathbf{X}$ and all scalars λ_1, λ_2. Such a linear transformation is uniquely determined on all of **X** by its action on any basis of **X**, that is, if $S = \{x_\alpha\}_{\alpha \in \mathcal{A}}$ is a basis of **X** and $y_\alpha = T(x_\alpha)$, then for any $x = \sum_{j=1}^m \lambda_j x_{\alpha_j}$ we have $T(x) = \sum_{j=1}^m \lambda_j y_{\alpha_j}$.

For a linear mapping it is common to omit parentheses and write Tx instead of $T(x)$, and we will always do so if it does not cause any confusion.

If T is such a linear mapping and $\mathbf{X}$ and $\mathbf{Y}$ are both of finite dimension, let us choose bases $\{x_1, x_2, \ldots, x_m\}$, $\{y_1, y_2, \ldots, y_n\}$ of $\mathbf{X}, \mathbf{Y}$, respectively. For $1 \le j \le m$ there must exist unique scalars a_{kj} such that $Tx_j = \sum_{k=1}^n a_{kj} y_k$ and it follows that

$$x = \sum_{j=1}^m \lambda_j x_j \Rightarrow Tx = \sum_{k=1}^n \mu_k y_k \qquad \text{where } \mu_k = \sum_{j=1}^m a_{kj} \lambda_j \tag{2.3.5}$$

For a given basis $\{x_1, x_2, \ldots, x_m\}$ of $\mathbf{X}$, if $x = \sum_{j=1}^m \lambda_j x_j$ we say that $\lambda_1, \lambda_2, \ldots, \lambda_m$ are the coordinates of x with respect to the given basis. The $n \times m$ matrix $A = [a_{kj}]$ thus maps the coordinates of x with respect to the basis $\{x_1, x_2, \ldots, x_m\}$ to the coordinates of Tx with respect to the basis $\{y_1, y_2, \ldots, y_n\}$, and therefore encodes all information about the linear mapping T.

If $T : \mathbf{X} \longmapsto \mathbf{Y}$ is linear, one-to-one and onto then we say T is an *isomorphism* between $\mathbf{X}$ and $\mathbf{Y}$, and the vector spaces $\mathbf{X}$ and $\mathbf{Y}$ are isomorphic whenever there exists an isomorphism between them. If T is such an isomorphism, and S is a basis of $\mathbf{X}$ then it is easy to check that the image set $T(S)$ is a basis of $\mathbf{Y}$. In particular, any two isomorphic vector spaces have the same finite dimension or are both infinite dimensional.

For any linear mapping $T : \mathbf{X} \to \mathbf{Y}$ we define the kernel, or null space, of T as

$$N(T) = \{x \in \mathbf{X} : Tx = 0\} \tag{2.3.6}$$

and the range of T as

$$R(T) = \{y \in \mathbf{Y} : y = Tx \text{ for some } x \in \mathbf{X}\} \tag{2.3.7}$$

It is immediate that $N(T)$ and $R(T)$ are subspaces of $\mathbf{X}, \mathbf{Y}$, respectively, and T is an isomorphism precisely if $N(T) = \{0\}$ and $R(T) = \mathbf{Y}$. If $\mathbf{X} = \mathbf{Y} = \mathbb{R}^N$ or $\mathbb{C}^N$, we learn in linear algebra that these two conditions are equivalent, but this is false in general.

2.4 EXERCISES

2.1. Using only the vector space axioms, show that the zero element in [V3] is unique.

2.2. Prove Propositions 2.2 and 2.3.

2.3. Show that the intersection of any family of subspaces of a vector space is also a subspace. Is the same true for the union of subspaces?

2.4. Show that $\mathcal{M}_{m \times n}$, the set of $m \times n$ matrices, with the usual definitions of addition and scalar multiplication, is a vector space of dimension mn. Show that the subset of symmetric matrices $n \times n$ matrices forms a subspace of $\mathcal{M}_{n \times n}$. What is its dimension?

2.5. Under what conditions on a measurable set $E \subset \mathbb{R}^N$ and $p \in (0, \infty]$ will it be true that $C(E)$ is a subspace of $L^p(E)$? Under what conditions is $L^p(E)$ a subspace of $L^q(E)$?

2.6. Let $u_j(t) = t^{\lambda_j}$ where $\lambda_1, \dots, \lambda_n$ are arbitrary unequal real numbers. Show that $\{u_1 \dots u_n\}$ are linearly independent functions on any interval $(a, b) \subset \mathbb{R}$. (Suggestion: If $\sum_{j=1}^{n} \alpha_j t^{\lambda_j} \equiv 0$, divide by t^{λ_1} and differentiate.)

2.7. A side condition for a differential equation is homogeneous if whenever two functions satisfy the side condition then so does any linear combination of the two functions. For example, the Dirichlet type boundary condition $u = 0$ for $x \in \partial\Omega$ is homogeneous. Now let $Lu = \sum_{|\alpha| \le m} a_\alpha(x) D^\alpha u$ denote any linear differential operator. Show that the set of functions satisfying $Lu = 0$ and any homogeneous side conditions is a vector space.

2.8. Consider the differential equation $u'' + u = 0$ on the interval $(0, \pi)$. What is the dimension of the vector space of solutions which satisfy the homogeneous boundary conditions (a) $u(0) = u(\pi)$, and (b) $u(0) = u(\pi) = 0$. Repeat the question if the interval $(0, \pi)$ is replaced by $(0, 1)$ and $(0, 2\pi)$.

2.9. Let $Df = f'$ for any differentiable function f on $\mathbb{R}$. For any $N \ge 0$ show that $D : \mathcal{P}_N \to \mathcal{P}_N$ is linear and find its null space and range.

2.10. If $\mathbf{X}$ and $\mathbf{Y}$ are vector spaces, then the Cartesian product of $\mathbf{X}$ and $\mathbf{Y}$, is defined as the set of ordered pairs

$$\mathbf{X} \times \mathbf{Y} = \{(x, y) : x \in \mathbf{X}, y \in \mathbf{Y}\} \tag{2.4.8}$$

Addition and scalar multiplication on $\mathbf{X} \times \mathbf{Y}$ are defined in the natural way,

$$(x, y) + (\widehat{x}, \widehat{y}) = (x + \widehat{x}, y + \widehat{y}) \qquad \lambda(x, y) = (\lambda x, \lambda y) \tag{2.4.9}$$

(a) Show that $\mathbf{X} \times \mathbf{Y}$ is a vector space.
(b) Show that $\mathbb{R} \times \mathbb{R}$ is isomorphic to $\mathbb{R}^2$.

2.11. If $\mathbf{X}, \mathbf{Y}$ are vector spaces of the same finite dimension, show $\mathbf{X}$ and $\mathbf{Y}$ are isomorphic.

2.12. Show that $L^p(0, 1)$ and $L^p(a, b)$ are isomorphic, for any $a, b \in \mathbb{R}$ and $p \in (0, \infty]$.

Chapter 3

Metric Spaces

3.1 AXIOMS OF A METRIC SPACE

The idea of a metric space is that of a set on which some natural notion of distance may be defined. The formal definition is as follows.

Definition 3.1. A metric space is a pair (X, d) where X is a set and d is a real valued mapping on $X \times X$, such that the following axioms hold.

[M1] $d(x, y) \geq 0$ for all $x, y \in X$
[M2] $d(x, y) = 0$ if and only if $x = y$
[M3] $d(x, y) = d(y, x)$ for all $x, y \in X$
[M4] $d(x, y) \leq d(x, z) + d(z, y)$ for all $x, y, z \in X$.

Here d is the metric on X, that is, $d(x, y)$ is regarded as the distance from x to y. Axiom [M4] is known as the triangle inequality. Although strictly speaking the metric space is the pair (X, d) it is a common practice to refer to X itself as being the metric space, if the metric d is understood from context. But as we will see in examples it is often possible to assign different metrics to the same set X.

If (X, d) is a metric space and $Y \subset X$ then it is clear that (Y, d) is also a metric space, and in this case we say that Y inherits the metric of X.

Example 3.1. If $X = \mathbb{R}^N$ then there are many choices of d for which $(\mathbb{R}^N, d)$ is a metric space. The most familiar is the ordinary Euclidean distance

$$d(x, y) = \left(\sum_{j=1}^{N} |x_j - y_j|^2 \right)^{1/2} \tag{3.1.1}$$

In general we may define

$$d_p(x, y) = \left(\sum_{j=1}^{N} |x_j - y_j|^p \right)^{1/p} \qquad 1 \leq p < \infty \tag{3.1.2}$$

and

$$d_\infty(x, y) = \max\left(|x_1 - y_1|, |x_2 - y_2|, \ldots, |x_n - y_n|\right) \tag{3.1.3}$$

Techniques of Functional Analysis for Differential and Integral Equations
http://dx.doi.org/10.1016/B978-0-12-811426-1.00003-9

The verification that $(\mathbb{R}^n, d_p)$ is a metric space for $1 \le p \le \infty$ is left to the exercises—the triangle inequality is the only nontrivial step. The same family of metrics may be used with $X = \mathbb{C}^N$. □

Example 3.2. To assign a metric to $C(E)$, the vector space of continuous functions on E, more specific assumptions must be made about E. If we assume, for example, that E is a closed and bounded[1] subset of $\mathbb{R}^N$ we may set

$$d_\infty(f,g) = \max_{x \in E} |f(x) - g(x)| \tag{3.1.4}$$

so that $d(f,g)$ is always finite by virtue of the well known theorem that a continuous function achieves its maximum on such a set. Other possibilities are

$$d_p(f,g) = \left(\int_E |f(x) - g(x)|^p \, dx \right)^{1/p} \qquad 1 \le p < \infty \tag{3.1.5}$$

Note the analogy with the definition of d_p in the case of $\mathbb{R}^N$ or $\mathbb{C}^N$.

For more arbitrary sets E there is in general no natural metric for $C(E)$. For example, if E is an open set, none of the metrics d_p can be used since there is no reason why $d_p(f,g)$ should be finite for $f, g \in C(E)$.

As in the case of vector spaces, some spaces of differentiable functions may also be made into metric spaces. For this we will assume a bit more about E, namely that E is the closure of a bounded open set $\mathcal{O} \subset \mathbb{R}^N$, and in this case will say that $D^\alpha f \in C(E)$ if the function $D^\alpha f$ defined in the usual pointwise sense on $\mathcal{O}$ has a continuous extension to E. We then can define

$$C^m(E) = \{f : D^\alpha f \in C(E) \text{ whenever } |\alpha| \le m\} \tag{3.1.6}$$

with metric

$$d(f,g) = \max_{|\alpha| \le m} \max_{x \in E} |D^\alpha (f - g)(x)| \tag{3.1.7}$$

which may be easily checked to satisfy [M1]–[M4].

We cannot define a metric on $C^\infty(E)$ in the obvious way just by letting $m \to \infty$ in the above definition, since there is no reason why the resulting maximum over m in Eq. (3.1.7) will be finite, even if $f \in C^m(E)$ for every m. See however Exercise 3.18. □

Example 3.3. Recall that if E is a measurable subset of $\mathbb{R}^N$, we have defined corresponding vector spaces $L^p(E)$ for $0 < p \le \infty$. To endow them with metric space structure let

$$d_p(f,g) = \left(\int_E |f(x) - g(x)|^p \, dx \right)^{1/p} \tag{3.1.8}$$

---

1. That is, E is compact in $\mathbb{R}^N$. Compactness is discussed in more detail below, and we avoid using the term until then.

for $1 \leq p < \infty$, and

$$d_\infty(f,g) = \text{ess sup}_{x\in E}|f(x) - g(x)| \tag{3.1.9}$$

(Recall the definition of ess sup is given just after Eq. 2.1.2.)

The validity of axioms [M1] and [M3] is clear, and the triangle inequality [M4] is an immediate consequence of the Minkowski inequality (A.2.23). But axiom [M2] does not appear to be satisfied here, since for example, two functions f, g agreeing except at a single point, or more generally agreeing except on a set of measure zero, would have $d_p(f,g) = 0$. It is necessary, therefore, to modify our point of view concerning $L^p(E)$ as follows. We define an equivalence relation $f \sim g$ if $f = g$ almost everywhere, that is, except on a set of measure zero. If $d_p(f,g) = 0$ we would be able to correctly conclude that $f \sim g$, in which case we will regard f and g as being the same element of $L^p(E)$. Thus strictly speaking, $L^p(E)$ is the set of equivalence classes of measurable functions, where the equivalence classes are defined by means of the above equivalence relation.

The distance $d_p([f],[g])$ between two equivalence classes $[f]$ and $[g]$ may be unambiguously determined by selecting a representative of each class and then evaluating the distance from Eq. (3.1.8) or (3.1.9). Likewise the vector space structure of $L^p(E)$ is maintained since, for example, we can define the sum of equivalence classes $[f] + [g]$ by selecting a representative of each class and observing that if $f_1 \sim f_2$ and $g_1 \sim g_2$ then $f_1 + g_1 \sim f_2 + g_2$. It is rarely necessary to make a careful distinction between a measurable function and the equivalence class it belongs to, and whenever it can cause no confusion we will follow the common practice of referring to members of $L^p(E)$ as functions rather than equivalence classes. The notation f may be used to stand for either a function or its equivalence class. An element $f \in L^p(E)$ will be said to be continuous if its equivalence class contains a continuous function, and in this way we can naturally regard $C(E) \cap L^p(E)$ as a subspace of $L^p(E)$. □

Although $L^p(E)$ is a vector space for $0 < p \leq \infty$, we cannot use the above definition of metric for $0 < p < 1$, since it turns out the triangle inequality is not satisfied (see Exercise 4.7 of Chapter 4) except in degenerate cases.

3.2 TOPOLOGICAL CONCEPTS

In a metric space, various concepts of point set topology may be introduced.

Definition 3.2. If (X, d) is a metric space then

1. $B(x,\epsilon) = \{y \in X : d(x,y) < \epsilon\}$ is the ball centered at x of radius ϵ.
2. A set $E \subset X$ is bounded if there exists some $x \in X$ and $R < \infty$ such that $E \subset B(x,R)$.
3. If $E \subset X$, then a point $x \in X$ is an interior point of E if there exists $\epsilon > 0$ such that $B(x,\epsilon) \subset E$.

4. If $E \subset X$, then a point $x \in X$ is a limit point of E if for any $\epsilon > 0$ there exists a point $y \in B(x, \epsilon) \cap E$, $y \neq x$.
5. A subset $E \subset X$ is open if every point of E is an interior point of E. By convention, the empty set is open.
6. A subset $E \subset X$ is closed if every limit point of E is in E.
7. The closure $\overline{E}$ of a set $E \subset X$ is the union of E and the limit points of E.
8. The interior E° of a set E is the set of all interior points of E.
9. A subset E is dense in X if $\overline{E} = X$.
10. X is separable if it contains a countable dense subset.
11. If $E \subset X$, we say that $x \in X$ is a boundary point of E if for any $\epsilon > 0$ the ball $B(x, \epsilon)$ contains at least one point of E and at least one point of the complement $E^c = \{x \in X : x \notin E\}$. The boundary of E is the set of boundary points, denoted ∂E.

The following proposition states a number of elementary but important properties. Proofs are essentially the same as in the more familiar special case when the metric space is a subset of $\mathbb{R}^N$, and will be left for the reader.

Proposition 3.1. *Let (X, d) be a metric space. Then*

1. *$B(x, \epsilon)$ is open for any $x \in X$ and $\epsilon > 0$.*
2. *$E \subset X$ is open if and only if its complement E^c is closed.*
3. *An arbitrary union or finite intersection of open sets is open.*
4. *An arbitrary intersection or finite union of closed sets is closed.*
5. *If $E \subset X$ then E° is the union of all open sets contained in E, E° is open, and E is open if and only if $E = E^\circ$.*
6. *$\overline{E}$ is the intersection of all closed sets containing E, $\overline{E}$ is closed, and E is closed if and only if $E = \overline{E}$.*
7. *If $E \subset X$ then $\partial E = \overline{E} \backslash E^\circ = \overline{E} \cap \overline{E^c}$.*

Next we study infinite sequences in (X, d).

Definition 3.3. We say that a sequence $\{x_n\}_{n=1}^\infty$ in X is convergent to x, that is, $\lim_{n\to\infty} x_n = x$, if for any $\epsilon > 0$ there exists $n_0 < \infty$ such that $d(x_n, x) < \epsilon$ whenever $n \geq n_0$.

Example 3.4. If $X = \mathbb{R}^N$ or $\mathbb{C}^N$, and d is any one of the metrics d_p defined in Example 3.1, then $x_n \to x$ if and only if each component sequence converges to the corresponding limit, that is, $x_{j,n} \to x_j$ as $n \to \infty$ in the ordinary sense of convergence in $\mathbb{R}$ or $\mathbb{C}$. (Here $x_{j,n}$ is the jth component of x_n.) □

Example 3.5. In the metric space $(C(E), d_\infty)$ of Example 3.2, $\lim_{n\to\infty} f_n = f$ is equivalent to the definition of uniform convergence on E. □

Definition 3.4. We say that a sequence $\{x_n\}_{n=1}^\infty$ in X is a *Cauchy sequence* if for any $\epsilon > 0$ there exists $n_0 < \infty$ such that $d(x_n, x_m) < \epsilon$ whenever $n, m \geq n_0$.

It is easy to see that a convergent sequence is always a Cauchy sequence, but the converse may be false.

Definition 3.5. A metric space X is said to be *complete* if every Cauchy sequence in X is convergent in X.

Example 3.6. Completeness is one of the fundamental properties of the real numbers $\mathbb{R}$, see, for example, Chapter 1 of [31], and which we take as a known result. If a sequence $\{x_n\}_{n=1}^{\infty}$ in $\mathbb{R}^N$ is Cauchy with respect to any of the metrics d_p, then each component sequence $\{x_{j,n}\}_{n=1}^{\infty}$ is a Cauchy sequence in $\mathbb{R}$, hence convergent in $\mathbb{R}$. It then follows immediately that $\{x_n\}_{n=1}^{\infty}$ is convergent in $\mathbb{R}^N$, again with any of the metrics d_p. The same conclusion holds for $\mathbb{C}^N$, so that $\mathbb{R}^N, \mathbb{C}^N$ are complete metric spaces, with respect to any one of these metrics. These spaces are also separable since the subset consisting of points with rational co-ordinates is countable and dense. A standard example of an incomplete metric space is the set of rational numbers with the metric inherited from $\mathbb{R}$. □

Most metric spaces used in this book, and indeed most metric spaces used in applied mathematics, are complete.

Proposition 3.2. *If $E \subset \mathbb{R}^N$ is closed and bounded, then the metric space $C(E)$ with metric $d = d_\infty$ is complete.*

Proof. Let $\{f_n\}_{n=1}^{\infty}$ be a Cauchy sequence in $C(E)$. If $\epsilon > 0$ we may then find n_0 such that

$$\max_{x \in E} |f_n(x) - f_m(x)| < \epsilon \tag{3.2.10}$$

whenever $n, m \geq n_0$. In particular the sequence of numbers $\{f_n(x)\}_{n=1}^{\infty}$ is Cauchy in $\mathbb{R}$ or $\mathbb{C}$ for each fixed $x \in E$, so we may define $f(x) := \lim_{n\to\infty} f_n(x)$. Letting $m \to \infty$ in Eq. (3.2.10) we obtain

$$|f_n(x) - f(x)| \leq \epsilon \qquad n \geq n_0 \quad x \in E \tag{3.2.11}$$

which means $d(f_n, f) \leq \epsilon$ for $n \geq n_0$. It remains to check that $f \in C(E)$. If we pick $x \in E$, then since $f_{n_0} \in C(E)$ there exists $\delta > 0$ such that $|f_{n_0}(x) - f_{n_0}(y)| < \epsilon$ if $|y - x| < \delta$. Thus for $|y - x| < \delta$ we have

$$|f(x) - f(y)| \leq |f(x) - f_{n_0}(x)| + |f_{n_0}(x) - f_{n_0}(y)| + |f_{n_0}(y) - f(y)| < 3\epsilon \tag{3.2.12}$$

Since ϵ is arbitrary, f is continuous at x, and since x is arbitrary $f \in C(E)$. Thus we have concluded that the Cauchy sequence $\{f_n\}_{n=1}^{\infty}$ is convergent in $C(E)$ to $f \in C(E)$, as needed. □

The final part of the previous proof should be recognized as the standard proof of the familiar fact that a uniform limit of continuous functions is continuous.

The spaces $C^m(E)$ can likewise be shown, again assuming that E is the closure of a bounded open set in $\mathbb{R}^N$, to be complete metric spaces with the metric defined in Eq. (3.1.7), see Exercise 3.19.

Example 3.7. If we were to choose the metric d_1 on $C(E)$ then the resulting metric space is not complete. To see this, choose, for example, $E = [-1, 1]$ and $f_n(x) = x^{1/(2n+1)}$ so that the pointwise limit of $f_n(x)$ is

$$f(x) = 1 \quad x > 0 \qquad f(x) = -1 \quad x < 0 \qquad f(0) = 0 \tag{3.2.13}$$

By a simple calculation

$$\int_{-1}^{1} |f_n(x) - f(x)| = \frac{1}{n+1} \tag{3.2.14}$$

so that $\{f_n\}_{n=1}^{\infty}$ must be Cauchy in $C(E)$ with metric d_1. On the other hand $\{f_n\}_{n=1}^{\infty}$ cannot be convergent in this space, since the only possible limit is f which does not belong to $C(E)$. □

The same example can be modified to show that $C(E)$ is not complete with any of the metrics d_p for $1 \le p < \infty$, and so d_∞ is in some sense the "natural" metric. For this reason $C(E)$ will always be assumed to supplied with the metric d_∞ unless otherwise stated.

We next summarize in the form of a theorem some especially important facts about the metric spaces $L^p(E)$, which may be found in any standard textbook on Lebesgue integration, for example, Chapter 3 of [32] or Chapter 8 of [40].

Theorem 3.1. *If $E \subset \mathbb{R}^N$ is measurable, then*

1. *$L^p(E)$ is complete for $1 \le p \le \infty$.*
2. *$L^p(E)$ is separable for $1 \le p < \infty$.*
3. *If $C_c(E)$ is the set of continuous functions of bounded support, that is,*

$$C_c(E) = \{f \in C(E) : \text{there exists } R < \infty \text{ such that } f(x) \equiv 0 \text{ for } |x| > R\} \tag{3.2.15}$$

then $C_c(E)$ is dense in $L^p(E)$ for $1 \le p < \infty$.

The completeness property of $L^p(E)$ is a significant result in measure theory, often known as the Riesz-Fischer theorem.

3.3 FUNCTIONS ON METRIC SPACES AND CONTINUITY

Next, suppose (X, d_X), (Y, d_Y) are two metric spaces.

Definition 3.6. Let $T : X \to Y$ be a mapping.

1. We say T is continuous at a point $x \in X$ if for any $\epsilon > 0$ there exists $\delta > 0$ such that $d_Y(T(x), T(\hat{x})) \le \epsilon$ whenever $d_X(x, \hat{x}) \le \delta$.
2. T is continuous on X if it is continuous at each point of X.
3. T is uniformly continuous on X if for any $\epsilon > 0$ there exists $\delta > 0$ such that $d_Y(T(x), T(\hat{x})) \le \epsilon$ whenever $d_X(x, \hat{x}) \le \delta$, $x, \hat{x} \in X$.

4. T is Lipschitz continuous on X if there exists L such that

$$d_Y(T(x), T(\widehat{x})) \le L d_X(x, \widehat{x}) \qquad x, \widehat{x} \in X \tag{3.3.16}$$

The infimum of all L's which work in this definition is called the Lipschitz constant of T.

Clearly we have the implications that T Lipschitz continuous implies T is uniformly continuous, which in turn implies that T is continuous.

T is one-to-one (or injective) if $T(x_1) = T(x_2)$ implies that $x_1 = x_2$, and onto (or surjective) if for every $y \in Y$ there exists some $x \in X$ such that $T(x) = y$. If T is both one-to-one and onto then we say it is bijective, and in this case there exists the inverse mapping $T^{-1} : Y \to X$.

For any mapping $T : X \to Y$ we define, for $E \subset X$ and $F \subset Y$

$$T(E) = \{y \in Y : y = T(x) \text{ for some } x \in E\} \tag{3.3.17}$$

the image of E in Y, and

$$T^{-1}(F) = \{x \in X : T(x) \in F\} \tag{3.3.18}$$

the preimage of F in X. Note that T is not required to be bijective in order that the preimage be defined.

The following theorem states two useful characterizations of continuity. Condition (b) is referred to as the sequential definition of continuity, for obvious reasons, while (c) is the topological definition, since it may be used to define continuity in much more general topological spaces.

Theorem 3.2. *Let X, Y be metric spaces and $T : X \to Y$. Then the following are equivalent:*

(a) *T is continuous on X.*
(b) *If $x_n \in X$ and $x_n \to x$ in X, then $T(x_n) \to T(x)$ in Y.*
(c) *If F is open in Y then $T^{-1}(F)$ is open in X.*

Proof. Assume T is continuous on X and let $x_n \to x$ in X. If $\epsilon > 0$ then there exists $\delta > 0$ such that $d_Y(T(\widehat{x}), T(x)) < \epsilon$ if $d_X(\widehat{x}, x) < \delta$. Choosing n_0 sufficiently large that $d_X(x_n, x) < \delta$ for $n \ge n_0$ we then must have $d_Y(T(x_n), T(x)) < \epsilon$ for $n \ge n_0$, so that $T(x_n) \to T(x)$. Thus (a) implies (b).

To see that (b) implies (c), suppose condition (b) holds, F is open in Y and $x \in T^{-1}(F)$. We must show that there exists $\delta > 0$ such that $\widehat{x} \in T^{-1}(F)$ whenever $d_X(\widehat{x}, x) < \delta$. If not then there exists a sequence $x_n \to x$ such that $x_n \notin T^{-1}(F)$, and by (b), $T(x_n) \to T(x)$. Since $y = T(x) \in F$ and F is open, there exists $\epsilon > 0$ such that $z \in F$ if $d_Y(z, y) < \epsilon$. Thus $T(x_n) \in F$ for sufficiently large n, that is, $x_n \in T^{-1}(F)$, a contradiction.

Finally, suppose (c) holds and fix $x \in X$. If $\epsilon > 0$ then corresponding to the open set $F = B(T(x), \epsilon)$ in Y there exists a ball $B(x, \delta)$ in X such that

$B(x,\delta) \subset T^{-1}(F)$. But this means precisely that if $d_X(\widehat{x},x) < \delta$ then $d_Y(T(\widehat{x}),T(x)) < \epsilon$, so that T is continuous at x. □

3.4 COMPACTNESS AND OPTIMIZATION

Another important topological concept is that of compactness.

Definition 3.7. If $E \subset X$ then a collection of open sets $\{G_\alpha\}_{\alpha\in A}$ is an open cover of E if $E \subset \cup_{\alpha\in A} G_\alpha$.

Here A is the *index set* and may be finite, countably or uncountably infinite.

Definition 3.8. $K \subset X$ is *compact* if any open cover of K has a finite subcover. More explicitly, K is compact if whenever $K \subset \cup_{\alpha\in A} G_\alpha$, where each G_α is open, there exists a finite number of indices $\alpha_1, \alpha_2, \ldots, \alpha_m \in A$ such that $K \subset \cup_{j=1}^m G_{\alpha_j}$. In addition, $E \subset X$ is *precompact* (or relatively compact) if $\overline{E}$ is compact. If X is a compact set, considered as a subset of itself, then we say X is a compact metric space.

Proposition 3.3. *A compact set is closed and bounded and a closed subset of a compact set is compact.*

Proof. Suppose that K is compact and pick $x \in K^c$. For any $r > 0$ let $G_r = \{y \in X : d(x,y) > r\}$. It is easy to see that each G_r is open and $K \subset \cup_{r>0} G_r$. Thus there exists $r_1, r_2, \ldots, r_m$ such that $K \subset \cup_{j=1}^m G_{r_j}$ and so $B(x,r) \subset K^c$ if $r < \min\{r_1, r_2, \ldots, r_m\}$. Thus K^c is open and so K is closed.

Obviously $\cup_{r>0} B(x,r)$ is an open cover of K for any fixed $x \in X$. If K is compact then there must exist $r_1, r_2, \ldots, r_m$ such that $K \subset \cup_{j=1}^m B(x,r_j)$ and so $K \subset B(x,R)$ where $R = \max\{r_1, r_2, \ldots, r_m\}$. Thus K is bounded.

Now suppose that $F \subset K$ where F is closed and K is compact. If $\{G_\alpha\}_{\alpha\in A}$ is an open cover of F then these sets together with the open set F^c are an open cover of K. Hence there exists $\alpha_1, \alpha_2, \ldots, \alpha_m$ such that $K \subset (\cup_{j=1}^m G_{\alpha_j}) \cup F^c$, from which we conclude that $F \subset \cup_{j=1}^m G_{\alpha_j}$. □

There will be frequent occasions for wanting to know if a certain set is compact, but it is rare to use the above definition directly. A useful equivalent condition is that of *sequential compactness*.

Definition 3.9. A set $K \subset X$ is sequentially compact if any infinite sequence in K has a subsequence convergent to a point of K.

Proposition 3.4. *A set $K \subset X$ is compact if and only if it is sequentially compact.*

We will not prove this result here, but instead refer to Theorem 16, Section 9.5 of [30] for details. It follows immediately that if $E \subset X$ is precompact then any infinite sequence in E has a convergent subsequence (the point being that the limit need not belong to E).

We point out that the concepts of compactness and sequential compactness are applicable in spaces even more general than metric spaces, and are not always equivalent in such situations. In the case that $X = \mathbb{R}^N$ or $\mathbb{C}^N$ we have the following even more explicit characterization of compactness, the well known Heine-Borel theorem, for which we refer to [31] for a proof.

Theorem 3.3. *$E \subset \mathbb{R}^N$ or $E \subset \mathbb{C}^N$ is compact if and only if it is closed and bounded.*

While we know from Proposition 3.3 that a compact set is always closed and bounded, the converse implication is definitely false in most function spaces we will be interested in.

In later chapters a great deal of attention will be paid to optimization problems in function spaces, that is, problems in the Calculus of Variations. A simple result along these lines that we can prove already is:

Theorem 3.4. *Let X be a compact metric space and $f : X \to \mathbb{R}$ be continuous. Then there exists $x_0 \in X$ such that*

$$f(x) \leq f(x_0) \quad \forall x \in X \tag{3.4.19}$$

Proof. Let $M = \sup_{x \in X} f(x)$ (which may be $+\infty$), so there exists a sequence $\{x_n\}_{n=1}^{\infty}$ such that $\lim_{n\to\infty} f(x_n) = M$. By sequential compactness there is a subsequence $\{x_{n_k}\}$ and $x_0 \in X$ such that $\lim_{k\to\infty} x_{n_k} = x_0$ and since f is continuous on X we must have $f(x_0) = \lim_{k\to\infty} f(x_{n_k}) = M$. Thus $M < \infty$ and Eq. (3.4.19) holds. ☐

A common notation expressing the same conclusion as Eq. (3.4.19) is[2]

$$x_0 = \text{argmax}(f(x)) \tag{3.4.20}$$

which is also useful in making the distinction between the maximum value of a function and the point(s) at which the maximum is achieved.

We emphasize here the distinction between maximum and supremum, which is an essential point in later discussion of optimization. If $E \subset \mathbb{R}$ then $M = \sup E$ if

- $x \leq M$ for all $x \in E$
- if $M' < M$ there exists $x \in E$ such that $x > M'$

Such a number M exists for any $E \subset \mathbb{R}$ if we allow the value $M = +\infty$; by convention $M = -\infty$ if E is the empty set. On the other hand $M = \max E$ if

- $x \leq M$ for all $x \in E$ and $M \in E$

in which case evidently the maximum is finite and equal to the supremum.

---

2. Since f may achieve its maximum value at more than one point it might be more appropriate to write this as $x_0 \in \text{argmax}(f(x))$, but we will use the above more common notation. Similarly we use the corresponding notation argmin for points where the minimum of f is achieved.

If $f : X \to \mathbb{C}$ is continuous on a compact metric space X, then we can apply Theorem 3.4 with f replaced by $|f|$, to obtain that there exists $x_0 \in X$ such that $|f(x)| \le |f(x_0)|$ for all $x \in X$. We can then also conclude, as in Example 3.2 and Proposition 3.2 that the following holds.

Proposition 3.5. *If X is a compact metric space, then*

$$C(X) = \{f : X \to \mathbb{C} : f \text{ is continuous at } x \text{ for every } x \in X\} \tag{3.4.21}$$

is a complete metric space with metric $d(f,g) = \max_{x\in X} |f(x) - g(x)|$.

In general $C(X)$, or even a bounded set in $C(X)$, is not itself precompact. A useful criteria for precompactness of a set of functions in $C(X)$ is given by the Arzela-Ascoli theorem, which we recall here (see, e.g., [31] for a proof).

Definition 3.10. We say a family of real or complex valued functions $\mathbb{F}$ defined on a metric space X is *uniformly bounded* if there exists a constant M such that

$$|f(x)| \le M \qquad \text{whenever } x \in X \quad f \in \mathbb{F} \tag{3.4.22}$$

and *equicontinuous* if for every $\epsilon > 0$ there exists $\delta > 0$ such that

$$|f(x) - f(y)| < \epsilon \qquad \text{whenever } x, y \in X \quad d(x,y) < \delta \quad f \in \mathbb{F} \tag{3.4.23}$$

We then have

Theorem 3.5. *(Arzela-Ascoli) If X is a compact metric space and $\mathbb{F} \subset C(X)$ is uniformly bounded and equicontinuous, then $\mathbb{F}$ is precompact in $C(X)$.*

Example 3.8. Let

$$\mathbb{F} = \{f \in C([0,1]) : |f'(x)| \le M \ \forall x \in (0,1), f(0) = 0\} \tag{3.4.24}$$

for some fixed M. Then for $f \in \mathbb{F}$ we have

$$f(x) = \int_0^x f'(s)\, ds \tag{3.4.25}$$

implying in particular that $|f(x)| \le \int_0^x M\, ds \le M$. Also

$$|f(x) - f(y)| = \left| \int_x^y f'(s)\, ds \right| \le M|x - y| \tag{3.4.26}$$

so that for any $\epsilon > 0$, $\delta = \epsilon/M$ works in the definition of equicontinuity. Thus by the Arzela-Ascoli theorem $\mathbb{F}$ is precompact in $C([0,1])$. □

If X is a compact subset of $\mathbb{R}^N$ then since uniform convergence implies L^p convergence, it follows that any set which is precompact in $C(X)$ is also precompact in $L^p(X)$. But there are also more refined, that is, less restrictive, criteria for precompactness in L^p spaces, which are known (see, e.g., [5, Section 4.5]).

3.5 CONTRACTION MAPPING THEOREM

One of the most important theorems about metric spaces, frequently used in applied mathematics, is the Contraction Mapping Theorem, which concerns fixed points of a mapping of X into itself.

Definition 3.11. A mapping $T : X \to X$ is a contraction on X if it is Lipschitz continuous with Lipschitz constant $\rho < 1$, that is, there exists $\rho \in [0, 1)$ such that

$$d(T(x), T(\widehat{x})) \le \rho d(x, \widehat{x}) \quad \forall x, \widehat{x} \in X \tag{3.5.27}$$

If $\rho = 1$ is allowed, we say T is nonexpansive.

Theorem 3.6. *If X is a complete metric space and T is a contraction on X then there exists a unique $x \in X$ such that $T(x) = x$.*

Proof. The uniqueness assertion is immediate, namely if $T(x_1) = x_1$ and $T(x_2) = x_2$ then $d(x_1, x_2) = d(T(x_1), T(x_2)) \le \rho d(x_1, x_2)$. Since $\rho < 1$ we must have $d(x_1, x_2) = 0$ so that $x_1 = x_2$.

To prove the existence of x, fix any point $x_1 \in X$ and define

$$x_{n+1} = T(x_n) \tag{3.5.28}$$

for $n = 1, 2, \ldots$. We first show that $\{x_n\}_{n=1}^{\infty}$ must be a Cauchy sequence.

Note that

$$d(x_3, x_2) = d(T(x_2), T(x_1)) \le \rho d(x_2, x_1) \tag{3.5.29}$$

and by induction

$$d(x_{n+1}, x_n) = d(T(x_n), T(x_{n-1}) \le \rho^{n-1} d(x_2, x_1) \tag{3.5.30}$$

Thus by the triangle inequality and the usual summation formula for a geometric series, if $m > n > 1$

$$d(x_m, x_n) \le \sum_{j=n}^{m-1} d(x_{j+1}, x_j) \le \sum_{j=n}^{m-1} \rho^{j-1} d(x_2, x_1) \tag{3.5.31}$$

$$= \frac{\rho^{n-1}(1 - \rho^{m-n})}{1 - \rho} d(x_2, x_1) \le \frac{\rho^{n-1}}{1 - \rho} d(x_2, x_1) \tag{3.5.32}$$

It follows immediately that $\{x_n\}_{n=1}^{\infty}$ is a Cauchy sequence, and since X is complete there exists $x \in X$ such that $\lim_{n\to\infty} x_n = x$. Since T is continuous $T(x_n) \to T(x)$ as $n \to \infty$ and so $x = T(x)$ must hold. □

The point x in the Contraction Mapping Theorem which satisfies $T(x) = x$ is called a fixed point of T, and the process (3.5.28) of generating the sequence $\{x_n\}_{n=1}^{\infty}$, is called fixed point iteration. Not only does the theorem show that T possesses a unique fixed point under the stated hypotheses, but the proof shows that the fixed point may be obtained by fixed point iteration starting from an arbitrary point of X.

As a simple application of the theorem, consider a second kind integral equation

$$u(x) - \int_\Omega K(x,y)u(y)\,dy = f(x) \tag{3.5.33}$$

with $\Omega \subset \mathbb{R}^N$ a bounded open set, a kernel function $K = K(x,y)$ defined and continuous for $(x,y) \in \overline{\Omega} \times \overline{\Omega}$ and $f \in C(\overline{\Omega})$. We can then define a mapping T on $X = C(\overline{\Omega})$ by

$$T(u)(x) = \int_\Omega K(x,y)u(y)\,dy + f(x) \tag{3.5.34}$$

so that Eq. (3.5.33) is equivalent to the fixed point problem $u = T(u)$ in X. Since K is uniformly continuous on $\overline{\Omega}\times\overline{\Omega}$ it is immediate that $Tu \in X$ whenever $u \in X$, and by elementary estimates we have

$$\begin{aligned} d(T(u),T(v)) &= \max_{x\in\overline{\Omega}} |T(u)(x) - T(v)(x)| \\ &= \max_{x\in\overline{\Omega}} \left| \int_\Omega K(x,y)(u(y) - v(y))\,dy \right| \le Ld(u,v) \end{aligned} \tag{3.5.35}$$

where $L := \max_{x\in\overline{\Omega}} \int_\Omega |K(x,y)|\,dy$. We therefore may conclude from the Contraction Mapping Theorem the following:

Proposition 3.6. *If*

$$\max_{x\in\overline{\Omega}} \int_\Omega |K(x,y)|\,dy < 1 \tag{3.5.36}$$

then Eq. (3.5.33) has a unique solution for every $f \in C(\overline{\Omega})$.

The condition (3.5.36) will be satisfied if either the maximum of $|K|$ is small enough or the size of the domain Ω is small enough. Eventually we will see that some such smallness condition is necessary for unique solvability of Eq. (3.5.33), but the exact conditions will be sharpened considerably.

If we consider instead the family of second kind integral equations

$$\lambda u(x) - \int_\Omega K(x,y)u(y)\,dy = f(x) \tag{3.5.37}$$

with the same conditions on K and f, then the above argument shows unique solvability for all sufficiently large λ, namely provided

$$\max_{x\in\overline{\Omega}} \int_\Omega |K(x,y)|\,dy < |\lambda| \tag{3.5.38}$$

As another example, consider the initial value problem for a first order ODE

$$\frac{du}{dt} = f(t,u) \quad u(t_0) = u_0 \tag{3.5.39}$$

where we assume at least that f is continuous on $[a,b] \times \mathbb{R}$ with $t_0 \in (a,b)$. If a classical solution u exists, then integrating both sides of the ODE from t_0 to t, and taking account of the initial condition we obtain

$$u(t) = u_0 + \int_{t_0}^{t} f(s, u(s))\, ds \tag{3.5.40}$$

Conversely, if $u \in C([a,b])$ and satisfies Eq. (3.5.40) then necessarily u' exists, is also continuous and Eq. (3.5.39) holds. Thus the problem of solving Eq. (3.5.39) is seen to be equivalent to that of finding a continuous solution of Eq. (3.5.40). In turn this can be viewed as the problem of finding a fixed point of the nonlinear mapping $T : C([a,b]) \to C([a,b])$ defined by

$$T(u)(t) = u_0 + \int_{t_0}^{t} f(s, u(s))\, ds \tag{3.5.41}$$

Now if we assume that f satisfies the Lipschitz condition with respect to u,

$$|f(t,u) - f(t,v)| \le L|u - v| \qquad u, v \in \mathbb{R} \quad t \in [a,b] \tag{3.5.42}$$

for some constant L, then

$$|T(u)(t) - T(v)(t)| \le L \int_{t_0}^{t} |u(s) - v(s)|\, ds \le L|b - a| \max_{a \le t \le b} |u(t) - v(t)| \tag{3.5.43}$$

or

$$d(T(u), T(v)) \le L|b - a| d(u, v) \tag{3.5.44}$$

where d is again the usual metric on $C([a,b])$. Thus the contraction mapping provides a unique local solution, that is, on any interval $[a,b]$ containing t_0 for which $(b - a) < 1/L$.

Instead of the requirement that the Lipschitz condition (3.5.44) be valid on the entire infinite strip $[a,b] \times \mathbb{R}$, it is actually only necessary to assume it holds on $[a,b] \times [c,d]$ where $u_0 \in (c,d)$. First order systems of ODEs (and thus scalar higher order equations) can be handled in essentially the same manner. Such generalizations may be found in standard ODE textbooks, for example, Chapter 1 of [7] or Chapter 3 of [4].

We conclude with a useful variant of the contraction mapping theorem. If $T : X \to X$ then we can define the (composition) powers of T by $T^2(x) = T(T(x))$, $T^3(x) = T(T^2(x))$, etc. Thus $T^n : X \to X$ for $n = 1, 2, 3, \ldots$.

Theorem 3.7. *If X is a complete metric space and there exists a positive integer n such that T^n is a contraction on X, then there exists a unique $x \in X$ such that $T(x) = x$.*

Proof. By Theorem 3.6 there exists a unique $x \in X$ such that $T^n(x) = x$. Applying T to both sides gives $T^n(T(x)) = T^{n+1}(x) = T(x)$ so that $T(x)$ is also a fixed point of T^n. By uniqueness, $T(x) = x$, that is, T has at least one fixed point. To see that the fixed point of T is unique, observe that any fixed point of T is also a fixed point of $T^2, T^3, \ldots$. In particular, if T has two distinct fixed points then so does T^n, which is a contradiction. □

3.6 EXERCISES

3.1. Verify that d_p defined in Example 3.1 is a metric on $\mathbb{R}^N$ or $\mathbb{C}^N$. (Suggestion: to prove the triangle inequality, use the finite dimensional version of the Minkowski inequality (A.2.28)).

3.2. If $(X, d_X), (Y, d_Y)$ are metric spaces, show that the Cartesian product

$$Z = X \times Y = \{(x, y) : x \in X, y \in Y\}$$

is a metric space with distance function

$$d_Z((x_1, y_1), (x_2, y_2)) = d_X(x_1, x_2) + d_Y(y_1, y_2)$$

3.3. Is $d(x, y) = |x - y|^2$ a metric on $\mathbf{R}$? What about $d(x, y) = \sqrt{|x - y|}$? Find reasonable conditions on a function $\phi : [0, \infty) \to [0, \infty)$ such that $d(x, y) = \phi(|x - y|)$ is a metric on $\mathbb{R}$.

3.4. If $K_1, K_2, \ldots$ are nonempty compact sets such that $K_{n+1} \subset K_n$ for all n, show that $\cap_{n=1}^{\infty} K_n$ is nonempty.

3.5. Let (X, d) be a metric space, $A \subset X$ be nonempty and define the distance from a point x to the set A to be

$$d(x, A) = \inf_{y \in A} d(x, y)$$

(a) Show that $|d(x, A) - d(y, A)| \le d(x, y)$ for $x, y \in X$ (i.e., $x \to d(x, A)$ is nonexpansive).

(b) Assume A is closed. Show that $d(x, A) = 0$ if and only if $x \in A$.

(c) Assume A is compact. Show that for any $x \in X$ there exists $z \in A$ such that $d(x, A) = d(x, z)$.

3.6. Suppose that F is closed and G is open in a metric space (X, d) and $F \subset G$. Show that there exists a continuous function $f : X \to \mathbb{R}$ such that

(i) $0 \le f(x) \le 1$ for all $x \in X$.

(ii) $f(x) = 1$ for $x \in F$.

(iii) $f(x) = 0$ for $x \in G^c$.

Hint: Consider

$$f(x) = \frac{d(x, G^c)}{d(x, G^c) + d(x, F)}$$

3.7. Two metrics $d, \hat{d}$ on a set X are said to be equivalent if there exist constants $0 < C < C^* < \infty$ such that

$$C \le \frac{d(x, y)}{\hat{d}(x, y)} \le C^* \qquad \forall x, y \in X$$

(a) If $d, \widehat{d}$ are equivalent, show that a sequence $\{x_k\}_{k=1}^{\infty}$ is convergent in (X, d) if and only if it is convergent in $(X, \widehat{d})$.
(b) Show that any two of the metrics d_p on $\mathbb{R}^n$ are equivalent.

3.8. Prove that $C([a,b])$ is separable (you may quote the Weierstrass approximation theorem) but $L^{\infty}(a,b)$ is not separable.

3.9. If X, Y are metric spaces, $f : X \to Y$ is continuous and K is compact in X, show that the image $f(K)$ is compact in Y.

3.10. Let

$$\mathbb{F} = \left\{ f \in C([0,1]) : |f(x) - f(y)| \le |x - y| \text{ for all } x, y, \int_0^1 f(x)\,dx = 0 \right\}$$

Show that $\mathbb{F}$ is compact in $C([0,1])$. (Suggestion: to prove that $\mathbb{F}$ is uniformly bounded, justify and use the fact that if $f \in \mathbb{F}$ then $f(x) = 0$ for some $x \in [0,1]$.)

3.11. Show that the set $\mathbb{F}$ in Example 3.8 is not closed.

3.12. From the proof of the contraction mapping it is clear that the smaller ρ is, the faster the sequence x_n converges to the fixed point x. With this in mind, explain why Newton's method

$$x_{n+1} = x_n - \frac{f(x_n)}{f'(x_n)}$$

is in general a very rapidly convergent method for approximating roots of $f : \mathbb{R} \to \mathbb{R}$, as long as the initial guess is close enough.

3.13. Let $f_n(x) = \sin^n x$ for $n = 1, 2, \ldots$.
(a) Is the sequence $\{f_n\}_{n=1}^{\infty}$ convergent in $C([0,\pi])$?
(b) Is the sequence convergent in $L^2(0,\pi)$?
(c) Is the sequence compact or precompact in either of these spaces?

3.14. Let X be a complete metric space and $T : X \to X$ satisfy

$$d(T(x), T(y)) < d(x,y) \qquad x, y \in X \quad x \ne y$$

Show that T can have at most one fixed point, but may have none. (Suggestion: for an example of nonexistence look at $T(x) = \sqrt{x^2+1}$ on $\mathbb{R}$.)

3.15. Let S denote the linear Volterra type integral operator

$$Su(x) = \int_a^x K(x,y)u(y)\,dy$$

where the kernel K is continuous and satisfies $|K(x,y)| \le M$ for $a \le y \le x$.
(a) Show that

$$|S^n u(x)| \le \frac{M^n (x-a)^n}{n!} \max_{a \le y \le x} |u(y)| \quad x > a \quad n = 1, 2, \ldots$$

(b) Deduce from this that for any $b > a$, there exists an integer n such that S^n is a contraction on $C([a,b])$.

(c) Show that for any $f \in C([a,b])$ the second kind Volterra integral equation

$$u(x) - \int_a^x K(x,y)u(y)\,dy = f(x) \quad a < x < b$$

has a unique solution $u \in C([a,b])$.

3.16. Show that for sufficiently small $|\lambda|$ there exists a unique solution of the boundary value problem

$$u'' + \lambda u = f(x) \quad 0 < x < 1 \qquad u(0) = u(1) = 0$$

for any $f \in C([0,1])$. (Suggestion: use the result of Chapter 1, Exercise 1.7 to transform the boundary value problem into a fixed point problem for an integral operator, then apply the Contraction Mapping Theorem.) Be as precise as you can about which values of λ are allowed.

3.17. Let $f = f(x,y)$ be continuously differentiable on $[0,1] \times \mathbb{R}$ and satisfy

$$0 < m \le \frac{\partial f}{\partial y}(x,y) \le M$$

Show that there exists a unique continuous function $\phi(x)$ such that

$$f(x, \phi(x)) = 0 \quad 0 < x < 1$$

(Suggestion: Define the transformation

$$(T\phi)(x) = \phi(x) - \lambda f(x, \phi(x))$$

and show that T is a contraction on $C([0,1])$ for some choice of λ. This is a special case of the implicit function theorem.)

3.18. Show that if we let

$$d(f,g) = \sum_{k=0}^{\infty} \frac{2^{-k} e_k}{1 + e_k}$$

where

$$e_k = \max_{x \in [a,b]} |f^{(k)}(x) - g^{(k)}(x)|$$

then $(C^\infty([a,b]), d)$ is a metric space, in which $f_n \to f$ if and only if $f_n^{(k)} \to f^{(k)}$ uniformly on $[a,b]$ for $k = 0, 1, \ldots$.

3.19. If $E \subset \mathbb{R}^N$ is the closure of a bounded open set, show that $C^1(E)$ is a complete metric space with the metric defined by Eq. (3.1.7).

Chapter 4

Banach Spaces

4.1 AXIOMS OF A NORMED LINEAR SPACE

In a normed linear space we combine vector space and a special kind of metric space structure.

Definition 4.1. A vector space $\mathbf{X}$ is said to be a normed linear space if for every $x \in \mathbf{X}$ there is defined a nonnegative real number $\|x\|$, the norm of x, such that the following axioms hold.

[N1] $\|x\| = 0$ if and only if $x = 0$
[N2] $\|\lambda x\| = |\lambda|\,\|x\|$ for any $x \in \mathbf{X}$ and any scalar λ.
[N3] $\|x + y\| \le \|x\| + \|y\|$ for any $x, y \in \mathbf{X}$.

As in the case of a metric space it is technically the pair $(\mathbf{X}, \|\cdot\|)$ which constitute a normed linear space, but the definition of the norm will usually be clear from the context. If two different normed spaces are needed we will use a notation such as $\|x\|_{\mathbf{X}}$ to indicate the space in which the norm is calculated.

Example 4.1. In the vector space $\mathbf{X} = \mathbb{R}^N$ or $\mathbb{C}^N$ we can define the family of norms

$$\|x\|_p = \left(\sum_{j=1}^N |x_j|^p\right)^{1/p} \qquad 1 \le p < \infty$$
$$\|x\|_\infty = \max_{1 \le j \le N} |x_j| \tag{4.1.1}$$

Axioms [N1] and [N2] are obvious, while axiom [N3] amounts to the Minkowski inequality (A.2.28). □

We obviously have $d_p(x, y) = \|x - y\|_p$ in the previous example, where d_p is the metric defined in the previous chapter, and this correspondence between norm and metric is a special case of the following general fact that a norm always gives rise to a metric. The proof is immediate from the definitions involved.

Techniques of Functional Analysis for Differential and Integral Equations
http://dx.doi.org/10.1016/B978-0-12-811426-1.00004-0

Proposition 4.1. *Let $(\mathbf{X}, \|\cdot\|)$ be a normed linear space. If we set $d(x,y) = \|x-y\|$ for $x, y \in \mathbf{X}$ then $(\mathbf{X}, d)$ is a metric space.*

Example 4.2. If $E \subset \mathbb{R}^N$ is closed and bounded then it is easy to verify that

$$\|f\| = \max_{x \in E} |f(x)| \tag{4.1.2}$$

defines a norm on $C(E)$, and the usual metric (3.1.4) on $C(E)$ amounts to $d(f,g) = \|f-g\|$. Likewise, the metrics (3.1.8), (3.1.9) on $L^p(E)$ may be viewed as coming from the corresponding L^p norms,

$$\|f\|_{L^p(E)} = \begin{cases} \left(\int_E |f(x)|^p\,dx\right)^{1/p} & 1 \le p < \infty \\ \text{ess sup}_{x\in E}|f(x)| & p = \infty \end{cases} \tag{4.1.3}$$

□

Note that for such a metric we must have $d(\lambda x, \lambda y) = |\lambda| d(x,y)$ so that if this property does not hold, the metric cannot arise from a norm in this way. For example,

$$d(x,y) = \frac{|x-y|}{1+|x-y|} \tag{4.1.4}$$

is a metric on $\mathbb{R}$ which does not come from a norm, since the above scaling property does not hold.

Since any normed linear space may now be regarded as metric space, all of the topological concepts defined for a metric space are meaningful in a normed linear space. Completeness holds in many situations of interest, so we have a special designation in that case.

Definition 4.2. A *Banach space* is a complete normed linear space.

Example 4.3. The spaces $\mathbb{R}^N, \mathbb{C}^N$ are vector spaces which are also complete metric spaces with any of the norms $\|\cdot\|_p$, hence they are Banach spaces. Similarly $C(E)$, $L^p(E)$ are Banach spaces with norms indicated above. □

Here are a few simple results we can prove already.

Proposition 4.2. *If $\mathbf{X}$ is a normed linear space then the norm is a Lipschitz continuous function on $\mathbf{X}$. If $E \subset \mathbf{X}$ is compact and $y \in \mathbf{X}$ then there exists $x_0 \in E$ such that*

$$\|y - x_0\| = \min_{x \in E} \|y - x\| \tag{4.1.5}$$

Proof. From the triangle inequality we get $|\|x_1\| - \|x_2\|| \le \|x_1 - x_2\|$ so that $f(x) = \|x\|$ is Lipschitz continuous (with Lipschitz constant 1) on $\mathbf{X}$. Similarly $f(x) = \|x - y\|$ is also continuous for any fixed y, so we may apply Theorem 3.4 with $\mathbf{X}$ replaced by the compact metric space E and $f(x) = -\|x-y\|$ to get the conclusion. □

Another topological point of interest is the following.

Theorem 4.1. *If M is a subspace of a normed linear space $\mathbf{X}$, and $\dim M < \infty$ then M is closed.*

Proof. The proof is by induction on the number of dimensions. Let $\dim(M) = 1$ so that $M = \{u = \lambda e : \lambda \in \mathbb{C}\}$ for some $e \in \mathbf{X}$, $\|e\| = 1$. If $u_n \in M$ then $u_n = \lambda_n e$ for some $\lambda_n \in \mathbb{C}$ and $u_n \to u$ in $\mathbf{X}$ implies, since $\|u_n - u_m\| = |\lambda_n - \lambda_m|$, that $\{\lambda_n\}$ is a Cauchy sequence in $\mathbb{C}$. Thus there exist $\lambda \in \mathbb{C}$ such that $\lambda_n \to \lambda$ so that $u_n \to u = \lambda e \in M$, as needed.

Now suppose we know that all N-dimensional subspaces are closed and $\dim M = N+1$, thus we can find $e_1, \dots, e_{N+1}$ linearly independent unit vectors such that $M = \text{span}(e_1, \dots, e_{N+1})$. Let $\widetilde{M} = \text{span}(e_1, \dots, e_N)$ which is closed by the induction assumption. If $u_n \in M$ there exists $\lambda_n \in \mathbb{C}$ and $v_n \in \widetilde{M}$ such that $u_n = v_n + \lambda_n e_{N+1}$. Suppose that $u_n \to u$ in $\mathbf{X}$. We claim first that $\{\lambda_n\}$ is bounded in $\mathbb{C}$. If not, there must exist λ_{n_k} such that $|\lambda_{n_k}| \to \infty$, and since u_n remains bounded in $\mathbf{X}$ we get $u_{n_k}/\lambda_{n_k} \to 0$. Since

$$e_{N+1} - \frac{u_{n_k}}{\lambda_{n_k}} = -\frac{v_{n_k}}{\lambda_{n_k}} \in \widetilde{M} \tag{4.1.6}$$

and $\widetilde{M}$ is closed, it would follow, upon letting $n_k \to \infty$, that $e_{N+1} \in \widetilde{M}$, which is impossible.

Thus we can find a subsequence $\lambda_{n_k} \to \lambda$ for some $\lambda \in \mathbb{C}$ and

$$v_{n_k} = u_{n_k} - \lambda_{n_k} e_{N+1} \to u - \lambda e_{N+1} \tag{4.1.7}$$

Again since $\widetilde{M}$ is closed it follows that $u - \lambda e_{N+1} \in \widetilde{M}$, so that $u \in M$ as needed. □

For an infinite dimensional subspace the theorem is false in general. For example, the Weierstrass approximation theorem states that if $f \in C([a,b])$ and $\epsilon > 0$ there exists a polynomial p such that $|p(x) - f(x)| \le \epsilon$ on $[a,b]$. Thus if we take $\mathbf{X} = C([a,b])$ and E to be the set of all polynomials on $[a,b]$ then clearly E is a subspace of $\mathbf{X}$ and every point of $\mathbf{X}$ is a limit point of E. Thus E cannot be closed since otherwise E would be equal to all of $\mathbf{X}$.

Recall that when $\overline{E} = \mathbf{X}$ as in this example, we say that E is a dense subspace of $\mathbf{X}$. Such subspaces play an important role in functional analysis. According to Theorem 4.1 a finite dimensional Banach space $\mathbf{X}$ has no dense subspace aside from $\mathbf{X}$ itself.

4.2 INFINITE SERIES

In a normed linear space we can study limits of sums, that is, infinite series. The basic definition is the same as for infinite series of numbers.

Definition 4.3. We say $\sum_{j=1}^{\infty} x_j$ is convergent in $\mathbf{X}$ to the limit $s \in \mathbf{X}$ if $\lim_{n\to\infty} s_n = s$, where $s_n = \sum_{j=1}^{n} x_j$ is the nth partial sum of the series.

A useful criterion for convergence can then be given, provided the space is also complete.

Proposition 4.3. *If* $\mathbf{X}$ *is a Banach space,* $x_j \in \mathbf{X}$ *for* $j = 1, 2, \ldots$ *and* $\sum_{j=1}^{\infty} \|x_j\| < \infty$ *then* $\sum_{j=1}^{\infty} x_j$ *is convergent to an element* $s \in \mathbf{X}$ *with* $\|s\| \leq \sum_{j=1}^{\infty} \|x_j\|$.

Proof. If $m > n$ we have $\|s_m - s_n\| = \| \sum_{j=n+1}^{m} x_j \| \leq \sum_{j=n+1}^{m} \|x_j\|$ by the triangle inequality. If $\sum_{j=1}^{\infty} \|x_j\|$ is convergent, its partial sums form a Cauchy sequence in $\mathbb{R}$, and hence $\{s_n\}$ is also Cauchy in $\mathbf{X}$. Since the space is complete $s = \lim_{n\to\infty} s_n$ exists. We also have $\|s_n\| \leq \sum_{j=1}^{n} \|x_j\|$ for any fixed n, and $\|s_n\| \to \|s\|$ by Proposition 4.2, so $\|s\| \leq \sum_{j=1}^{\infty} \|x_j\|$ must hold. □

The concepts of linear combination, linear independence and basis may now be extended to allow for infinite sums in an obvious way: We say a countably infinite set of vectors $\{x_n\}_{n=1}^{\infty}$ is linearly independent if

$$\sum_{n=1}^{\infty} \lambda_n x_n = 0 \quad \text{if and only if } \lambda_n = 0 \text{ for all } n \tag{4.2.8}$$

and $x \in \operatorname{span}(\{x_n\}_{n=1}^{\infty})$ provided $x = \sum_{n=1}^{\infty} \lambda_n x_n$ for some scalars $\{\lambda_n\}_{n=1}^{\infty}$. A basis of $\mathbf{X}$ is then a linearly independent spanning set, or equivalently $\{x_n\}_{n=1}^{\infty}$ is a basis of $\mathbf{X}$ if for any $x \in \mathbf{X}$ there exist unique scalars $\{\lambda_n\}_{n=1}^{\infty}$ such that $x = \sum_{n=1}^{\infty} \lambda_n x_n$.

We emphasize that this definition of basis is not the same as that given in Definition 2.4 for a basis of a vector space, the difference being that the sum there is required to always be finite. The term *Schauder basis* is sometimes used for the definition just given if the distinction needs to be made. Throughout the remainder of this text, the term basis will always mean Schauder basis unless otherwise stated.

A Banach space $\mathbf{X}$ which contains a Schauder basis $\{x_n\}_{n=1}^{\infty}$ is always separable, since then the set of all finite linear combinations of the x_n's with rational coefficients is easily seen to be countable and dense. It is known that not every separable Banach space has a Schauder basis (recall there must exist a Hamel basis), see, for example, Section 1.1 of [41].

4.3 LINEAR OPERATORS AND FUNCTIONALS

We have previously defined what it means for a mapping $T : \mathbf{X} \longmapsto \mathbf{Y}$ between vector spaces to be linear. When the spaces $\mathbf{X}, \mathbf{Y}$ are normed linear spaces we usually refer to such a mapping T as a *linear operator*. We say that T is *bounded* if there exists a finite constant C such that $\|Tx\| \leq C\|x\|$ for every $x \in \mathbf{X}$, and we may then define the norm of T as the smallest such C, or equivalently

$$\|T\| = \sup_{x \neq 0} \frac{\|Tx\|}{\|x\|} \tag{4.3.9}$$

The following simple proposition includes in particular the fact that boundedness is equivalent to continuity of T.

Proposition 4.4. *If* $\mathbf{X}, \mathbf{Y}$ *are normed linear spaces and* $T : \mathbf{X} \to \mathbf{Y}$ *is linear then the following conditions are equivalent.*

(a) T *is bounded.*
(b) T *is continuous.*
(c) *There exists* $x_0 \in \mathbf{X}$ *such that* T *is continuous at* x_0.
(d) T *is continuous at* 0.

Proof. If $x_0, x \in \mathbf{X}$ then

$$\|T(x) - T(x_0)\| = \|T(x - x_0)\| \leq \|T\| \, \|x - x_0\| \tag{4.3.10}$$

Thus if T is bounded then it is (Lipschitz) continuous at any point of $\mathbf{X}$. The implication that (b) implies (c) is trivial. Assuming that T is continuous at x_0 and $x_n \to 0$ then since $x_0 - x_n \to x_0$ and $T(0) = 0$, the linearity implies that $T(x_n) = T(x_0) - T(x_0 - x_n) \to 0$. Finally suppose T is continuous at 0. For any $\epsilon > 0$ there must exist $\delta > 0$ such that $\|T(z)\| = \|T(z) - T(0)\| \leq \epsilon$ if $\|z\| \leq \delta$. For any $x \neq 0$, choose $z = \delta \frac{x}{\|x\|}$ to get

$$\|T\left(\delta \frac{x}{\|x\|}\right)\| \leq \epsilon \tag{4.3.11}$$

or equivalently, using the linearity of T, $\|Tx\| \leq C\|x\|$ with $C = \epsilon/\delta$. Thus T is bounded. □

A continuous linear operator is therefore the same as a bounded linear operator, and the two terms are used interchangeably. When the range space $\mathbf{Y}$ is the scalar field $\mathbb{R}$ or $\mathbb{C}$ the convention is to use the terminology *linear functional* instead of linear operator, and correspondingly T is a bounded (or continuous) linear functional if $|Tx| \leq C\|x\|$ for some finite constant C and all $x \in \mathbf{X}$.

We introduce the notation

$$\mathcal{B}(\mathbf{X}, \mathbf{Y}) = \{T : \mathbf{X} \to \mathbf{Y} : T \text{ is linear and bounded}\} \tag{4.3.12}$$

and the special cases

$$\mathcal{B}(\mathbf{X}) = \mathcal{B}(\mathbf{X}, \mathbf{X}) \qquad \mathbf{X}^* = \mathcal{B}(\mathbf{X}, \mathbb{C}) \tag{4.3.13}$$

The space of linear functionals $\mathbf{X}^*$ (also commonly denoted by $\mathbf{X}'$ and referred to as the *dual space of* $\mathbf{X}$) especially will play a very significant role later on. A number of examples of linear operators will be given in Chapter 9. For now we just give two simple cases.

Example 4.4. If $\mathbf{X} = \mathbb{C}^N, \mathbf{Y} = \mathbb{C}^M$ and A is an $M \times N$ complex matrix with entries a_{kj}, then

$$y_k = \sum_{j=1}^{N} a_{kj}x_j \quad k = 1, \ldots, M \tag{4.3.14}$$

defines a linear mapping, and according to the discussion of Section 2.3 any linear mapping of $\mathbb{C}^N$ to $\mathbb{C}^M$ can be regarded as being of this form. It is not hard to check that T is always bounded, assuming that we use any of the norms $\|\cdot\|_p$ in $\mathbf{X}$ and in $\mathbf{Y}$, see more details in Example 9.1. Evidently T is a linear functional if $M = 1$. □

Example 4.5. If $E \subset \mathbb{R}^N$ is compact and $\mathbf{X} = C(E)$ pick $x_0 \in E$ and set $T(f) = f(x_0)$ for $f \in \mathbf{X}$. Clearly T is a linear functional and $|Tf| \leq \|f\|$ so that $\|T\| \leq 1$. □

We mention here, without proof, one further property of the dual space, see Chapter 5 of [32] or Chapter 1 of [5] for details.

Proposition 4.5. *If $\mathbf{X}$ is a Banach space and $x \in \mathbf{X}$ then*

$$\|x\| = \max_{\phi \in \mathbf{X}^*, \|\phi\|=1} |\phi(x)| \tag{4.3.15}$$

In particular, if $\phi(x) = 0$ for every $\phi \in \mathbf{X}^*$, then $x = 0$.

4.4 CONTRACTION MAPPINGS IN A BANACH SPACE

If the Contraction Mapping theorem, Theorem 3.6, is specialized to a Banach space, the resulting statement is that if $\mathbf{X}$ is a Banach space and $F : \mathbf{X} \to \mathbf{X}$ satisfies

$$\|F(x) - F(y)\| \leq L\|x - y\| \quad x, y \in \mathbf{X} \tag{4.4.16}$$

for some $L < 1$, then F has a unique fixed point in $\mathbf{X}$.

A particular case which arises frequently in applications is when the mapping F has the form $F(x) = Tx + b$ for some $b \in \mathbf{X}$ and bounded linear operator T on $\mathbf{X}$, in which case the contraction condition (4.4.16) simply amounts to the requirement that $\|T\| < 1$. If we then initialize the fixed point iteration process (3.5.28) with $x_1 = b$, the successive iterates are

$$x_2 = F(x_1) = F(b) = Tb + b \tag{4.4.17}$$

$$x_3 = F(x_2) = Tx_2 + b = T^2b + Tb + b \tag{4.4.18}$$

etc., the general pattern being

$$x_n = \sum_{j=0}^{n-1} T^j b \quad n = 1, 2, \ldots \tag{4.4.19}$$

with $T^0 = I$ as usual. If $\|T\| < 1$ we already know that this sequence must converge, but it could also be checked directly from Proposition 4.3 using the obvious inequality $\|T^j b\| \leq \|T\|^j \|b\|$. In fact we know that $x_n \to x$, the unique fixed point of F, so

$$x = \sum_{j=0}^{\infty} T^j b \tag{4.4.20}$$

is an explicit solution formula for the linear, inhomogeneous equation $x - Tx = b$. The right-hand side of Eq. (4.4.20) is known as the *Neumann series* for $x = (I - T)^{-1} b$, and symbolically we may write

$$(I - T)^{-1} = \sum_{j=0}^{\infty} T^j \tag{4.4.21}$$

Note the formal similarity to the usual geometric series formula for $(1 - z)^{-1}$ if $z \in \mathbb{C}$, $|z| < 1$. If T and b are such that $\|T^j b\| << \|Tb\|$ for $j \geq 2$, then truncating the series after two terms we get the *Born approximation* formula $x \approx b + Tb$.

4.5 EXERCISES

4.1. Give the proof of Proposition 4.1.

4.2. Show that any two norms on a finite dimensional normed linear space are equivalent. That is to say, if $(\mathbf{X}, \|\cdot\|_1)$, $(\mathbf{X}, \|\cdot\|_2)$ are both normed linear spaces and $\dim(\mathbf{X}) < \infty$, then there exist constants $0 < c < C < \infty$ such that

$$c \leq \frac{\|x\|_2}{\|x\|_1} \leq C \qquad \text{for all } x \in \mathbf{X}, x \neq 0$$

4.3. If $\mathbf{X}$ is a normed linear space and $\mathbf{Y}$ is a Banach space, show that $\mathcal{B}(\mathbf{X}, \mathbf{Y})$ is a Banach space, with the norm given by Eq. (4.3.9).

4.4. For the linear functional in Example 4.5 show that $\|T\| = 1$.

4.5. If T is a linear integral operator, $Tu(x) = \int_\Omega K(x, y)u(y)\, dy$, then T^2 is also a linear integral operator. What is the kernel for T^2?

4.6. If $\mathbf{X}$ is a normed linear space and E is a subspace of $\mathbf{X}$, show that $\overline{E}$ is also a subspace of $\mathbf{X}$.

4.7. If $p \in (0, 1)$ show that $\|f\|_p = \left(\int_\Omega |f(x)|^p\, dx\right)^{1/p}$ does not define a norm.

4.8. The simple initial value problem

$$u' = u \qquad u(0) = 1$$

is equivalent to the integral equation

$$u(x) = 1 + \int_0^x u(s)\, ds$$

which may be viewed as a fixed point problem of the special type discussed in Section 4.4. Find the Neumann series for the solution u. Where does it converge?

4.9. If $Tf = f(0)$, show that T is not a bounded linear functional on $L^p(-1,1)$ for $1 \le p < \infty$.

4.10. Let $\mathbf{X}$ be a Banach space and $A \in \mathcal{B}(\mathbf{X})$.

(a) Show that

$$\exp(A) = e^A := \sum_{n=0}^{\infty} \frac{A^n}{n!} \tag{4.5.22}$$

is defined in $\mathcal{B}(\mathbf{X})$.

(b) If also $B \in \mathcal{B}(\mathbf{X})$ and $AB = BA$ show that $\exp(A+B) = \exp(A)\exp(B)$.

(c) Show that $\exp((t+s)A) = \exp(tA)\exp(sA)$ for any $t, s \in \mathbb{R}$.

(d) Show that the conclusion in (b) is false, in general, if A and B do not commute. (Suggestion: a counterexample can be found in $\mathbf{X} = \mathbb{R}^2$.)

4.11. Find an integral equation of the form $u = Tu + f$, T linear, which is equivalent to the initial value problem

$$u'' + u = x^2 \quad x > 0 \qquad u(0) = 1 \quad u'(0) = 2 \tag{4.5.23}$$

Calculate the Born approximation to the solution u and compare to the exact solution.

Chapter 5

Hilbert Spaces

5.1 AXIOMS OF AN INNER PRODUCT SPACE

We now add one further structural ingredient to our spaces, generalizing the calculus concept of inner product (also commonly called the scalar product or dot product).

Definition 5.1. A vector space $\mathbf{X}$ is said to be an inner product space if for every $x, y \in \mathbf{X}$ there is defined a scalar $\langle x, y\rangle$, the inner product of x and y, such that the following axioms hold.

[H1] $\langle x, x\rangle \geq 0$ for all $x \in \mathbf{X}$.
[H2] $\langle x, x\rangle = 0$ if and only if $x = 0$.
[H3] $\langle \lambda x, y\rangle = \lambda\langle x, y\rangle$ for any $x, y \in \mathbf{X}$ and any scalar λ.
[H4] $\langle x, y\rangle = \overline{\langle y, x\rangle}$ for any $x, y \in \mathbf{X}$.
[H5] $\langle x + y, z\rangle = \langle x, z\rangle + \langle y, z\rangle$ for any $x, y, z \in \mathbf{X}$.

Unless otherwise stated, the scalar field will now be taken to be $\mathbb{C}$, but occasionally it will be convenient to allow only real scalars. Certain obvious simplifications in the rules then occur, for example, [H4] becomes $\langle x, y\rangle = \langle y, x\rangle$ for all x, y. Properties derived in the remainder of this chapter are valid for either choice of scalar field.

Note that from axioms [H3] and [H4] it follows that

$$\langle x, \lambda y\rangle = \overline{\langle \lambda y, x\rangle} = \overline{\lambda\langle y, x\rangle} = \bar{\lambda}\overline{\langle y, x\rangle} = \bar{\lambda}\langle x, y\rangle \tag{5.1.1}$$

Example 5.1. The vector space $\mathbb{C}^N$ is an inner product space if we define

$$\langle x, y\rangle = \sum_{j=1}^{N} x_j \bar{y}_j \tag{5.1.2}$$

In the case of $\mathbb{R}^N$ of course this simplifies to

$$\langle x, y\rangle = \sum_{j=1}^{N} x_j y_j \tag{5.1.3}$$

which we recognize as the usual dot product formula from vector calculus. □

Techniques of Functional Analysis for Differential and Integral Equations
http://dx.doi.org/10.1016/B978-0-12-811426-1.00005-2

Example 5.2. For the vector space $L^2(\Omega)$, with $\Omega \subset \mathbb{R}^N$, we may define

$$\langle f, g \rangle = \int_\Omega f(x)\overline{g(x)}\, dx \tag{5.1.4}$$

Note the formal analogy with the inner product in the case of $\mathbb{R}^N$ or $\mathbb{C}^N$. The finiteness of $\langle f, g \rangle$ is guaranteed by the Hölder inequality (A.2.19), and the validity of [H1]–[H5] is clear. □

Example 5.3. Another important inner product space which we introduce at this point is the sequence space

$$\ell^2 = \left\{ x = \{x_j\}_{j=1}^\infty : \sum_{j=1}^\infty |x_j|^2 < \infty \right\} \tag{5.1.5}$$

with inner product

$$\langle x, y \rangle = \sum_{j=1}^\infty x_j \bar{y}_j \tag{5.1.6}$$

The fact that $\langle x, y \rangle$ is finite for any $x, y \in \ell^2$ follows now from Eq. (A.2.27), the discrete form of the Hölder inequality. More generally, if A denotes any index set then

$$\ell^2(A) = \left\{ x = \{x_\alpha\}_{\alpha \in A} : \sum_{\alpha \in A} |x_\alpha|^2 < \infty \right\} \tag{5.1.7}$$

It is possible to allow for uncountable index sets A, but will we only make use of the case that A is a finite or countably infinite set. □

5.2 NORM IN A HILBERT SPACE

Proposition 5.1. *If* $\mathbf{X}$ *is an inner product space and* $x, y \in \mathbf{X}$*, then*

$$|\langle x, y \rangle|^2 \le \langle x, x \rangle \langle y, y \rangle \tag{5.2.8}$$

Proof. For any $z \in \mathbf{X}$ we have

$$0 \le \langle x - z, x - z \rangle = \langle x, x \rangle - \langle x, z \rangle - \langle z, x \rangle + \langle z, z \rangle \tag{5.2.9}$$

$$= \langle x, x \rangle + \langle z, z \rangle - 2\mathrm{Re}\, \langle x, z \rangle \tag{5.2.10}$$

and hence

$$2\mathrm{Re}\, \langle z, x \rangle \le \langle x, x \rangle + \langle z, z \rangle \tag{5.2.11}$$

If $y = 0$ there is nothing to prove, otherwise choose $z = (\langle x, y \rangle / \langle y, y \rangle) y$ to get

$$2 \frac{|\langle x, y \rangle|^2}{\langle y, y \rangle} \le \langle x, x \rangle + \frac{|\langle x, y \rangle|^2}{\langle y, y \rangle} \tag{5.2.12}$$

The conclusion (5.2.8) now follows upon rearrangement. □

Theorem 5.1. *If* $\mathbf{X}$ *is an inner product space and if we set* $\|x\| = \sqrt{\langle x, x\rangle}$ *then* $\|\cdot\|$ *is a norm on* $\mathbf{X}$.

Proof. By axiom [H1], $\|x\|$ is defined as a nonnegative real number for every $x \in \mathbf{X}$, and axiom [H2] implies the corresponding axiom [N1] of norm. If λ is any scalar then $\|\lambda x\|^2 = \langle \lambda x, \lambda x\rangle = \lambda\bar{\lambda}\langle x, x\rangle = |\lambda|^2\|x\|^2$ so that [N2] also holds. Finally, if $x, y \in \mathbf{X}$ then

$$\|x+y\|^2 = \langle x+y, x+y\rangle = \|x\|^2 + 2\mathrm{Re}\,\langle x, y\rangle + \|y\|^2 \tag{5.2.13}$$

$$\leq \|x\|^2 + 2|\langle x, y\rangle| + \|y\|^2 \leq \|x\|^2 + 2\|x\|\,\|y\| + \|y\|^2 \tag{5.2.14}$$

$$= (\|x\| + \|y\|)^2 \tag{5.2.15}$$

so that the triangle inequality [N3] is also valid. □

The inequality (5.2.8) may now be restated as

$$|\langle x, y\rangle| \leq \|x\|\,\|y\| \tag{5.2.16}$$

for any $x, y \in \mathbf{X}$, and in this form is usually called the Schwarz or Cauchy-Schwarz inequality.

Corollary 5.1. *If* $x_n \to x$ *in* $\mathbf{X}$ *then* $\langle x_n, y\rangle \to \langle x, y\rangle$ *for any* $y \in \mathbf{X}$.

Proof. We have that

$$|\langle x_n, y\rangle - \langle x, y\rangle| = |\langle x_n - x, y\rangle| \leq \|x_n - x\|\,\|y\| \to 0 \tag{5.2.17}$$

□

Another immediate consequence of the axioms is that

$$\|x+y\|^2 = \langle x+y, x+y\rangle = \|x\|^2 + 2\mathrm{Re}\,\langle x, y\rangle + \|y\|^2 \tag{5.2.18}$$

If we replace y by $-y$ and add the resulting identities we obtain the so-called Parallelogram Law

$$\|x+y\|^2 + \|x-y\|^2 = 2\|x\|^2 + 2\|y\|^2 \tag{5.2.19}$$

This is thus a necessary condition for a norm which is defined in terms of an inner product.

By Theorem 5.1 an inner product space may always be regarded as a normed linear space, and analogously to the definition of Banach space we have

Definition 5.2. A Hilbert space is a complete inner product space.

Example 5.4. The spaces $\mathbb{R}^N$ and $\mathbb{C}^N$ are Hilbert spaces, as is $L^2(\Omega)$ on account of the completeness property mentioned in Theorem 3.1 of Chapter 3. On the other hand if we consider $C(E)$ with inner product $\langle f, g\rangle = \int_E f(x)\overline{g(x)}\,dx$, then it is an inner product space which is *not* a Hilbert space, since as previously observed, $C(E)$ is not complete with the metric corresponding to the $L^2(\Omega)$ norm. Any sequence space $\ell^2(A)$ is also a Hilbert space, see Exercise 5.7. □

5.3 ORTHOGONALITY

Recall from elementary calculus that in $\mathbb{R}^N$ the inner product allows one to calculate the angle between two vectors, namely

$$\langle x, y\rangle = \|x\|\,\|y\|\cos\theta \tag{5.3.20}$$

where θ is the angle between x and y. In particular x and y are perpendicular if and only if $\langle x, y\rangle = 0$. The concept of perpendicularity, or as it will henceforth be referred to, orthogonality, is fundamental in Hilbert space analysis, even if the geometric picture is less clear.

Definition 5.3. If $\mathbf{X}$ is an inner product space and $x, y \in \mathbf{X}$, we say x, y are *orthogonal* if $\langle x, y\rangle = 0$.

From Eq. (5.2.18) we obtain immediately the "Pythagorean Theorem" that if x and y are orthogonal then

$$\|x + y\|^2 = \|x\|^2 + \|y\|^2 \tag{5.3.21}$$

A collection of vectors $\{x_1, x_2, \ldots, x_n\}$ is called an *orthogonal set* if x_j and x_k are orthogonal whenever $j \neq k$, and for such a set we have

$$\left\|\sum_{j=1}^{n} x_j\right\|^2 = \sum_{j=1}^{n} \|x_j\|^2 \tag{5.3.22}$$

The same terminology is used for countably infinite sets, with Eq. (5.3.22) still valid provided that the series on the right is convergent. The set is called *orthonormal* if in addition $\|x_j\| = 1$ for every j.

Example 5.5. If $f_n(x) = \sin nx$ then $\{f_n\}_{n=1}^{\infty}$ is orthogonal in $L^2(0, \pi)$. This is seen simply by evaluating the necessary inner products $\langle f_n, f_m\rangle = \int_0^{\pi} \sin nx \sin mx\, dx$. It is an orthonormal set if each f_n is replaced by $f_n/\sqrt{2\pi}$. □

We also use the notation $x \perp y$ if x, y are orthogonal, and if $E \subset \mathbf{X}$ we define the orthogonal complement of E to be

$$E^{\perp} = \{x \in \mathbf{X} : \langle x, y\rangle = 0 \text{ for all } y \in E\}$$

If $E = \{x\}$, a set containing a single point, then we use the simpler notation $x^{\perp}$. Obviously we have $0^{\perp} = \mathbf{X}$ and $\mathbf{X}^{\perp} = \{0\}$ also, since if $x \in \mathbf{X}^{\perp}$ then $\langle x, x\rangle = 0$ so that $x = 0$.

Proposition 5.2. *If $E \subset \mathbf{X}$ then $E^{\perp}$ is a closed subspace of $\mathbf{X}$ and E is a closed subspace if and only if $E = E^{\perp\perp}$.*

We leave the proof as an exercise. Here $E^{\perp\perp}$ of course means $(E^{\perp})^{\perp}$, the orthogonal complement of the orthogonal complement.

Example 5.6. If $\mathbf{X} = \mathbb{R}^3$ and $E = \{x = (x_1, x_2, x_3) : x_1 = x_2 = 0\}$ then $E^{\perp} = \{x \in \mathbb{R}^3 : x_3 = 0\}$. □

Example 5.7. If $\mathbf{X} = L^2(\Omega)$ with Ω a bounded open set in $\mathbb{R}^N$, let E denote the set of constant functions. Then $f \in E^\perp$ if and only if $\langle f, 1\rangle = \int_\Omega f(x)\,dx = 0$. Thus $E^\perp$ is the set of functions in $L^2(\Omega)$ with mean value zero. □

5.4 PROJECTIONS

If $\mathbf{X}$ is a normed linear space, $E \subset \mathbf{X}$ and $x \in \mathbf{X}$, then the *projection of x onto E*, denoted $P_E x$, is defined to be the unique element of E closest to x, if such an element exists. That is, $y = P_E(x)$ if y is the unique solution of the minimization problem

$$\min_{z\in E} \|x - z\| \tag{5.4.23}$$

Of course the minimization problem may not possess any solution, and may not be unique if it does exist. In a Hilbert space, however, we will see that the projection is well defined provided E is closed and convex.

Definition 5.4. If $\mathbf{X}$ is a vector space and $E \subset \mathbf{X}$, we say E is convex if $\lambda x + (1-\lambda)y \in E$ whenever $x, y \in E$ and $\lambda \in [0,1]$.

Example 5.8. If $\mathbf{X}$ is a vector space then any subspace of $\mathbf{X}$ is convex. If $\mathbf{X}$ is a normed linear space then any ball $B(x,R) \subset \mathbf{X}$ is convex. □

Theorem 5.2. *Let $\mathbf{H}$ be a Hilbert space, $E \subset \mathbf{H}$ closed and convex, and $x \in \mathbf{H}$. Then $y = P_E x$ exists. Furthermore, $y = P_E x$ if and only if*

$$y \in E \qquad \mathrm{Re}\,\langle x - y, z - y\rangle \le 0 \quad \text{for all } z \in E \tag{5.4.24}$$

Proof. Set $d = \inf_{z\in E}\|x - z\|$ so that there exists a sequence $z_n \in E$ such that $\|x - z_n\| \to d$. We wish to show that $\{z_n\}$ is a Cauchy sequence. From the Parallelogram Law (5.2.19) applied to $z_n - x, z_m - x$ we have

$$\|z_n - z_m\|^2 = 2\|z_n - x\|^2 + 2\|z_m - x\|^2 - 4\left\|\frac{z_n + z_m}{2} - x\right\|^2 \tag{5.4.25}$$

Since E is convex, $(z_n + z_m)/2 \in E$ so that $\|\frac{z_n+z_m}{2} - x\| \ge d$, and it follows that

$$\|z_n - z_m\|^2 \le 2\|z_n - x\|^2 + 2\|z_m - x\|^2 - 4d^2 \tag{5.4.26}$$

Letting $n, m \to \infty$ the right-hand side tends to zero, so that $\{z_n\}$ is Cauchy. Since the space is complete there exists $y \in \mathbf{H}$ such that $\lim_{n\to\infty} z_n = y$, and $y \in E$ since E is closed. It follows that $\|y - x\| = \lim_{n\to\infty}\|z_n - x\| = d$ so that $\min_{z\in E}\|z - x\|$ is achieved at y.

For the uniqueness assertion, suppose $\|y - x\| = \|\widehat{y} - x\| = d$ with $y, \widehat{y} \in E$. Then Eq. (5.4.26) holds with z_n, z_m replaced by $y, \widehat{y}$ giving

$$\|y - \widehat{y}\|^2 \le 2\|y - x\|^2 + 2\|\widehat{y} - x\|^2 - 4d^2 = 0 \tag{5.4.27}$$

so that $y = \widehat{y}$. Thus $y = P_E x$ exists.

Finally, to obtain the characterization (5.4.24), note that for any $z \in E$

$$f(t) = \|x - (y + t(z - y))\|^2 \tag{5.4.28}$$

has its minimum value with respect to the interval $[0, 1]$ at $t = 0$, since $y + t(z - y) = tz + (1 - t)y \in E$. We explicitly calculate

$$f(t) = \|x - y\|^2 - 2t \operatorname{Re} \langle x - y, z - y\rangle + t^2 \|z - y\|^2 \tag{5.4.29}$$

By elementary calculus considerations, the minimum of this quadratic occurs at $t = 0$ only if $f'(0) = -2 \operatorname{Re} \langle x-y, z-y\rangle \geq 0$ which is equivalent to Eq. (5.4.24). If, on the other hand, Eq. (5.4.24) holds, then for any $z \in E$ we must have

$$\|z - x\|^2 = f(1) \geq f(0) = \|x - y\|^2 \tag{5.4.30}$$

so that $\min_{z\in E} \|z - x\|$ must occur at y, that is, $y = P_E x$. □

The most important special case of the previous theorem is when E is a closed subspace of the Hilbert space $\mathbf{H}$ (recall a subspace is always convex), in which case we have

Theorem 5.3. *If $E \subset \mathbf{H}$ is a closed subspace of a Hilbert space $\mathbf{H}$ and $x \in \mathbf{H}$ then $y = P_E x$ if and only if $y \in E$ and $x - y \in E^\perp$. Furthermore*

1. $x - y = x - P_E x = P_{E^\perp} x$
2. *We have that*

$$x = y + (x - y) = P_E x + P_{E^\perp} x \tag{5.4.31}$$

is the unique decomposition of x as the sum of an element of E and an element of $E^\perp$.

3. *P_E is a linear operator on $\mathbf{H}$ with $\|P_E\| = 1$ except for the trivial case $E - \{0\}$.*

Proof. If $y = P_E x$ then for any $w \in E$ we also have $y \pm w \in E$, and choosing $z = y \pm w$ in Eq. (5.4.24) gives $\pm \operatorname{Re} \langle x - y, w\rangle \leq 0$. Thus $\operatorname{Re} \langle x - y, w\rangle = 0$, and repeating the same argument with $z = y \pm iw$ gives $\operatorname{Re} \langle x - y, iw\rangle = \operatorname{Im} \langle x - y, w\rangle = 0$ also. We conclude that $\langle x - y, w\rangle = 0$ for all $w \in E$, that is, $x - y \in E^\perp$. The converse statement may be proved in a similar manner.

Recall that $E^\perp$ is always a closed subspace of $\mathbf{H}$. The statement that $x - y = P_{E^\perp} x$ is then equivalent, by the previous paragraph, to $x - y \in E^\perp$ and $\langle x - (x - y), w\rangle = \langle y, w\rangle = 0$ for every $w \in E^\perp$, which is evidently true since $y \in E$.

Next, if $x = y_1 + z_1 = y_2 + z_2$ with $y_1, y_2 \in E$ and $z_1, z_2 \in E^\perp$ then $y_1 - y_2 = z_2 - z_1$ implying that $y = y_1 - y_2$ belongs to both E and $E^\perp$. But then $y \perp y$, that is, $\langle y, y\rangle = 0$, must hold so that $y = 0$ and hence $y_1 = y_2, z_1 = z_2$. We leave the proof of the final statement to the exercises. □

If we denote by I the identity mapping, we have just proved that $P_{E^\perp} = I - P_E$. We also obtain that

$$\|x\|^2 = \|P_E x\|^2 + \|P_{E^\perp} x\|^2 \tag{5.4.32}$$

for any $x \in \mathbf{H}$.

Example 5.9. In the Hilbert space $L^2(-1,1)$ let E denote the subspace of even functions, that is, $f \in E$ if $f(x) = f(-x)$ for almost every $x \in (-1,1)$. We claim that $E^\perp$ is the subspace of odd functions on $(-1,1)$. The fact that any odd function belongs to $E^\perp$ is clear, since if f is even and g is odd then $f\overline{g}$ is odd and so $\langle f, g\rangle = \int_{-1}^{1} f(x)\overline{g(x)}\,dx = 0$. Conversely, if $g \perp E$ then for any $f \in E$ we have

$$0 = \langle g, f\rangle = \int_{-1}^{1} g(x)\overline{f(x)}\,dx = \int_0^1 (g(x) + g(-x))\overline{f(x)}\,dx \tag{5.4.33}$$

by an obvious change of variables. Choosing $f(x) = g(x) + g(-x)$ we see that

$$\int_0^1 |g(x) + g(-x)|^2\,dx = 0 \tag{5.4.34}$$

so that $g(x) = -g(-x)$ for almost every $x \in (0,1)$ and hence for almost every $x \in (-1,1)$. Thus any element of $E^\perp$ is an odd function on $(-1,1)$.

Any function $f \in L^2(-1,1)$ thus has the unique decomposition $f = P_E f + P_{E^\perp} f$, a sum of an even and an odd function. Since one such splitting is

$$f(x) = \frac{f(x) + f(-x)}{2} + \frac{f(x) - f(-x)}{2} \tag{5.4.35}$$

we conclude from the uniqueness property that these two term are the projections, that is,

$$P_E f(x) = \frac{f(x) + f(-x)}{2} \qquad P_{E^\perp} f(x) = \frac{f(x) - f(-x)}{2} \tag{5.4.36}$$

□

Example 5.10. Let $\{x_1, x_2, \dots, x_n\}$ be an orthogonal set of nonzero elements in a Hilbert space $\mathbf{X}$ and $E = \text{span}\{x_1, x_2, \dots, x_n\}$. Let us compute P_E for this closed subspace E. If $y = P_E x$ then $y = \sum_{j=1}^n \lambda_j x_j$ for some scalars $\lambda_1, \dots, \lambda_n$ since $y \in E$. From Theorem 5.3 we also have that $x - y \perp E$ which is equivalent to $x - y \perp x_k$ for each k. Thus $\langle x, x_k\rangle = \langle y, x_k\rangle = \lambda_k \langle x_k, x_k\rangle$ using the orthogonality assumption. Therefore we conclude that

$$y = P_E x = \sum_{j=1}^n \frac{\langle x, x_j\rangle}{\langle x_j, x_j\rangle} x_j \tag{5.4.37}$$

□

5.5 GRAM-SCHMIDT METHOD

The projection formula (5.4.37) provides an explicit and very convenient expression for the solution y of the best approximation problem (5.4.23) *provided E* is a subspace spanned by mutually orthogonal vectors $\{x_1, x_2, \ldots, x_n\}$. If instead $E = \text{span}\{x_1, x_2, \ldots, x_n\}$ is a subspace but $\{x_1, x_2, \ldots, x_n\}$ are not orthogonal vectors, we can still use Eq. (5.4.37) to compute $y = P_E x$ if we can find a set of orthogonal vectors $\{y_1, y_2, \ldots, y_m\}$ such that $E = \text{span}\{x_1, x_2, \ldots, x_n\} = \text{span}\{y_1, y_2, \ldots, y_m\}$, that is, if we can find an orthogonal basis of E. This may always be done by adapting the Gram-Schmidt orthogonalization procedure from linear algebra, which we now recall.

Assume for the moment that $\{x_1, x_2, \ldots, x_n\}$ are linearly independent, so that $m = n$ must hold. First set $y_1 = x_1$. If orthogonal vectors $y_1, y_2, \ldots, y_k$ have been chosen for some $1 \le k < n$ such that $E_k := \text{span}\{y_1, y_2, \ldots, y_k\} = \text{span}\{x_1, x_2, \ldots, x_k\}$ then define $y_{k+1} = x_{k+1} - P_{E_k} x_{k+1}$. Clearly $\{y_1, y_2, \ldots, y_{k+1}\}$ are orthogonal since y_{k+1} is the projection of x_{k+1} onto $E_k^{\perp}$. Also since y_{k+1}, x_{k+1} differ by an element of E_k it is evident that $\text{span}\{x_1, x_2, \ldots, x_{k+1}\} = \text{span}\{y_1, y_2, \ldots, y_{k+1}\}$. Thus after n steps we obtain an orthogonal set $\{y_1, y_2, \ldots, y_n\}$ which spans E. If the original set $\{x_1, x_2, \ldots, x_n\}$ were not linearly independent then some of the y_k's will be zero. After discarding these and relabeling, we obtain $\{y_1, y_2, \ldots, y_m\}$ for some $m < n$, again an orthogonal basis for E. Note that we may compute y_{k+1} using Eq. (5.4.37), namely

$$y_{k+1} = x_{k+1} - \sum_{j=1}^{k} \frac{\langle x_{k+1}, y_j \rangle}{\langle y_j, y_j \rangle} y_j \tag{5.5.38}$$

In practice the Gram-Schmidt method is often modified to produce an *orthonormal* basis of E by normalizing y_k to be a unit vector at each step, or else discarding it if it is already a linear combination of $\{y_1, y_2, \ldots, y_{k-1}\}$. More explicitly:

- Set $y_1 = \frac{x_1}{\|x_1\|}$
- If orthonormal vectors $\{y_1, y_2, \ldots, y_k\}$ have been chosen, set

$$\tilde{y}_{k+1} = x_{k+1} - \sum_{j=1}^{k} \langle x_{k+1}, y_j \rangle y_j \tag{5.5.39}$$

If $\tilde{y}_{k+1} = 0$ discard it, otherwise set $y_{k+1} = \frac{\tilde{y}_{k+1}}{\|\tilde{y}_{k+1}\|}$.

The reader may easily check that $\{y_1, y_2, \ldots, y_m\}$ constitutes an orthonormal basis of E, and consequently $P_E x = \sum_{j=1}^{m} \langle x, y_j \rangle y_j$ for any $x \in \mathbf{H}$.

5.6 BESSEL'S INEQUALITY AND INFINITE ORTHOGONAL SEQUENCES

The formula (5.4.37) for P_E may be adapted for use in infinite dimensional subspaces E. If $\{x_n\}_{n=1}^{\infty}$ is a countable orthogonal set in $\mathbf{H}$, $x_n \neq 0$ for all n, we formally expect, simply by letting $n \to \infty$ in Eq. (5.4.37), that if $E = \text{span}\,(\{x_n\}_{n=1}^{\infty})$ then

$$P_E x = \sum_{n=1}^{\infty} \frac{\langle x, x_n \rangle}{\langle x_n, x_n \rangle} x_n \tag{5.6.40}$$

To verify that this is correct, we must in particular show that the infinite series in Eq. (5.6.40) is guaranteed to be convergent in **H**.

First of all, let us set

$$e_n = \frac{x_n}{\|x_n\|} \qquad c_n = \langle x, e_n \rangle \qquad E_N = \text{span}\{x_1, x_2, \ldots, x_N\} \tag{5.6.41}$$

so that $\{e_n\}_{n=1}^{\infty}$ is an orthonormal set, and

$$P_{E_N} x = \sum_{n=1}^{N} c_n e_n \tag{5.6.42}$$

From Eq. (5.4.32) we have

$$\sum_{n=1}^{N} |c_n|^2 = \|P_{E_N} x\|^2 \leq \|x\|^2 \tag{5.6.43}$$

Letting $N \to \infty$ we obtain *Bessel's inequality*

$$\sum_{n=1}^{\infty} |c_n|^2 = \sum_{n=1}^{\infty} |\langle x, e_n \rangle|^2 \leq \|x\|^2 \tag{5.6.44}$$

The immediate consequence that $\lim_{n\to\infty} c_n = 0$ is sometimes called the Riemann-Lebesgue lemma.

Proposition 5.3. *(Riesz-Fischer) Let $\{e_n\}_{n=1}^{\infty}$ be an orthonormal set in **H**, $E =$ span $(\{e_n\}_{n=1}^{\infty})$, $x \in$ **H** and $c_n = \langle x, e_n \rangle$. Then the infinite series $\sum_{n=1}^{\infty} c_n e_n$ is convergent in **H** to $P_E x$.*

Proof. First we note that the series $\sum_{n=1}^{\infty} c_n e_n$ is Cauchy in **H** since if $M > N$

$$\left\| \sum_{n=N}^{M} c_n e_n \right\|^2 = \sum_{n=N}^{M} |c_n|^2 \tag{5.6.45}$$

which is less than any prescribed $\epsilon > 0$ for $M > N$ sufficiently large, since $\sum_{n=1}^{\infty} |c_n|^2 < \infty$. Thus $y = \sum_{n=1}^{\infty} c_n e_n$ exists in **H**, and clearly $y \in E$. Since $\langle \sum_{n=1}^{N} c_n e_n, e_m \rangle = c_m$ if $N > m$ it follows easily that $\langle y, e_m \rangle = c_m = \langle x, e_m \rangle$. Thus $y - x \perp e_m$ for any m which implies $y - x \in E^{\perp}$. From Theorem 5.3 we conclude that $y = P_E x$. □

5.7 CHARACTERIZATION OF A BASIS OF A HILBERT SPACE

Now suppose we have an orthogonal set $\{x_n\}_{n=1}^{\infty}$ and we wish to determine whether or not it is a basis of the Hilbert space **H**. There are a number of interesting and useful ways to answer this question, summarized in Theorem 5.4. First we must make some more definitions.

Definition 5.5. A collection of vectors $\{x_n\}_{n=1}^\infty$ is *closed* in **H** if the set of all finite linear combinations of $\{x_n\}_{n=1}^\infty$ is dense in **H**

A collection of vectors $\{x_n\}_{n=1}^\infty$ is *complete* in **H** if there is no nonzero vector orthogonal to all of them, that is, $\langle x, x_n\rangle = 0$ for all n if and only if $x = 0$.

An orthonormal set $\{x_n\}_{n=1}^\infty$ in **H** is a *maximal* orthonormal set if it is not contained in any strictly larger orthonormal set.

Theorem 5.4. *Let $\{e_n\}_{n=1}^\infty$ be an orthonormal set in a Hilbert space* **H**. *Then the following are equivalent.*

(a) $\{e_n\}_{n=1}^\infty$ *is a basis of* **H**.
(b) $x = \sum_{n=1}^\infty \langle x, e_n\rangle e_n$ *for every* $x \in \mathbf{H}$.
(c) $\langle x, y\rangle = \sum_{n=1}^\infty \langle x, e_n\rangle \langle e_n, y\rangle$ *for every* $x, y \in \mathbf{H}$.
(d) $\|x\|^2 = \sum_{n=1}^\infty |\langle x, e_n\rangle|^2$ *for every* $x \in \mathbf{H}$.
(e) $\{e_n\}_{n=1}^\infty$ *is a maximal orthonormal set.*
(f) $\{e_n\}_{n=1}^\infty$ *is closed in* **H**.
(g) $\{e_n\}_{n=1}^\infty$ *is complete in* **H**.

Proof. *(a) implies (b):* If $\{e_n\}_{n=1}^\infty$ is a basis of **H** then for any $x \in \mathbf{H}$ there exist unique constants d_n such that $x = \lim_{N\to\infty} S_N$ where $S_N = \sum_{n=1}^N d_n e_n$. Since $\langle S_N, e_m\rangle = d_m$ if $N > m$ it follows for such N that

$$|d_m - \langle x, e_m\rangle| = |\langle S_N - x, e_m\rangle| \le \|S_N - x\| \, \|e_m\| \to 0 \tag{5.7.46}$$

as $N \to \infty$, using the Schwarz inequality. Hence

$$x = \sum_{n=1}^\infty d_n e_n = \sum_{n=1}^\infty \langle x, e_n\rangle e_n \tag{5.7.47}$$

(b) implies (c): For any $x, y \in \mathbf{H}$ we have

$$\langle x, y\rangle = \left\langle x, \lim_{N\to\infty} \sum_{n=1}^N \langle y, e_n\rangle e_n \right\rangle \tag{5.7.48}$$

$$= \lim_{N\to\infty} \left\langle x, \sum_{n=1}^N \langle y, e_n\rangle e_n \right\rangle = \lim_{N\to\infty} \sum_{n=1}^N \overline{\langle y, e_n\rangle} \langle x, e_n\rangle \tag{5.7.49}$$

$$= \sum_{n=1}^\infty \langle x, e_n\rangle \overline{\langle y, e_n\rangle} = \sum_{n=1}^\infty \langle x, e_n\rangle \langle e_n, y\rangle \tag{5.7.50}$$

Here we have used Corollary 5.1 in the second equality.

(c) implies (d): We simply choose $x = y$ in the identity stated in (c).

(d) implies (e): If $\{e_n\}_{n=1}^\infty$ is not maximal then there exists $e \in \mathbf{H}$ such that

$$\{e_n\}_{n=1}^\infty \cup \{e\} \tag{5.7.51}$$

is orthonormal. Since $\langle e, e_n \rangle = 0$ but $\|e\| = 1$ (d) cannot be true for the case $x = e$.

(e) implies (f): Let E denote the set of finite linear combinations of the e_n's. If $\{e_n\}_{n=1}^{\infty}$ is not closed then $\overline{E} \neq \mathbf{H}$ so there must exist $x \notin \overline{E}$. If we let $y = x - P_{\overline{E}}x$ then $y \neq 0$ and $y \perp E$. If $e = y/\|y\|$ we would then have that $\{e_n\}_{n=1}^{\infty} \cup \{e\}$ is orthonormal so that $\{e_n\}_{n=1}^{\infty}$ could not be maximal.

(f) implies (g): Assume that $\langle x, e_n \rangle = 0$ for all n. If $\{e_n\}_{n=1}^{\infty}$ is closed then for any $\epsilon > 0$ there exists N and $\lambda_1, \dots, \lambda_N$ such that $\|x - \sum_{n=1}^{N} \lambda_n e_n\|^2 < \epsilon$. But then $\|x\|^2 + \sum_{n=1}^{N} |\lambda_n|^2 < \epsilon$ and in particular $\|x\|^2 < \epsilon$. Thus $x = 0$ so $\{e_n\}_{n=1}^{\infty}$ is complete.

(g) implies (a): Let $E = \text{span}\,(\{e_n\}_{n=1}^{\infty})$. If $x \in \mathbf{H}$ and $y = \sum_{n=1}^{\infty} \langle x, e_n \rangle e_n$ then as in the proof of Proposition 5.3, $y = P_E x$ and $\langle y, e_n \rangle = \langle x, e_n \rangle$. Since $\{e_n\}_{n=1}^{\infty}$ is complete it follows that $x = y \in E$ so that span $\{e_n\}_{n=1}^{\infty} = \mathbf{H}$. Since an orthonormal set is obviously linearly independent it follows that $\{e_n\}_{n=1}^{\infty}$ is a basis of $\mathbf{H}$. □

Because of the equivalence of the conditions stated in this theorem, the phrases "complete orthonormal set," "maximal orthonormal set," and "closed orthonormal set" are often used interchangeably with "orthonormal basis" in a Hilbert space setting. The identity in (d) is called the Bessel equality (recall the corresponding inequality (5.6.44) is valid whether or not the orthonormal set $\{e_n\}_{n=1}^{\infty}$ is a basis), while the identity in (c) is the Parseval equality. For reasons which should become more clear in Chapter 7 the infinite series $\sum_{n=1}^{\infty} \langle x, e_n \rangle e_n$ is often called the generalized Fourier series of x with respect to the orthonormal basis $\{e_n\}_{n=1}^{\infty}$, and $\langle x, e_n \rangle$ is the nth generalized Fourier coefficient.

Theorem 5.5. *Every separable Hilbert space has an orthonormal basis.*

Proof. If $\{x_n\}_{n=1}^{\infty}$ is a countable dense sequence in $\mathbf{H}$ and we carry out the Gram-Schmidt procedure, we obtain an orthonormal sequence $\{e_n\}_{n=1}^{\infty}$. This sequence must be complete, since any vector orthogonal to every e_n must also be orthogonal to every x_n, so must be zero, since $\{x_n\}_{n=1}^{\infty}$ is dense. Therefore by Theorem 5.4 $\{e_n\}_{n=1}^{\infty}$ (or $\{e_1, e_2, \dots, e_N\}$ in the finite dimensional case) is an orthonormal basis of $\mathbf{H}$. □

The same conclusion is actually correct in a nonseparable Hilbert space also, but needs more explanation. See, for example, Chapter 4 of [32].

5.8 ISOMORPHISMS OF A HILBERT SPACE

There are two interesting and important isomorphisms of every separable Hilbert space, one is to its so-called dual space, and the second is to the sequence space ℓ^2. In this section we explain both of these facts.

Recall that in Chapter 4 we have already introduced the so-called dual space $\mathbf{X}^* = \mathcal{B}(\mathbf{X}, \mathbb{C})$, the space of continuous linear functionals on the normed linear space $\mathbf{X}$. It is itself always a Banach space (see Exercise 4.3 of Chapter 4).

Example 5.11. If $\mathbf{H}$ is a Hilbert space and $y \in \mathbf{H}$, define $\phi(x) = \langle x, y \rangle$. Then $\phi : \mathbf{H} \to \mathbb{C}$ is clearly linear, and $|\phi(x)| \leq \|y\| \, \|x\|$ by the Schwarz inequality, hence $\phi \in \mathbf{H}^*$, with $\|\phi\| \leq \|y\|$. □

The following fundamental theorem asserts that every element of the dual space $\mathbf{H}^*$ arises in this way.

Theorem 5.6. *(Riesz representation theorem) If $\mathbf{H}$ is a Hilbert space and $\phi \in \mathbf{H}^*$ then there exists a unique $y \in \mathbf{H}$ such that $\phi(x) = \langle x, y \rangle$. Furthermore $\|y\| = \|\phi\|$.*

Proof. Let $M = \{x \in \mathbf{H} : \phi(x) = 0\}$, which is clearly a closed subspace of $\mathbf{H}$. If $M = \mathbf{H}$ then ϕ is the zero functional, so $y = 0$ has the required properties. Otherwise, there must exist $e \in M^\perp$ such that $\|e\| = 1$. For any $x \in \mathbf{H}$ let $z = \phi(x)e - \phi(e)x$ and observe that $\phi(z) = 0$ so $z \in M$, and in particular $z \perp e$. It then follows that

$$0 = \langle z, e \rangle = \phi(x)\langle e, e \rangle - \phi(e)\langle x, e \rangle \tag{5.8.52}$$

Thus $\phi(x) = \langle x, y \rangle$ with $y := \overline{\phi(e)}e$, for every $x \in \mathbf{H}$. As in the previous example, $\|\phi\| \leq \|y\|$, and from $\phi(y) = \|y\|^2$ we get that $\|\phi\| = \|y\|$.

The uniqueness property is even easier to show. If $\phi(x) = \langle x, y_1 \rangle = \langle x, y_2 \rangle$ for every $x \in \mathbf{H}$ then necessarily $\langle x, y_1 - y_2 \rangle = 0$ for all x, and choosing $x = y_1 - y_2$ we get $\|y_1 - y_2\|^2 = 0$, that is, $y_1 = y_2$. □

We view the element $y \in \mathbf{H}$ as "representing" the linear functional $\phi \in \mathbf{H}^*$, hence the name of the theorem. There are actually several theorems one may encounter, all called the Riesz representation theorem, and what they all have in common is that the dual space of some other space is characterized. The Hilbert space version here is by the far the easiest of these theorems.

If we define the mapping $R : \mathbf{H} \to \mathbf{H}^*$ (the *Riesz map*) by the condition $R(y) = \phi$, with ϕ, y related as mentioned previously, then Theorem 5.6 implies that R is one to one and onto. Since it is easy to check that R is also linear, it follows that R is an isomorphism from $\mathbf{H}$ to $\mathbf{H}^*$. Because of the property $\|R(y)\| = \|y\|$ for all y, we say R is an *isometric isomorphism.*

Next, suppose that $\mathbf{H}$ is an infinite dimensional separable Hilbert space. According to Theorem 5.5 there exists an orthonormal basis of $\mathbf{H}$ which cannot be finite, and so may be written as $\{e_n\}_{n=1}^\infty$. Associate with any $x \in \mathbf{H}$ the corresponding sequence of generalized Fourier coefficients $\{c_n\}_{n=1}^\infty$, where $c_n = \langle x, e_n \rangle$, and let Λ denote this mapping, that is, $\Lambda(x) = \{c_n\}_{n=1}^\infty$.

We know by Theorem 5.4 that $\sum_{n=1}^\infty |c_n|^2 < \infty$, that is, $\Lambda(x) \in \ell^2$. On the other hand, suppose $\sum_{n=1}^\infty |c_n|^2 < \infty$ and let $x = \sum_{n=1}^\infty c_n e_n$. This series is Cauchy, hence convergent in $\mathbf{H}$, by precisely the same argument as used in the beginning of the proof of Proposition 5.3. Since $\{e_n\}_{n=1}^\infty$ is a basis, we must have $c_n = \langle x, e_n \rangle$, thus $\Lambda(x) = \{c_n\}_{n=1}^\infty$, and consequently $\Lambda : \mathbf{H} \to \ell^2$ is onto. It is also one-to-one, since $\Lambda(x_1) = \Lambda(x_2)$ means that $\langle x_1 - x_2, e_n \rangle = 0$ for every n, hence $x_1 - x_2 = 0$ by the completeness property of a basis. Finally it is straightforward to check that Λ is linear, so that Λ is an isomorphism. Like the Riesz map, the isomorphism Λ is also isometric, $\|\Lambda(x)\| = \|x\|$, on account of the Bessel equality. By the above considerations we have then established the following theorem.

Theorem 5.7. *If **H** is an infinite dimensional separable Hilbert space, then **H** is isometrically isomorphic to* ℓ^2.

Since all such Hilbert spaces are isometrically isomorphic to ℓ^2, they are then obviously isometrically isomorphic to each other. If **H** is a Hilbert space of dimension N, the same arguments show that **H** is isometrically isomorphic to the Hilbert space $\mathbb{R}^N$ or $\mathbb{C}^N$, depending on whether real or complex scalars are allowed. Finally, see Theorem 4.17 of [32] for the nonseparable case.

5.9 EXERCISES

5.1. Prove Proposition 5.2.

5.2. In the Hilbert space $L^2(-1,1)$ find $M^\perp$ if
(a) $M = \{u : u(x) = u(-x) \text{ a.e.}\}$.
(b) $M = \{u : u(x) = 0 \text{ a.e. for } -1 < x < 0\}$.
Give an explicit formula for the projection onto M in each case.

5.3. If E is a closed subspace of the Hilbert space **H**, show that P_E is a linear operator on **H** with norm $\|P_E\| = 1$ except in the trivial case when $E = \{0\}$. Suggestion: If $x = c_1x_1 + c_2x_2$ first show that

$$P_Ex - c_1P_Ex_1 - c_2P_Ex_2 = -P_{E^\perp}x + c_1P_{E^\perp}x_1 + c_2P_{E^\perp}x_2$$

5.4. Show that the parallelogram law fails in $L^\infty(\Omega)$, so there is no choice of inner product which can give rise to the norm in $L^\infty(\Omega)$. (The same is true in $L^p(\Omega)$ for any $p \neq 2$.)

5.5. If $(X, \langle\cdot,\cdot\rangle)$ is an inner product space prove the *polarization identity*

$$\langle x, y\rangle = \frac{1}{4}\left(\|x+y\|^2 - \|x-y\|^2 + i\|x+iy\|^2 - i\|x-iy\|^2\right)$$

Thus, in any normed linear space, there can exist at most one inner product giving rise to the norm.

5.6. Let M be a closed subspace of a Hilbert space **H**, and P_M be the corresponding projection. Show that
(a) $P_M^2 = P_M$.
(b) $\langle P_Mx, y\rangle = \langle P_Mx, P_My\rangle = \langle x, P_My\rangle$ for any $x, y \in$ **H**.

5.7. Show that ℓ^2 is a Hilbert space. (Discussion: The only property you need to check is completeness, and you may freely use the fact that $\mathbb{C}$ is complete. A Cauchy sequence in this case is a sequence of sequences, so use a notation like

$$x^{(n)} = \{x_1^{(n)}, x_2^{(n)}, \dots\}$$

where $x_j^{(n)}$ denotes the jth term of the nth sequence $x^{(n)}$. Given a Cauchy sequence $\{x^{(n)}\}_{n=1}^\infty$ in ℓ^2 you'll first find a sequence x such that $\lim_{n\to\infty} x_j^{(n)} = x_j$ for each fixed j. You then must still show that $x \in \ell^2$, and one good way to do this is by first showing that $x - x^{(n)} \in \ell^2$ for some n.)

5.8. Let $\mathbf{H}$ be a Hilbert space.
(a) If $x_n \to x$ in $\mathbf{H}$ show that $\{x_n\}_{n=1}^\infty$ is bounded in $\mathbf{H}$.
(b) If $x_n \to x, y_n \to y$ in $\mathbf{H}$ show that $\langle x_n, y_n\rangle \to \langle x, y\rangle$.

5.9. Compute orthogonal polynomials of degree 0,1,2,3 on $[-1,1]$ and on $[0,1]$ by applying the Gram-Schmidt procedure to $1, x, x^2, x^3$ in $L^2(-1,1)$ and $L^2(0,1)$. (In the case of $L^2(-1,1)$, you are finding so-called Legendre polynomials.)

5.10. Use the result of Exercise 5.9 and the projection formula (5.6.40) to compute the best polynomial approximations of degrees 0,1,2, and 3 to $u(x) = e^x$ in $L^2(-1,1)$. Feel free to use any symbolic calculation tool you know to compute the necessary integrals, but give exact coefficients, not calculator approximations. If possible, produce a graph displaying u and the four approximations.

5.11. Let $\Omega \subset \mathbb{R}^N$, ρ be a measurable function on Ω, and $\rho(x) > 0$ a.e. on Ω. Let $\mathbf{X}$ denote the set of measurable functions u for which $\int_\Omega |u(x)|^2 \rho(x)\, dx$ is finite. We can then define the *weighted* inner product

$$\langle u, v\rangle_\rho = \int_\Omega u(x)\overline{v(x)}\rho(x)\, dx$$

and corresponding norm $\|u\|_\rho = \sqrt{\langle u, u\rangle_\rho}$ on $\mathbf{X}$. The resulting inner product space is complete, often denoted $L^2_\rho(\Omega)$. (As in the case of $\rho(x) = 1$ we regard any two functions which agree a.e. as being the same element, so $L^2_\rho(\Omega)$ is again really a set of equivalence classes.)
(a) Verify that all of the inner product axioms are satisfied.
(b) Suppose that there exist constants C_1, C_2 such that $0 < C_1 \le \rho(x) \le C_2$ a.e. Show that $u_n \to u$ in $L^2_\rho(\Omega)$ if and only if $u_n \to u$ in $L^2(\Omega)$.

5.12. More classes of orthogonal polynomials may be derived by applying the Gram-Schmidt procedure to $\{1, x, x^2, \dots\}$ in $L^2_\rho(a,b)$ for various choices of ρ, a, b, two of which occur in Exercise 5.9. Another class is the Laguerre polynomials, corresponding to $a = 0, b = \infty$ and $\rho(x) = e^{-x}$. Find the first four Laguerre polynomials.

5.13. Show that equality holds in the Schwarz inequality (5.2.8) if and only if x, y are linearly dependent.

5.14. Show by examples that the best approximation problem (5.4.23) may not have a solution if E is either not closed or not convex.

5.15. If Ω is a compact subset of $\mathbb{R}^N$, show that $C(\Omega)$ is a subspace of $L^2(\Omega)$ which isn't closed.

5.16. Show that

$$\left\{\frac{1}{\sqrt{2}}, \cos n\pi x, \sin n\pi x\right\}_{n=1}^{\infty} \tag{5.9.53}$$

is an orthonormal set in $L^2(-1,1)$. (Completeness of this set will be shown in Chapter 7.)

5.17. For nonnegative integers n define

$$v_n(x) = \cos(n \cos^{-1} x)$$

(a) Show that $v_{n+1}(x) + v_{n-1}(x) = 2xv_n(x)$ for $n = 1, 2, \ldots$

(b) Show that v_n is a polynomial of degree n (the so-called Chebyshev polynomials).

(c) Show that $\{v_n\}_{n=1}^{\infty}$ are orthogonal in $L^2_\rho(-1,1)$ where the weight function is $\rho(x) = \frac{1}{\sqrt{1-x^2}}$.

5.18. If $\mathbf{H}$ is a Hilbert space we say a sequence $\{x_n\}_{n=1}^{\infty}$ converges weakly to x (notation: $x_n \xrightarrow{w} x$) if $\langle x_n, y\rangle \to \langle x, y\rangle$ for every $y \in \mathbf{H}$.

(a) Show that if $x_n \to x$ then $x_n \xrightarrow{w} x$.

(b) Prove that the converse is false, as long as $\dim(\mathbf{H}) = \infty$, by showing that if $\{e_n\}_{n=1}^{\infty}$ is any orthonormal sequence in $\mathbf{H}$ then $e_n \xrightarrow{w} 0$, but $\lim_{n\to\infty} e_n$ doesn't exist.

(c) Prove that if $x_n \xrightarrow{w} x$ then $\|x\| \le \liminf_{n\to\infty} \|x_n\|$.

(d) Prove that if $x_n \xrightarrow{w} x$ and $\|x_n\| \to \|x\|$ then $x_n \to x$.

5.19. Let M_1, M_2 be closed subspaces of a Hilbert space $\mathbf{H}$ and suppose $M_1 \perp M_2$. Show that

$$M_1 \oplus M_2 = \{x \in \mathbf{H} : x = y + z, y \in M_1, z \in M_2\}$$

is also a closed subspace of H.

Chapter 6

Distribution Spaces

In this chapter we will introduce and study the concept of *distribution*, also sometimes known as *generalized function*. Commonly occurring sets of distributions form vector spaces, which are fundamental in the modern theory of differential equations. To motivate this study we first mention two examples.

Example 6.1. As was discussed in Section 1.3.3, the wave equation $u_{tt} - u_{xx} = 0$ has the general solution $u(x,t) = F(x+t) + G(x-t)$ where F, G must be in $C^2(\mathbb{R})$ in order that u be a classical solution. However from a physical point of view there is no apparent reason why such smoothness restrictions on F, G should be needed. Indeed the two terms represent waves of fixed shape moving to the left and right respectively with speed one, and it ought to be possible to allow the shape functions F, G to even have discontinuities. The calculus of distributions will allow us to regard u as a solution of the wave equation in a well defined sense even for such irregular F, G. □

Example 6.2. In physics and engineering one frequently encounters the so-called *Dirac delta function* $\delta(x)$, which has the properties

$$\delta(x) = 0 \quad x \neq 0 \qquad \int_{-\infty}^{\infty} \delta(x)\, dx = 1 \tag{6.0.1}$$

representing, for example, the idealized limit of a sequence of functions all with integral equal to one, and supported in smaller and smaller intervals centered at $x = 0$. Unfortunately these properties are inconsistent for ordinary functions—any function which is zero except at a single point must have integral zero. The theory of distributions will allow us to give a precise mathematical meaning to the delta function and in so doing justify formal calculations with it. □

Roughly speaking a distribution is a mathematical object whose unique identity is specified by *how it acts on all test functions*. It is in a sense quite analogous to a function in the ordinary sense, whose unique identity is specified by how it acts (i.e., how it maps) all points in its domain. As we will see, most ordinary functions may viewed as a special kind of distribution, which explains the "generalized function" terminology. In addition, there is a well

Techniques of Functional Analysis for Differential and Integral Equations
http://dx.doi.org/10.1016/B978-0-12-811426-1.00006-4

defined calculus of distributions which we will start to make extensive use of. We now start to give precise meaning to these concepts.

6.1 THE SPACE OF TEST FUNCTIONS

For any real or complex valued function f defined on some domain in $\mathbb{R}^N$, the *support* of f, denoted $\operatorname{supp} f$, is the closure of the set $\{x : f(x) \neq 0\}$.

Definition 6.1. If Ω is any open set in $\mathbb{R}^N$ the space of test functions on Ω is

$$C_0^\infty(\Omega) = \{\phi \in C^\infty(\Omega) : \operatorname{supp} \phi \text{ is compact in } \Omega\} \tag{6.1.2}$$

This function space is also commonly denoted $\mathcal{D}(\Omega)$, which is the notation we will use from now on. Clearly $\mathcal{D}(\Omega)$ is a vector space, but it may not be immediately evident that it contains any function other than $\phi \equiv 0$.

Example 6.3. Define

$$\phi(x) = \begin{cases} e^{1/(x^2-1)} & |x| < 1 \\ 0 & |x| \geq 1 \end{cases} \tag{6.1.3}$$

Then $\phi \in \mathcal{D}(\Omega)$ with $\Omega = \mathbb{R}$. To see this one only needs to check that $\lim_{x\to 1-} \phi^{(k)}(x) = 0$ for $k = 0, 1, \ldots$, and similarly at $x = -1$. Once we have one such function then many others can be derived from it by dilation ($\phi(x) \to \phi(\alpha x)$), translation ($\phi(x) \to \phi(x - \alpha)$), scaling ($\phi(x) \to \alpha\phi(x)$), differentiation ($\phi(x) \to \phi^{(k)}(x)$), or any linear combination of such terms. See also Exercise 6.1. Test functions of more than one variable may be found similarly. □

Next, we define convergence in the test function space.

Definition 6.2. If $\phi_n \in \mathcal{D}(\Omega)$ then we say $\phi_n \to 0$ in $\mathcal{D}(\Omega)$ if

(i) There exists a compact set $K \subset \Omega$ such that $\operatorname{supp} \phi_n \subset K$ for every n
(ii) $\lim_{n\to\infty} \max_{x\in\Omega} |D^\alpha \phi_n(x)| = 0$ for every multiindex α

We also say that $\phi_n \to \phi$ in $\mathcal{D}(\Omega)$ provided $\phi_n - \phi \to 0$ in $\mathcal{D}(\Omega)$. By specifying what convergence of a sequence in $\mathcal{D}(\Omega)$ means, we are partly, but not completely, specifying a topology on $\mathcal{D}(\Omega)$. We will have no need of further details about this topology, but see Chapter 6 of [33] for more on this point.

6.2 THE SPACE OF DISTRIBUTIONS

We come now to the basic definition—a distribution is a continuous linear functional on $\mathcal{D}(\Omega)$. More precisely

Definition 6.3. A linear mapping $T : \mathcal{D}(\Omega) \to \mathbb{C}$ is a *distribution* on Ω if $T(\phi_n) \to T(\phi)$ whenever $\phi_n \to \phi$ in $\mathcal{D}(\Omega)$. The set of all distributions on Ω is denoted $\mathcal{D}'(\Omega)$.

Recall we have earlier defined the dual space $\mathbf{X}^*$ of any normed linear space $\mathbf{X}$ to be $\mathcal{B}(\mathbf{X}, \mathbb{C})$, the vector space of continuous linear functionals on $\mathbf{X}$. Here $\mathcal{D}(\Omega)$ is not a normed linear space, but the dual space concept is still meaningful, as long as convergence of a sequence in $\mathbf{X}$ has been defined. That is to say, in such a case, a linear map $T : \mathbf{X} \to \mathbb{C}$ is in the dual space of $\mathbf{X}$ if $T(x_n) \to T(x)$ whenever $x_n \to x$ in $\mathbf{X}$. The distribution space $\mathcal{D}'(\Omega)$ is another example of a dual space, and we use the more common notation $\mathcal{D}'(\Omega)$ in place of $\mathcal{D}(\Omega)^*$.

Many more examples of dual spaces will be discussed later on. We emphasize that the distribution T is defined solely in terms of the values it assigns to test functions ϕ, in particular two distributions T_1, T_2 are equal precisely if $T_1(\phi) = T_2(\phi)$ for every $\phi \in \mathcal{D}(\Omega)$.

To further clarify the distribution concept, let us discuss a number of examples.

Example 6.4. If $f \in L^1(\Omega)$ define

$$T(\phi) = \int_\Omega f(x)\phi(x)\,dx \tag{6.2.4}$$

Obviously $|T(\phi)| \le \|f\|_{L^1(\Omega)}\|\phi\|_{L^\infty(\Omega)}$, so that $T : \mathcal{D}(\Omega) \to \mathbb{C}$ and is also linear. If $\phi_n \to \phi$ in $\mathcal{D}(\Omega)$ then by the same token

$$|T(\phi_n) - T(\phi)| \le \|f\|_{L^1(\Omega)}\|\phi_n - \phi\|_{L^\infty(\Omega)} \to 0 \tag{6.2.5}$$

so that T is continuous. Thus $T \in \mathcal{D}'(\Omega)$.

Because of the fact that ϕ must have compact support in Ω one does not really need f to be in $L^1(\Omega)$ but only in $L^1(K)$ for any compact subset K of Ω. For any $1 \le p \le \infty$ let us therefore define

$$L^p_{loc}(\Omega) = \{f : f \in L^p(K) \text{ for any compact set } K \subset \Omega\} \tag{6.2.6}$$

Thus a function in $L^p_{loc}(\Omega)$ can become infinite arbitrarily rapidly at the boundary of Ω. We say that $f_n \to f$ in $L^p_{loc}(\Omega)$ if $f_n \to f$ in $L^p(K)$ for every compact subset $K \subset \Omega$. Functions in L^1_{loc} are said to be *locally integrable* on Ω.

Now if we let $f \in L^1_{loc}(\Omega)$ the definition (6.2.4) still produces a finite value, since

$$|T(\phi)| = \int_\Omega f(x)\phi(x)\,dx = \int_K f(x)\phi(x)\,dx \le \|f\|_{L^1(K)}\|\phi\|_{L^\infty(K)} < \infty \tag{6.2.7}$$

if $K = \operatorname{supp}\phi$. Similarly if $\phi_n \to \phi$ in $\mathcal{D}(\Omega)$ we can choose a fixed compact set $K \subset \Omega$ containing $\operatorname{supp}\phi$ and $\operatorname{supp}\phi_n$ for every n, hence again

$$|T(\phi_n) - T(\phi)| \le \|f\|_{L^1(K)}\|\phi_n - \phi\|_{L^\infty(K)} \to 0 \tag{6.2.8}$$

so that $T \in \mathcal{D}'(\Omega)$.

When convenient, we will denote the distribution in Eq. (6.2.4) by T_f. The correspondence $f \to T_f$ allows us to think of $L^1_{loc}(\Omega)$ as a special subspace of $\mathcal{D}'(\Omega)$, that is, corresponding to any locally integrable function f is the distribution T_f. From this point of view f is thought of as the mapping

$$\phi \to \int_\Omega f\phi\, dx \tag{6.2.9}$$

instead of the more conventional

$$x \to f(x) \tag{6.2.10}$$

In fact for L^1_{loc} functions the former is in some sense more natural since it doesn't require us to make special arrangements for sets of measure zero. A distribution of the form $T = T_f$ for some $f \in L^1_{loc}(\Omega)$ is sometimes referred to as a regular distribution, while any distribution not of this type is a singular distribution. □

The correspondence $f \to T_f$ is also one-to-one in the following sense.

Theorem 6.1. *Two distributions T_{f_1}, T_{f_2} on Ω are equal if and only if $f_1 = f_2$ almost everywhere on Ω.*

This is a slightly technical result in measure theory which we leave for the exercises, for those with the necessary background. See also Theorem 2, Chapter II of [34].

Example 6.5. Fix a point $x_0 \in \Omega$ and define

$$T(\phi) = \phi(x_0) \tag{6.2.11}$$

We observe that T is defined and linear on $\mathcal{D}(\Omega)$ and if $\phi_n \to \phi$ in $\mathcal{D}(\Omega)$ then

$$|T(\phi_n) - T(\phi)| = |\phi_n(x_0) - \phi(x_0)| \to 0 \tag{6.2.12}$$

since $\phi_n \to \phi$ uniformly on Ω. We claim that T is not of the form T_f for any $f \in L^1_{loc}(\Omega)$ (i.e. f is not a regular distribution). To see this, suppose some such f existed. We would then have

$$\int_\Omega f(x)\phi(x)\, dx = 0 \tag{6.2.13}$$

for any test function ϕ with $\phi(x_0) = 0$. In particular if $\Omega' = \Omega\backslash\{x_0\}$ and $\phi \in \mathcal{D}(\Omega')$ then defining $\phi(x_0) = 0$ we have $\phi \in \mathcal{D}(\Omega)$ and $T(\phi) = 0$. Hence $f = 0$ a.e. on Ω' and so also on Ω, by Theorem 6.1. On the other hand we must also have, for any $\phi \in \mathcal{D}(\Omega)$ that

$$\phi(x_0) = T(\phi) = \int_\Omega f(x)\phi(x)\, dx \tag{6.2.14a}$$

$$= \int_\Omega f(x)(\phi(x) - \phi(x_0))\, dx + \phi(x_0)\int_\Omega f(x)\, dx = \phi(x_0)\int_\Omega f(x)\, dx \tag{6.2.14b}$$

since $f = 0$ a.e. on Ω, and therefore $\int_\Omega f(x)\, dx = 1$ a contradiction. Note that $f(x) = 0$ for a.e. $x \in \Omega$ and $\int_\Omega f(x)\, dx = 1$ are precisely the formal properties of the delta function mentioned in Example 6.2.

We define T to be the *Dirac delta distribution* with singularity at x_0, usually denoted δ_{x_0}, or simply δ in the case $x_0 = 0$. By an acceptable abuse of notation, pretending that δ is an actual function, we may write a formula like

$$\int_\Omega \delta(x)\phi(x)\,dx = \phi(0) \tag{6.2.15}$$

but we emphasize that this is simply a formal expression of Eq. (6.2.11), and any rigorous arguments must make use of Eq. (6.2.11) directly. In the same formal sense $\delta_{x_0}(x) = \delta(x - x_0)$ so that

$$\int_\Omega \delta(x - x_0)\phi(x)\,dx = \phi(x_0) \tag{6.2.16}$$

□

Example 6.6. Fix a point $x_0 \in \Omega$, a multiindex α and define

$$T(\phi) = (D^\alpha\phi)(x_0) \tag{6.2.17}$$

One may show, as in the previous example, that $T \in \mathcal{D}'(\Omega)$. □

Example 6.7. Let Σ be a sufficiently smooth hypersurface in Ω of dimension $m \le n - 1$ and define

$$T(\phi) = \int_\Sigma \phi(x)\,ds(x) \tag{6.2.18}$$

where ds is the surface area element on Σ. Then T is a distribution on Ω sometimes referred to as the delta distribution concentrated on Σ, which we denote as δ_Σ. □

Example 6.8. Let $\Omega = \mathbb{R}$ and define

$$T(\phi) = \lim_{\epsilon\to 0+} \int_{|x|>\epsilon} \frac{\phi(x)}{x}\,dx \tag{6.2.19}$$

As we'll show below, the indicated limit always exists and is finite for $\phi \in \mathcal{D}(\Omega)$ (even for $\phi \in C_0^1(\Omega)$). The expression (6.2.19) appears formally to be the same as $T_f(\phi)$ when $f(x) = 1/x$, but since $f \notin L^1_{loc}(\mathbb{R})$ this is not a suitable definition.

In general, a limit of the form

$$\lim_{\epsilon\to 0+} \int_{\Omega\cap|x-a|>\epsilon} f(x)\,dx \tag{6.2.20}$$

when it exists, is called the *Cauchy principal value* of $\int_\Omega f(x)\,dx$, which may be finite even when $\int_\Omega f(x)\,dx$ is divergent in the ordinary sense. For example, $\int_{-1}^1 \frac{dx}{x}$ is divergent, regarded as either a Lebesgue integral or an improper Riemann integral, but

$$\lim_{\epsilon\to 0+} \int_{1>|x|>\epsilon} \frac{dx}{x} = 0 \tag{6.2.21}$$

To distinguish the principal value meaning of the integral, the notation

$$\text{pv}\int_{\Omega} f(x)\,dx \tag{6.2.22}$$

may be used instead of Eq. (6.2.20), where the point a in question must be clear from context.

Let us now check that Eq. (6.2.19) defines a distribution. If $\operatorname{supp}\phi \subset [-M,M]$ then since

$$\begin{aligned}\int_{|x|>\epsilon} \frac{\phi(x)}{x}\,dx &= \int_{M>|x|>\epsilon} \frac{\phi(x)}{x}\,dx \\ &= \int_{M>|x|>\epsilon} \frac{\phi(x)-\phi(0)}{x}\,dx + \phi(0)\int_{M>|x|>\epsilon} \frac{1}{x}\,dx \end{aligned}\tag{6.2.23}$$

and the last term on the right is zero, we have

$$T(\phi) = \lim_{\epsilon\to 0+} \int_{M>|x|>\epsilon} \psi(x)\,dx \tag{6.2.24}$$

where $\psi(x) = (\phi(x)-\phi(0))/x$. It now follows from the mean value theorem that

$$|T(\phi)| \le \int_{|x|<M} |\psi(x)|\,dx \le 2M\|\phi'\|_{L^\infty} \tag{6.2.25}$$

so $T(\phi)$ is defined and finite for all test functions. Linearity of T is clear, and if $\phi_n \to \phi$ in $\mathcal{D}(\Omega)$ then

$$|T(\phi_n) - T(\phi)| \le 2M\|\phi_n' - \phi'\|_{L^\infty} \to 0 \tag{6.2.26}$$

where M is chosen so that $\operatorname{supp}\phi_n$, $\operatorname{supp}\phi \subset [-M,M]$, and it follows that T is continuous.

The distribution T is often denoted $\text{pv}\frac{1}{x}$, so, for example, $\text{pv}\frac{1}{x}(\phi)$ means the same thing as the right-hand side of Eq. (6.2.19). For reasons which will become more clear later, it may also be referred to as $\text{pf}\frac{1}{x}$, pf standing for *pseudofunction* (also *partie finie* in French). □

6.3 ALGEBRA AND CALCULUS WITH DISTRIBUTIONS

6.3.1 Multiplication of Distributions

It is clear that distributions can be added and multiplied by scalars, that is, $\mathcal{D}'(\Omega)$ is a vector space. In general it is not possible to multiply together arbitrary distributions, for example, $\delta^2 = \delta\cdot\delta$ cannot be defined in any consistent way. We may, however, always multiply a distribution by a C^∞ function. More precisely, if $a \in C^\infty(\Omega)$ and $T \in \mathcal{D}'(\Omega)$ then we may define the product aT as a distribution via

Definition 6.4. $aT(\phi) = T(a\phi) \qquad \phi \in \mathcal{D}(\Omega)$

We emphasize that in order for this to be a valid definition of a new distribution aT, it must be checked that the map $\phi \to T(a\phi)$ satisfies the basic Definition 6.3 of a distribution. Since $a\phi \in \mathcal{D}(\Omega)$ the right-hand side is well defined, and it is straightforward to check that aT satisfies the necessary linearity and continuity conditions. One should also note that if $T = T_f$ then this definition is consistent with ordinary pointwise multiplication of the functions f and a.

6.3.2 Convergence of Distributions

An appropriate definition of convergence of a sequence of distributions is as follows.

Definition 6.5. If $T, T_n \in \mathcal{D}'(\Omega)$ for $n = 1, 2, \dots$ then we say $T_n \to T$ in $\mathcal{D}'(\Omega)$ (or in the sense of distributions) if $T_n(\phi) \to T(\phi)$ for every $\phi \in \mathcal{D}(\Omega)$.

It is an interesting fact, which we shall not prove here, that it is not necessary to assume that the limit T belongs to $\mathcal{D}'(\Omega)$, that is to say, if $T(\phi) := \lim_{n\to\infty} T_n(\phi)$ exists for every $\phi \in \mathcal{D}(\Omega)$ then necessarily $T \in \mathcal{D}'(\Omega)$ (see Theorem 6.17 of [33]).

Example 6.9. If $f_n \in L^1_{loc}(\Omega)$ and $f_n \to f$ in $L^1_{loc}(\Omega)$ then the corresponding distribution $T_{f_n} \to T_f$ in the sense of distributions, since

$$|T_{f_n}(\phi) - T_f(\phi)| \le \int_K |f_n - f||\phi|\,dx \le \|f_n - f\|_{L^1(K)} \|\phi\|_{L^\infty(\Omega)} \tag{6.3.27}$$

where K is the support of ϕ. Because of the one-to-one correspondence $f \leftrightarrow T_f$, we will usually write instead that $f_n \to f$ in the sense of distributions, in place of the more cumbersome $T_{f_n} \to T_f$. □

Example 6.10. Define

$$f_n(x) = \begin{cases} n & 0 < x < \frac{1}{n} \\ 0 & \text{otherwise} \end{cases} \tag{6.3.28}$$

We claim that $f_n \to \delta$ in the sense of distributions. We see this by first observing that

$$|T_{f_n}(\phi) - \delta(\phi)| = \left| n\int_0^{1/n} \phi(x)\,dx - \phi(0) \right| = \left| n\int_0^{1/n} (\phi(x) - \phi(0))\,dx \right| \tag{6.3.29}$$

By the continuity of ϕ, if $\epsilon > 0$ there exists $\delta > 0$ such that $|\phi(x) - \phi(0)| \le \epsilon$ whenever $|x| \le \delta$. Thus if we choose $n > \frac{1}{\delta}$ there follows

$$n\int_0^{1/n} |\phi(x) - \phi(0)|\,dx \le n\epsilon \int_0^{1/n} dx = \epsilon \tag{6.3.30}$$

from which the conclusion follows. Note that the formal properties of the δ function, $\delta(x) = 0, x \ne 0$, $\delta(0) = +\infty$, $\int \delta(x)\,dx = 1$, are reflected in the pointwise limit of the sequence f_n, but it is only the distributional definition that is mathematically satisfactory. □

Sequences converging to δ play a very large role in methods of applied mathematics, especially in the theory of differential and integral equations. The following theorem includes many cases of interest.

Theorem 6.2. *Suppose $f_n \in L^1(\mathbb{R}^N)$ for $n = 1, 2, \ldots$ and assume*

(a) $\int_{\mathbb{R}^N} f_n(x)\,dx = 1$ *for all n.*
(b) *There exists a constant C such that* $\|f_n\|_{L^1(\mathbb{R}^N)} \le C$ *for all n.*
(c) $\lim_{n\to\infty} \int_{|x|>\delta} |f_n(x)|\,dx = 0$ *for all* $\delta > 0$.

If ϕ is bounded on $\mathbb{R}^N$ and continuous at $x = 0$ then

$$\lim_{n\to\infty} \int_{\mathbb{R}^N} f_n(x)\phi(x)\,dx = \phi(0) \tag{6.3.31}$$

and in particular $f_n \to \delta$ in $\mathcal{D}'(\mathbb{R}^N)$.

Proof. For any ϕ satisfying the stated conditions, we have

$$\int_{\mathbb{R}^N} f_n(x)\phi(x)\,dx - \phi(0) = \int_{\mathbb{R}^N} f_n(x)(\phi(x) - \phi(0))\,dx \tag{6.3.32}$$

and so we will be done if we show that the integral on the right tends to zero as $n \to \infty$. Fix $\epsilon > 0$ and choose $\delta > 0$ such that $|\phi(x) - \phi(0)| \le \epsilon$ whenever $|x| < \delta$. Write the integral on the right in Eq. (6.3.32) as the sum $A_{n,\delta} + B_{n,\delta}$ where

$$A_{n,\delta} = \int_{|x|\le\delta} f_n(x)(\phi(x) - \phi(0))\,dx \qquad B_{n,\delta} = \int_{|x|>\delta} f_n(x)(\phi(x) - \phi(0))\,dx \tag{6.3.33}$$

We then have, by obvious estimations, that

$$|A_{n,\delta}| \le \epsilon \int_{\mathbb{R}^N} |f_n(x)| \le C\epsilon \tag{6.3.34}$$

while

$$\limsup_{n\to\infty} |B_{n,\delta}| \le \limsup_{n\to\infty} 2\|\phi\|_{L^\infty} \int_{|x|>\delta} |f_n(x)|\,dx = 0 \tag{6.3.35}$$

Thus

$$\limsup_{n\to\infty} \left| \int_{\mathbb{R}^N} f_n(x)\phi(x)\,dx - \phi(0) \right| \le C\epsilon \tag{6.3.36}$$

and the conclusion follows since $\epsilon > 0$ is arbitrary. □

We will refer to any sequence satisfying the assumptions of Theorem 6.2 as a *delta sequence*. It is often the case that $f_n \ge 0$ for all n, in which case assumption (b) follows automatically from (a) with $C = 1$. A common way to construct such a sequence is to pick any $f \in L^1(\mathbb{R}^N)$ with $\int_{\mathbb{R}^N} f(x)\,dx = 1$ and set

$$f_n(x) = n^N f(nx) \tag{6.3.37}$$

The verification of this is left to the exercises. If, for example, we choose $f(x) = \chi_{[0,1]}(x)$, then the resulting sequence $f_n(x)$ is the same as is defined in Eq. (6.3.28). Since we can certainly choose such an f in $\mathcal{D}(\mathbb{R}^N)$ we also have

Corollary 6.1. *There exists a sequence* $\{f_n\}_{n=1}^{\infty}$ *such that* $f_n \in \mathcal{D}(\mathbb{R}^N)$ *and* $f_n \to \delta$ *in* $\mathcal{D}'(\mathbb{R}^N)$.

6.3.3 Derivative of a Distribution

Next we explain how to define the derivative of an arbitrary distribution. For the moment, suppose $(a,b) \subset \mathbb{R}$, $f \in C^1(a,b)$ and $T = T_f$ is the corresponding distribution. We then have from integration by parts that

$$T_{f'}(\phi) = \int_a^b f'(x)\phi(x)\,dx = -\int_a^b f(x)\phi'(x)\,dx = -T_f(\phi') \tag{6.3.38}$$

This suggests defining

$$T'(\phi) = -T(\phi') \qquad \phi \in C_0^\infty(a,b) \tag{6.3.39}$$

whenever $T \in \mathcal{D}'(a,b)$. The previous equation shows that this definition is consistent with the ordinary concept of differentiability for C^1 functions. Clearly, $T'(\phi)$ is always defined, since ϕ' is a test function whenever ϕ is, linearity of T' is obvious, and if $\phi_n \to \phi$ in $C_0^\infty(a,b)$ then $\phi_n' \to \phi'$ also in $C_0^\infty(a,b)$ so that

$$T'(\phi_n) = -T(\phi_n') \to -T(\phi') = T'(\phi) \tag{6.3.40}$$

Thus, $T' \in \mathcal{D}'(a,b)$.

Example 6.11. Consider the case of the Heaviside (unit step) function $H(x)$

$$H(x) = \begin{cases} 0 & x < 0 \\ 1 & x > 0 \end{cases} \tag{6.3.41}$$

If we seek the derivative of H (i.e., of T_H) according to the preceding distributional definition, then we compute

$$H'(\phi) = -H(\phi') = -\int_{-\infty}^{\infty} H(x)\phi'(x)\,dx = -\int_0^{\infty} \phi'(x)\,dx = \phi(0) \tag{6.3.42}$$

(where we use the natural notation H' in place of T_H'). This means that $H'(\phi) = \delta(\phi)$ for any test function ϕ, and so $H' = \delta$ in the sense of distributions. This relationship captures the fact that $H' = 0$ at all points where the derivative exists in the classical sense, since we think of the delta function as being zero on any interval not containing the origin. Since H is not differentiable at the origin, the distributional derivative is itself a distribution which is not a function.

Since δ is again a distribution, it will itself have a derivative, namely

$$\delta'(\phi) = -\delta(\phi') = -\phi'(0) \tag{6.3.43}$$

a distribution of the type discussed in Example 6.6, often referred to as the *dipole distribution*, which of course we may regard as the second derivative of H. □

For an arbitrary domain $\Omega \subset \mathbb{R}^N$ and sufficiently smooth function f we have the similar integration by parts formula (see Eq. A.3.31)

$$\int_\Omega \frac{\partial f}{\partial x_i}\phi\, dx = -\int_\Omega f\frac{\partial \phi}{\partial x_i}\, dx \tag{6.3.44}$$

motivating the following definition.

Definition 6.6. If $T \in \mathcal{D}'(\Omega)$,

$$\frac{\partial T}{\partial x_i}(\phi) = -T\left(\frac{\partial \phi}{\partial x_i}\right) \qquad \phi \in \mathcal{D}(\Omega) \tag{6.3.45}$$

As in the one-dimensional case we easily check that $\frac{\partial T}{\partial x_i}$ belongs to $\mathcal{D}'(\Omega)$ whenever T does. This has the far reaching consequence that *every distribution is infinitely differentiable in the sense of distributions.* Furthermore we have the general formula, obtained by repeated application of the basic definition, that

$$(D^\alpha T)(\phi) = (-1)^{|\alpha|}T(D^\alpha\phi) \tag{6.3.46}$$

for any multiindex α. If $T = T_f$ is a regular distribution we will allow an alternative notation such as $D^\alpha f(\phi)$ in place of $T_{D^\alpha f}(\phi)$.

A simple and useful property is

Proposition 6.1. *If $T_n \to T$ in $\mathcal{D}'(\Omega)$ then $D^\alpha T_n \to D^\alpha T$ in $\mathcal{D}'(\Omega)$ for any multiindex α.*

Proof. $D^\alpha T_n(\phi) = (-1)^{|\alpha|}T_n(D^\alpha\phi) \to (-1)^{|\alpha|}T(D^\alpha\phi) = D^\alpha T(\phi)$ for any test function ϕ. □

Next we consider a more generic one-dimensional situation. Let $x_0 \in \mathbb{R}$ and consider a function f which is C^∞ on $(-\infty, x_0)$ and on (x_0, ∞), and for which $f^{(k)}$ has finite left- and right-hand limits at $x = x_0$, for any k. Thus, at the point $x = x_0$, f or any of its derivatives may have a jump discontinuity, and we denote

$$\Delta^k f = \lim_{x\to x_0+} f^{(k)}(x) - \lim_{x\to x_0-} f^{(k)}(x) \tag{6.3.47}$$

(and by convention $\Delta f = \Delta^0 f$.) Define also

$$[f^{(k)}](x) = \begin{cases} f^{(k)}(x) & x \ne x_0 \\ \text{undefined} & x = x_0 \end{cases} \tag{6.3.48}$$

which we'll refer to as the *pointwise kth derivative*. The notation $f^{(k)}$ will always be understood to mean the distributional derivative unless otherwise stated. The distinction between $f^{(k)}$ and $[f^{(k)}]$ is crucial, for example, if $f(x) = H(x)$, the Heaviside function, then $H' = \delta$ but $[H'] = 0$ for $x \neq 0$, and is undefined for $x = 0$.

For f as just described we now proceed to calculate the distributional derivative. If $\phi \in C_0^\infty(\mathbb{R})$ we have

$$\int_{-\infty}^{\infty} f(x)\phi'(x)\,dx = \int_{-\infty}^{x_0} f(x)\phi'(x)\,dx + \int_{x_0}^{\infty} f(x)\phi'(x)\,dx \tag{6.3.49a}$$

$$= f(x)\phi(x)|_{-\infty}^{x_0} - \int_{-\infty}^{x_0} f'(x)\phi(x)\,dx + f(x)\phi(x)|_{x_0}^{\infty} - \int_{x_0}^{\infty} f'(x)\phi(x)\,dx \tag{6.3.49b}$$

$$= -\int_{-\infty}^{\infty} [f'(x)]\phi(x)\,dx + (f(x_0-) - f(x_0+))\phi(x_0) \tag{6.3.49c}$$

It follows that

$$f'(\phi) = \int_{-\infty}^{\infty} [f'(x)]\phi(x)\,dx + (\Delta f)\phi(x_0) \tag{6.3.50}$$

or

$$f' = [f'] + (\Delta f)\delta(x - x_0) \tag{6.3.51}$$

Note in particular that $f' = [f']$ if and only if f is continuous at x_0.

The function $[f']$ satisfies all of the same assumptions as f itself, with $\Delta f' = \Delta[f']$, thus we can differentiate again in the distribution sense to obtain

$$f'' = [f']' + (\Delta f)\delta'(x - x_0) = [f''] + (\Delta^1 f)\delta(x - x_0) + (\Delta f)\delta'(x - x_0) \tag{6.3.52}$$

Here we use the evident fact that the distributional derivative of $\delta(x - x_0)$ is $\delta'(x - x_0)$.

A similar calculation can be carried out for higher derivatives of f, leading to the general formula

$$f^{(k)} = [f^{(k)}] + \sum_{j=0}^{k-1} (\Delta^j f)\delta^{(k-1-j)}(x - x_0) \tag{6.3.53}$$

One can also obtain a similar formula if f is allowed to have any finite number of such singular points, or even a countably infinite number of isolated singular points.

Example 6.12. Let

$$f(x) = \begin{cases} x & x < 0 \\ \cos x & x > 0 \end{cases} \tag{6.3.54}$$

We see that f satisfies all of the assumptions mentioned previously with $x_0 = 0$, and

$$[f'](x) = \begin{cases} 1 & x < 0 \\ -\sin x & x > 0 \end{cases} \tag{6.3.55}$$

$$[f''](x) = \begin{cases} 0 & x < 0 \\ -\cos x & x > 0 \end{cases} \tag{6.3.56}$$

so that $\Delta f = 1, \Delta^1 f = -1$. Thus

$$f' = [f'] + \delta \qquad f'' = [f''] - \delta + \delta' \tag{6.3.57}$$

□

Here is one more instructive example in the one-dimensional case.

Example 6.13. Let

$$f(x) = \begin{cases} \log x & x > 0 \\ 0 & x \le 0 \end{cases} \tag{6.3.58}$$

Since $f \in L^1_{loc}(\mathbb{R})$ we may regard it as a distribution on $\mathbb{R}$, but its pointwise derivative $H(x)/x$ is not locally integrable, so does not have an obvious distributional meaning. Nevertheless f' must exist in the sense of $\mathcal{D}'(\mathbb{R})$, which we may anticipate is still related to $H(x)/x$ somehow. To find it, we use the definition above,

$$f'(\phi) = -f(\phi') = -\int_0^\infty \phi'(x) \log x \, dx \tag{6.3.59}$$

$$= -\lim_{\epsilon \to 0+} \int_\epsilon^\infty \phi'(x) \log x \, dx \tag{6.3.60}$$

$$= \lim_{\epsilon \to 0+} \left[\phi(\epsilon) \log \epsilon + \int_\epsilon^\infty \frac{\phi(x)}{x} dx \right] \tag{6.3.61}$$

$$= \lim_{\epsilon \to 0+} \left[\phi(0) \log \epsilon + \int_\epsilon^\infty \frac{\phi(x)}{x} dx \right] \tag{6.3.62}$$

where the final equality is valid because the difference between it and the previous line is $\lim_{\epsilon \to 0}(\phi(\epsilon) - \phi(0)) \log \epsilon = 0$. The functional defined by the final expression above will be denoted[1] as $\text{pf}\left(\frac{H(x)}{x}\right)$, that is,

$$\text{pf}\left(\frac{H(x)}{x}\right)(\phi) = \lim_{\epsilon \to 0+} \left[\phi(0) \log \epsilon + \int_\epsilon^\infty \frac{\phi(x)}{x} dx \right] \tag{6.3.63}$$

Since we have already established that the derivative of a distribution is also a distribution, it follows that $\text{pf}\left(\frac{H(x)}{x}\right) \in \mathcal{D}'(\mathbb{R})$ and in particular the limit here

1. Recall the pf notation was mentioned earlier in Section 6.2.

always exists for $\phi \in \mathcal{D}(\mathbb{R})$. It should be emphasized that if $\phi(0) \neq 0$ then neither of the two terms on the right-hand side in Eq. (6.3.63) will have a finite limit separately, but the sum always will. For a test function ϕ with support disjoint from the singularity at $x = 0$, the action of the distribution pf $\left(\frac{H(x)}{x}\right)$ coincides with that of the ordinary function $H(x)/x$, as we might expect. □

Next we turn to examples involving partial derivatives.

Example 6.14. Let $F \in L^1_{loc}(\mathbb{R})$ and set $u(x,t) = F(x+t)$. We claim that $u_{tt} - u_{xx} = 0$ in $\mathcal{D}'(\mathbb{R}^2)$. Recall that this is the point that was raised in the first example at the beginning of this chapter. A similar argument works for $F(x-t)$. To verify this claim, first observe that for any $\phi \in \mathcal{D}(\mathbb{R}^2)$

$$(u_{tt}-u_{xx})(\phi) = u(\phi_{tt}-\phi_{xx}) = \iint_{\mathbb{R}^2} F(x+t)(\phi_{tt}(x,t)-\phi_{xx}(x,t))\,dxdt \quad (6.3.64)$$

Make the change of coordinates

$$\xi = x-t \quad \eta = x+t \quad (6.3.65)$$

to obtain

$$\begin{aligned}(u_{tt} - u_{xx})(\phi) &= 2\int_{-\infty}^{\infty} F(\eta)\left[\int_{-\infty}^{\infty} \phi_{\xi\eta}(\xi,\eta)\,d\xi\right] d\eta \\ &= 2\int_{-\infty}^{\infty} F(\eta)\left(\phi_\eta(\xi,\eta)|_{\xi=-\infty}^{\infty}\right) d\eta = 0 \end{aligned} \quad (6.3.66)$$

since ϕ has compact support. □

Example 6.15. Let $N \geq 3$ and define

$$u(x) = \frac{1}{|x|^{N-2}} \quad (6.3.67)$$

We claim that

$$\Delta u = C_N\delta \qquad \text{in } \mathcal{D}'(\mathbb{R}^N) \quad (6.3.68)$$

where $C_N = (2-N)\Omega_{N-1}$ and Ω_{N-1} is the surface area[2] of the unit sphere in $\mathbb{R}^N$. First note that for any R we have

$$\int_{|x|<R} |u(x)|\,dx = \Omega_{N-1}\int_0^R \frac{1}{r^{N-2}} r^{N-1}\,dr < \infty \quad (6.3.69)$$

(using, e.g., Eq. A.4.37) so $u \in L^1_{loc}(\mathbb{R}^N)$ and in particular $u \in \mathcal{D}'(\mathbb{R}^N)$.

It is natural here to use spherical coordinates in $\mathbb{R}^N$, see Section A.4 for a review. In particular the expression for the Laplacian in spherical coordinates

2. The usual notation is to use $N-1$ rather than N as the subscript because the sphere is a surface of dimension $N-1$.

may be derived from the chain rule, as was done in Eq. (1.3.102) for the two-dimensional case. When applied to a function depending only on $r = |x|$, such as u, the result is

$$\Delta u = u_{rr} + \frac{N-1}{r} u_r \tag{6.3.70}$$

(see Exercise 1.17 of Chapter 1) and it follows that $\Delta u(x) = 0$ for $x \neq 0$.

We may use Green's identity (A.3.34) to obtain, for any $\phi \in \mathcal{D}(\mathbb{R}^N)$

$$\Delta u(\phi) = u(\Delta\phi) = \int_{\mathbb{R}^N} u(x)\Delta\phi(x)\,dx \tag{6.3.71}$$

$$= \lim_{\epsilon\to 0+} \int_{|x|>\epsilon} u(x)\Delta\phi(x)\,dx \tag{6.3.72}$$

$$= \lim_{\epsilon\to 0+} \left[\int_{|x|>\epsilon} \Delta u(x)\phi(x)\,dx + \int_{|x|=\epsilon} \left(u(x)\frac{\partial\phi}{\partial n}(x) - \phi(x)\frac{\partial u}{\partial n}(x)\right) dS(x)\right] \tag{6.3.73}$$

Since $\Delta u = 0$ for $x \neq 0$ and $\frac{\partial}{\partial n} = -\frac{\partial}{\partial r}$ on $\{x : |x| = \epsilon\}$ this simplifies to

$$\Delta u(\phi) = \lim_{\epsilon\to 0+} \int_{|x|=\epsilon} \left(\frac{2-N}{\epsilon^{N-1}}\phi(x) - \frac{1}{\epsilon^{N-2}}\frac{\partial\phi}{\partial r}(x)\right) dS(x) \tag{6.3.74}$$

We next observe that

$$\lim_{\epsilon\to 0+} \int_{|x|=\epsilon} \frac{2-N}{\epsilon^{N-1}}\phi(x)\,dS(x) = (2-N)\Omega_{N-1}\phi(0) \tag{6.3.75}$$

since the average of ϕ over the sphere of radius ϵ converges to $\phi(0)$ as $\epsilon \to 0$. Finally, the second integral tends to zero, since

$$\left|\int_{|x|=\epsilon} \frac{1}{\epsilon^{N-2}}\frac{\partial\phi}{\partial r}(x)\,dS(x)\right| \le \frac{\Omega_{N-1}\epsilon^{N-1}}{\epsilon^{N-2}}\|\nabla\phi\|_{L^\infty} \to 0 \tag{6.3.76}$$

Thus Eq. (6.3.68) holds. When $N = 2$ an analogous calculation shows that if $u(x) = \log|x|$ then $\Delta u = 2\pi\delta$ in $\mathcal{D}'(\mathbb{R}^2)$. □

6.4 CONVOLUTION AND DISTRIBUTIONS

If f, g are locally integrable functions on $\mathbb{R}^N$ the classical convolution of f and g is defined to be

$$(f * g)(x) = \int_{\mathbb{R}^N} f(x-y)g(y)\,dy \tag{6.4.77}$$

whenever the integral is defined. By an obvious change of variable we see that convolution is commutative, that is, $f * g = g * f$.

Proposition 6.2. *If $f \in L^p(\mathbb{R}^N)$ and $g \in L^q(\mathbb{R}^N)$ then $f * g \in L^r(\mathbb{R}^N)$ if $1 + \frac{1}{r} = \frac{1}{p} + \frac{1}{q}$, so in particular $f * g$ is defined almost everywhere. Furthermore*

$$\|f * g\|_{L^r(\mathbb{R}^N)} \le \|f\|_{L^p(\mathbb{R}^N)} \|g\|_{L^q(\mathbb{R}^N)} \tag{6.4.78}$$

The inequality (6.4.78) is *Young's convolution inequality*, and we refer to [40, Theorem 9.2] for a proof. In the case $r = \infty$ it can actually be shown that $f * g \in C(\mathbb{R}^N)$.

Our goal here is to generalize the definition of convolution in such a way that at least one of the two factors can be a distribution. Let us introduce the notations for translation and reflection of a function f,

$$(\tau_h f)(x) = f(x - h) \tag{6.4.79}$$

$$\check{f}(x) = f(-x) \tag{6.4.80}$$

so that $f(x - y) = (\tau_x \check{f})(y)$. If $f \in \mathcal{D}(\mathbb{R}^N)$ then so is $(\tau_x \check{f})$ so that $(f * g)(x)$ may be regarded as $T_g(\tau_x \check{f})$, that is, the value obtained when the distribution corresponding to the locally integrable function g acts on the test function $(\tau_x \check{f})$. This motivates the following definition.

Definition 6.7. If $T \in \mathcal{D}'(\mathbb{R}^N)$ and $\phi \in \mathcal{D}(\mathbb{R}^N)$ then $(T * \phi)(x) = T(\tau_x \check{\phi})$.

By this definition $(T * \phi)(x)$ exists and is finite for every $x \in \mathbb{R}^N$ but other smoothness or decay properties of $T * \phi$ may not be apparent.

Example 6.16. If $T = \delta$ then

$$(T * \phi)(x) = \delta(\tau_x \check{\phi}) = (\tau_x \check{\phi})(y)|_{y=0} = \phi(x - y)|_{y=0} = \phi(x) \tag{6.4.81}$$

Thus, δ is the "convolution identity," $\delta * \phi = \phi$ at least for $\phi \in \mathcal{D}(\mathbb{R}^N)$. Formally this corresponds to the widely used formula

$$\int_{\mathbb{R}^N} \delta(x - y)\phi(y)\, dy = \phi(x) \tag{6.4.82}$$

If $T_n \to \delta$ in $\mathcal{D}'(\mathbb{R}^N)$ then likewise

$$(T_n * \phi)(x) = T_n(\tau_x \check{\phi}) \to \delta(\tau_x \check{\phi}) = \phi(x) \tag{6.4.83}$$

for any fixed $x \in \mathbb{R}^N$. □

A key property of convolution is that in computing a derivative $D^\alpha(T * \phi)$, the derivative may be applied to either factor in the convolution. More precisely we have the following theorem.

Theorem 6.3. *If $T \in \mathcal{D}'(\mathbb{R}^N)$ and $\phi \in \mathcal{D}(\mathbb{R}^N)$ then $T * \phi \in C^\infty(\mathbb{R}^N)$ and for any multiindex α*

$$D^\alpha(T * \phi) = D^\alpha T * \phi = T * D^\alpha \phi \tag{6.4.84}$$

Proof. First observe that

$$(-1)^{|\alpha|} D^\alpha(\tau_x \check{\phi}) = \tau_x((D^\alpha \phi)\check{\ }) \tag{6.4.85}$$

and applying T to these identical test functions we get the right-hand equality in Eq. (6.4.84). We refer to Theorem 6.30 of [33] for the proof of the left-hand equality. □

When f, g are continuous functions of compact support it is elementary to see that $\text{supp}\,(f * g) \subset \text{supp} f + \text{supp}\, g$. The same property holds for $T * \phi$ if $T \in \mathcal{D}'(\mathbb{R}^N)$ and $\phi \in \mathcal{D}(\mathbb{R}^N)$, once a proper definition of the support of a distribution is given, which we now proceed to do.

If $\omega \subset \Omega$ is an open set we say that $T = 0$ in ω if $T(\phi) = 0$ whenever $\phi \in \mathcal{D}(\Omega)$ and $\text{supp}\,(\phi) \subset \omega$. If W denotes the largest open subset of Ω on which $T = 0$ (equivalently the union of all open subsets of Ω on which $T = 0$) then the support of T is the complement of W in Ω. In other words, $x \notin \text{supp}\, T$ if there exists $\epsilon > 0$ such that $T(\phi) = 0$ whenever ϕ is a test function with support in $B(x, \epsilon)$. One can easily verify that the support of a distribution is closed, and agrees with the usual notion of support of a function, up to sets of measure zero. The set of distributions of compact support in Ω forms a vector subspace of $\mathcal{D}'(\Omega)$ which is denoted $\mathcal{E}'(\Omega)$. This notation is appropriate because $\mathcal{E}'(\Omega)$ turns out to be precisely the dual space of $C^\infty(\Omega) =: \mathcal{E}(\Omega)$ when a suitable definition of convergence is given, see, for example, Chapter II, section 5 of [34].

If now $T \in \mathcal{E}'(\mathbb{R}^N)$ and $\phi \in \mathcal{D}(\mathbb{R}^N)$, we observe that

$$\text{supp}\,(\tau_x \check{\phi}) = x - \text{supp}\,\phi \tag{6.4.86}$$

Thus

$$(T * \phi)(x) = T(\tau_x \check{\phi}) = 0 \tag{6.4.87}$$

unless there is a nonempty intersection of $\text{supp}\, T$ and $x -\text{supp}\,\phi$, in other words, $x \in \text{supp}\, T + \text{supp}\,\phi$. Thus from these remarks and Theorem 6.3 we have

Proposition 6.3. *If $T \in \mathcal{E}'(\mathbb{R}^N)$ and $\phi \in \mathcal{D}(\mathbb{R}^N)$ then*

$$\text{supp}\,(T * \phi) \subset \text{supp}\, T + \text{supp}\,\phi \tag{6.4.88}$$

and in particular $T * \phi \in \mathcal{D}(\mathbb{R}^N)$.

Convolution provides an extremely useful and convenient way to approximate functions and distributions by very smooth functions, the exact sense in which the approximation takes place being dependent on the object being approximated. We will next discuss several results of this type.

Theorem 6.4. *Let $f \in C(\mathbb{R}^N)$ with $\text{supp} f$ compact in $\mathbb{R}^N$. Pick $\phi \in \mathcal{D}(\mathbb{R}^N)$, with $\int_{\mathbb{R}^N} \phi(x)\, dx = 1$, set $\phi_n(x) = n^N \phi(nx)$ and $f_n = f * \phi_n$. Then $f_n \in \mathcal{D}(\mathbb{R}^N)$ and $f_n \to f$ uniformly on $\mathbb{R}^N$.*

Proof. The fact that $f_n \in \mathcal{D}(\mathbb{R}^N)$ is immediate from Proposition 6.3. Fix $\epsilon > 0$. By the assumption that f is continuous and of compact support it must be uniformly continuous on $\mathbb{R}^N$ so there exists $\delta > 0$ such that $|f(x) - f(z)| < \epsilon$ if $|x - z| < \delta$. Now choose n_0 such that $\text{supp}\,\phi_n \subset B(0, \delta)$ for $n > n_0$. We then have, for $n > n_0$ that

$$|f_n(x) - f(x)| = \left| \int_{\mathbb{R}^N} (f(x-y) - f(x))\phi_n(y)\, dy \right| \tag{6.4.89}$$

$$\leq \int_{|y|<\delta} |f(x-y) - f(x)||\phi_n(y)|\, dy \leq \epsilon \|\phi\|_{L^1(\mathbb{R}^N)} \tag{6.4.90}$$

and the conclusion follows. □

If f is not assumed continuous then of course it is not possible for there to exist $f_n \in \mathcal{D}(\mathbb{R}^N)$ converging uniformly to f. However the following can be shown.

Theorem 6.5. *Let $f \in L^p(\mathbb{R}^N)$, $1 \leq p < \infty$. Pick $\phi \in \mathcal{D}(\mathbb{R}^N)$, with $\int_{\mathbb{R}^N} \phi(x)\, dx = 1$, set $\phi_n(x) = n^N \phi(nx)$ and $f_n = f * \phi_n$. Then $f_n \in C^\infty(\mathbb{R}^N) \cap L^p(\mathbb{R}^N)$ and $f_n \to f$ in $L^p(\mathbb{R}^N)$.*

Proof. If $\epsilon > 0$ we can find $g \in C(\mathbb{R}^N)$ of compact support such that $\|f - g\|_{L^p(\mathbb{R}^N)} < \epsilon$. If $g_n = g * \phi_n$ then

$$\|f - f_n\|_{L^p(\mathbb{R}^N)} \leq \|f - g\|_{L^p(\mathbb{R}^N)} + \|g - g_n\|_{L^p(\mathbb{R}^N)} + \|f_n - g_n\|_{L^p(\mathbb{R}^N)} \tag{6.4.91}$$

$$\leq C\|f - g\|_{L^p(\mathbb{R}^N)} + \|g - g_n\|_{L^p(\mathbb{R}^N)} \tag{6.4.92}$$

Here we have used Young's convolution inequality (6.4.78) to obtain

$$\|f_n - g_n\|_{L^p(\mathbb{R}^N)} \leq \|\phi_n\|_{L^1(\mathbb{R}^N)} \|f - g\|_{L^p(\mathbb{R}^N)} = \|\phi\|_{L^1(\mathbb{R}^N)} \|f - g\|_{L^p(\mathbb{R}^N)} \tag{6.4.93}$$

Since $g_n \to g$ uniformly by Theorem 6.4 and $g - g_n$ has support in a fixed compact set independent of n, it follows that $\|g - g_n\|_{L^p(\mathbb{R}^N)} \to 0$, and so $\limsup_{n\to\infty} \|f - f_n\|_{L^p(\mathbb{R}^N)} \leq C\epsilon$. □

Further refinements and variants of these results can be proved, see, for example, Section C.4 of [11]. We state explicitly one such case of particular importance, see, for example, Theorem 2.19 in [1], or Corollary 4.23 in [5].

Theorem 6.6. *If $\Omega \subset \mathbb{R}^N$ is open and $1 \leq p < \infty$ then $\mathcal{D}(\Omega)$ is dense in $L^p(\Omega)$.*

Finally we consider the possibility of approximating a general $T \in \mathcal{D}'(\mathbb{R}^N)$ by smooth functions. As in Corollary 6.1 we can choose $\psi_n \in \mathcal{D}(\mathbb{R}^N)$ such that $\psi_n \to \delta$ in $\mathcal{D}'(\mathbb{R}^N)$. Set $T_n = T * \psi_n$, so that $T_n \in C^\infty(\mathbb{R}^N)$. If $\phi \in \mathcal{D}(\mathbb{R}^N)$ we than have

$$T_n(\phi) = (T_n * \check{\phi})(0) = ((T * \psi_n) * \check{\phi})(0) \tag{6.4.94}$$

$$= (T * (\psi_n * \check{\phi}))(0) = T((\psi_n * \check{\phi})\check{}) \tag{6.4.95}$$

It may be checked that $\psi_n * \check{\phi} \to \check{\phi}$ in $\mathcal{D}(\mathbb{R}^N)$, thus $T_n(\phi) \to T(\phi)$ for all $\phi \in \mathcal{D}(\mathbb{R}^N)$, that is, $T_n \to T$ in $\mathcal{D}'(\mathbb{R}^N)$.

In the above derivation we used associativity of convolution. This property is not completely obvious, and in fact is false in a more general setting in which convolution of two distributions is defined. For example, if we were to assume that convolution of distributions was always defined and that Theorem 6.3 holds, we would have $1 * (\delta' * H) = 1 * H' = 1 * \delta = 1$, but $(1 * \delta') * H = 0 * H = 0$. Nevertheless, associativity is correct in the case we have just used it, and we refer to [33, Theorem 6.30(c)], for the proof.

The pattern of the results just stated is that $T * \psi_n$ converges to T in the topology appropriate to the space that T itself belongs to, but this cannot be true in all situations which may be encountered. For example, it cannot be true that if $f \in L^\infty$ then $f * \psi_n$ converges to f in L^∞ since this would amount to uniform convergence of a sequence of continuous functions, which is impossible if f itself is not continuous.

6.5 EXERCISES

6.1. Construct a test function $\phi \in C_0^\infty(\mathbb{R})$ with the following properties: $0 \le \phi(x) \le 1$ for all $x \in \mathbb{R}$, $\phi(x) \equiv 1$ for $|x| < 1$ and $\phi(x) \equiv 0$ for $|x| > 2$. (Suggestion: think about what ϕ' would have to look like.)

6.2. Show that

$$T(\phi) = \sum_{n=1}^{\infty} \phi^{(n)}(n)$$

defines a distribution $T \in \mathcal{D}'(\mathbb{R})$.

6.3. If $\phi \in \mathcal{D}(\mathbb{R})$ show that $\psi(x) = (\phi(x) - \phi(0))/x$ (this function appeared in Example 6.8) belongs to $C^\infty(\mathbb{R})$. (Suggestion: first prove $\psi(x) = \int_0^1 \phi'(xt)\,dt$.)

6.4. Find the distributional derivative of $f(x) = [x]$, the greatest integer function.

6.5. Find the distributional derivatives up through order four of $f(x) = |x| \sin x$.

6.6. (For readers familiar with the concept of absolute continuity.) If f is absolutely continuous on (a, b) and $f' = g$ a.e., show that $f' = g$ in the sense of distributions on (a, b).

6.7. Let $\lambda_n > 0$, $\lambda_n \to +\infty$ and set

$$f_n(x) = \sin \lambda_n x \qquad g_n(x) = \frac{\sin \lambda_n x}{\pi x}$$

(a) Show that $f_n \to 0$ in $\mathcal{D}'(\mathbb{R})$ as $n \to \infty$.

(b) Show that $g_n \to \delta$ in $\mathcal{D}'(\mathbb{R})$ as $n \to \infty$.

(You may use without proof the fact that the value of the improper integral $\int_{-\infty}^{\infty} \frac{\sin x}{x}\,dx$ is π.)

6.8. Let $\phi \in C_0^\infty(\mathbb{R})$ and $f \in L^1(\mathbb{R})$.

(a) If $\psi_n(x) = n(\phi(x + \frac{1}{n}) - \phi(x))$, show that $\psi_n \to \phi'$ in $C_0^\infty(\mathbb{R})$. (Suggestion: use the mean value theorem over and over again.)

(b) If $g_n(x) = n(f(x + \frac{1}{n}) - f(x))$, show that $g_n \to f'$ in $\mathcal{D}'(\mathbb{R})$.

6.9. Let $T = \text{pv}\,\frac{1}{x}$. Find a formula analogous to Eq. (6.3.62) for the distributional derivative of T.

6.10. Find $\lim_{n\to\infty} \sin^2 nx$ in $\mathcal{D}'(\mathbb{R})$, or show that it doesn't exist.

6.11. Define the distribution

$$T(\phi) = \int_{-\infty}^{\infty} \phi(x,x)\,dx$$

for $\phi \in C_0^\infty(\mathbb{R}^2)$. Show that T satisfies the wave equation $u_{xx} - u_{yy} = 0$ in the sense of distributions on $\mathbb{R}^2$. Does it make sense to regard T as being a special case of the general solution (1.3.77)?

6.12. Let $\Omega \subset \mathbb{R}^N$ be a bounded open set and $K \subset\subset \Omega$. Show that there exists $\phi \in C_0^\infty(\Omega)$ such that $0 \le \phi(x) \le 1$ and $\phi(x) \equiv 1$ for $x \in K$. (Hint: approximate the characteristic function of Σ by convolution, where Σ satisfies $K \subset\subset \Sigma \subset\subset \Omega$. Use Proposition 6.3 for the needed support property.)

6.13. If $a \in C^\infty(\Omega)$ and $T \in \mathcal{D}'(\Omega)$ prove the product rule

$$\frac{\partial}{\partial x_j}(aT) = a\frac{\partial T}{\partial x_j} + \frac{\partial a}{\partial x_j}T$$

6.14. Let $T \in \mathcal{D}'(\mathbb{R}^N)$. We may then regard $\phi \longmapsto A\phi = T * \phi$ as a linear mapping from $C_0^\infty(\mathbb{R}^n)$ into $C^\infty(\mathbb{R}^n)$. Show that A commutes with translations, that is, $\tau_h A\phi = A\tau_h\phi$ for any $\phi \in C_0^\infty(\mathbb{R}^N)$. (The following interesting converse statement can also be proved: If $A : C_0^\infty(\mathbb{R}^N) \longmapsto C(\mathbb{R}^N)$ is continuous and commutes with translations then there exists a unique $T \in \mathcal{D}'(\mathbb{R}^N)$ such that $A\phi = T * \phi$. An operator commuting with translations is also said to be *translation invariant*.)

6.15. If $f \in L^1(\mathbb{R}^N)$, $\int_{\mathbb{R}^N} f(x)\,dx = 1$, and $f_n(x) = n^N f(nx)$, use Theorem 6.2 to show that $f_n \to \delta$ in $\mathcal{D}'(\mathbb{R}^N)$.

6.16. Prove Theorem 6.1.

6.17. If $T \in \mathcal{D}'(\Omega)$ prove the equality of mixed partial derivatives

$$\frac{\partial^2 T}{\partial x_i \partial x_j} = \frac{\partial^2 T}{\partial x_j \partial x_i} \qquad (6.5.96)$$

in the sense of distributions, and discuss why there is no contradiction with known examples from calculus showing that the mixed partial derivatives need not be equal.

6.18. Show that the expression

$$T(\phi) = \int_{-1}^{1} \frac{\phi(x) - \phi(0)}{|x|}\,dx + \int_{|x|>1} \frac{\phi(x)}{|x|}\,dx$$

defines a distribution on $\mathbb{R}$. Show also that $xT = \operatorname{sgn} x$.

6.19. If f is a function defined on $\mathbb{R}^N$ and $\lambda > 0$, let $f_\lambda(x) = f(\lambda x)$. We say that f is homogeneous of degree α if $f_\lambda = \lambda^\alpha f$ for any $\lambda > 0$. If T is a distribution on $\mathbb{R}^N$ we say that T is homogeneous of degree α if

$$T(\phi_\lambda) = \lambda^{-\alpha-N} T(\phi)$$

(a) Show that these two definitions are consistent, that is, if $T = T_f$ for some $f \in L^1_{loc}(\mathbb{R}^N)$ then T is homogeneous of degree α if and only if f is homogeneous of degree α.

(b) Show that the delta function is homogeneous of degree $-N$.

6.20. Show that $u(x) = \frac{1}{2\pi} \log|x|$ satisfies $\Delta u = \delta$ in $\mathcal{D}'(\mathbb{R}^2)$.

6.21. Without appealing to Theorem 6.3, give a direct proof of the fact that $T * \phi$ is a continuous function of x, for $T \in \mathcal{D}'(\mathbb{R}^N)$ and $\phi \in \mathcal{D}(\mathbb{R}^N)$.

6.22. Let

$$f(x) = \begin{cases} \log^2 x & x > 0 \\ 0 & x < 0 \end{cases}$$

Show that $f \in \mathcal{D}'(\mathbb{R})$ and find the distributional derivative f'.

6.23. If $a \in C^\infty(\mathbb{R})$, show that

$$a\delta' = a(0)\delta' - a'(0)\delta$$

6.24. If $T \in \mathcal{E}'(\Omega)$, show that $T(\phi)$ is defined in an unambiguous way for any $\phi \in C^\infty(\mathbb{R}^N) =: \mathcal{E}(\mathbb{R}^N)$. (Suggestion: write $\phi = \psi\phi + (1 - \psi)\phi$ where $\psi \in \mathcal{D}(\mathbb{R}^N)$ satisfies $\psi \equiv 1$ on the support of T.)

Chapter 7

Fourier Analysis

In this chapter we present some of the elements of Fourier analysis, with special attention to those aspects connected to the theory of distributions. Fourier analysis is often viewed as made up of two parts, one being a collection of topics relating to Fourier series, and the second being those concerning the Fourier transform. The essential distinction is that the former focuses on periodic functions while the latter is concerned with functions defined on all of $\mathbb{R}^N$. In either case the central question is that of how we may represent fairly arbitrary functions, or even distributions, as combinations of particularly simple periodic functions.

We will begin with Fourier series, and restrict attention to the one-dimensional case. See, for example, [28] for treatment of multidimensional Fourier series.

7.1 FOURIER SERIES IN ONE SPACE DIMENSION

The fundamental point at the foundation of the theory of Fourier series is that if $u_n(x) = e^{inx}$ then the set of functions $\{u_n\}_{n=-\infty}^{\infty}$ is an orthogonal basis of the Hilbert space $\mathbf{H} = L^2(-\pi, \pi)$. It will then follow from the general considerations of Chapter 5 that any $f \in \mathbf{H}$ may be expressed as a linear combination

$$f(x) = \sum_{n=-\infty}^{\infty} c_n e^{inx} \tag{7.1.1}$$

where

$$c_n = \frac{\langle f, u_n \rangle}{\langle u_n, u_n \rangle} = \frac{1}{2\pi} \int_{-\pi}^{\pi} f(y) e^{-iny}\, dy \tag{7.1.2}$$

The right-hand side of Eq. (7.1.1) is a Fourier series for f, and Eq. (7.1.2) is a formula for the nth Fourier coefficient of f. It must be understood that the equality in Eq. (7.1.1) is meant only in the sense of L^2 convergence of the partial sums, and need not be true at any particular point. From the theory of Lebesgue integration it follows that there is a subsequence of the partial

Techniques of Functional Analysis for Differential and Integral Equations
http://dx.doi.org/10.1016/B978-0-12-811426-1.00007-6

sums converging almost everywhere on $(-\pi,\pi)$, and more will be said about pointwise convergence properties later in the chapter.

Any finite sum $\sum_{n=-N}^{N} \gamma_n e^{inx}$ is called a *trigonometric polynomial*, so keeping in mind Theorem 5.4, this basis property amounts to showing that trigonometric polynomials are dense in $\mathbf{H}$.

Let us set

$$e_n(x) = \frac{1}{\sqrt{2\pi}} e^{inx} \qquad n = 0, \pm 1, \pm 2, \ldots \tag{7.1.3}$$

$$D_n(x) = \frac{1}{2\pi} \sum_{k=-n}^{n} e^{ikx} \tag{7.1.4}$$

$$K_N(x) = \frac{1}{N+1} \sum_{n=0}^{N} D_n(x) \tag{7.1.5}$$

It is immediate from checking the necessary integrals that $\{e_n\}_{n=-\infty}^{\infty}$ is an orthonormal set in $\mathbf{H}$. The main goal now is to prove that $\{e_n\}_{n=-\infty}^{\infty}$ is actually an orthonormal basis of this space.

For the rest of this section, the inner product symbol $\langle f, g\rangle$ and norm $\|\cdot\|$ refer to the inner product and norm in $\mathbf{H} = L^2(-\pi,\pi)$ unless otherwise stated. In the context of Fourier analysis, D_n, K_N are known as the *Dirichlet kernel* and *Féjer kernel* respectively. Note that

$$\int_{-\pi}^{\pi} D_n(x)\, dx = \int_{-\pi}^{\pi} K_N(x)\, dx = 1 \tag{7.1.6}$$

for any n, N.

If $f \in \mathbf{H}$, let

$$s_n(x) = \sum_{k=-n}^{n} c_k e^{ikx} \tag{7.1.7}$$

where c_k is given by Eq. (7.1.2) and

$$\sigma_N(x) = \frac{1}{N+1} \sum_{n=0}^{N} s_n(x) \tag{7.1.8}$$

Since

$$s_n(x) = \sum_{k=-n}^{n} \langle f, e_k\rangle e_k(x) \tag{7.1.9}$$

it follows that the partial sum s_n is also the projection of f onto the span of $\{e_k\}_{k=-n}^{n}$ and so in particular the Bessel inequality

$$\|s_n\| = \sqrt{\sum_{k=-n}^{n} |c_k|^2} \leq \|f\| \tag{7.1.10}$$

holds for all n. In particular, $\lim_{n\to\infty}\langle f, e_n\rangle = 0$, which is the Riemann-Lebesgue lemma for the Fourier coefficients of $f \in \mathbf{H}$.

Next observe that by substitution of Eq. (7.1.2) into Eq. (7.1.7) we obtain

$$s_n(x) = \int_{-\pi}^{\pi} f(y) D_n(x-y)\, dy \tag{7.1.11}$$

We can therefore regard s_n as being given by the convolution $D_n * f$ if we let $f(x) = 0$ outside of the interval $(-\pi, \pi)$. We can also express D_n in the following alternative and useful way:

$$D_n(x) = \frac{1}{2\pi} e^{-inx} \sum_{k=0}^{2n} e^{ikx} = \frac{1}{2\pi} e^{-inx} \left(\frac{1 - e^{(2n+1)ix}}{1 - e^{ix}} \right) \tag{7.1.12}$$

for $x \neq 0$. Multiplying top and bottom of the fraction by $e^{-ix/2}$ then yields

$$D_n(x) = \frac{1}{2\pi} \frac{\sin\,(n+\frac{1}{2})x}{\sin\frac{x}{2}} \qquad x \neq 0 \tag{7.1.13}$$

and obviously $D_n(0) = (2n+1)/2\pi$.

An alternative viewpoint of the convolutional relation (7.1.11), which is in some sense more natural, starts by defining the unit circle as $\mathbb{T} = \mathbb{R} \bmod 2\pi\mathbb{Z}$, that is, we identify any two points of $\mathbb{R}$ differing by an integer multiple of 2π. Any 2π periodic function, such as e_n, D_n, s_n, etc. may be regarded as a function on $\mathbb{T}$, and if f is originally given as a function on $(-\pi, \pi)$ then it may be extended in a 2π periodic manner to all of $\mathbb{R}$ and so also viewed as a function on the circle $\mathbb{T}$. With f, D_n both 2π periodic, the integral (7.1.11) could be written as

$$s_n(x) = \int_{\mathbb{T}} f(y) D_n(x-y)\, dy \tag{7.1.14}$$

since Eq. (7.1.11) simply amounts to using one natural parametrization of the independent variable. By the same token

$$s_n(x) = \int_a^{a+2\pi} f(y) D_n(x-y)\, dy \tag{7.1.15}$$

for any convenient choice of a. A 2π periodic function is continuous on $\mathbb{T}$ if it is continuous on $[-\pi, \pi]$ and $f(\pi) = f(-\pi)$, and so the space $C(\mathbb{T})$ may simply be regarded as

$$C(\mathbb{T}) = \{f \in C([-\pi, \pi]) : f(\pi) = f(-\pi)\} \tag{7.1.16}$$

a closed subspace of $C([-\pi, \pi])$, so is itself a Banach space with maximum norm. Likewise we can define

$$C^m(\mathbb{T}) = \{f \in C^m([-\pi, \pi]) : f^{(j)}(\pi) = f^{(j)}(-\pi), j = 0, 1, \ldots, m\} \tag{7.1.17}$$

a Banach space with the analogous norm.

Next let us make some corresponding observations about K_N.

Proposition 7.1. *There holds*

$$\sigma_N(x) = \int_{\mathbb{T}} K_N(x-y)f(y)\,dy \tag{7.1.18}$$

and

$$K_N(x) = \frac{1}{2\pi}\sum_{k=-N}^{N}\left(1-\frac{|k|}{N+1}\right)e^{ikx} = \frac{1}{2\pi(N+1)}\left(\frac{\sin\left(\frac{(N+1)x}{2}\right)}{\sin\left(\frac{x}{2}\right)}\right)^2 \quad x \neq 0 \tag{7.1.19}$$

Proof. The identity (7.1.18) is immediate from Eq. (7.1.14) and the definition of K_N, and the first identity in Eq. (7.1.19) is left as an exercise. To complete the proof we observe that

$$2\pi\sum_{n=0}^{N} D_n(x) = \frac{\sum_{n=0}^{N}\sin\left(n+\frac{1}{2}\right)x}{\sin\frac{x}{2}} \tag{7.1.20}$$

$$= \frac{\operatorname{Im}\left(e^{i\frac{x}{2}}\sum_{n=0}^{N}e^{inx}\right)}{\sin\frac{x}{2}} \tag{7.1.21}$$

$$= \frac{\operatorname{Im}\left(e^{i\frac{x}{2}}\left(\frac{1-e^{i(N+1)x}}{1-e^{ix}}\right)\right)}{\sin\frac{x}{2}} \tag{7.1.22}$$

$$= \frac{\operatorname{Im}\left(\frac{1-\cos(N+1)x-i\sin(N+1)x}{-2i\sin\frac{x}{2}}\right)}{\sin\frac{x}{2}} \tag{7.1.23}$$

$$= \frac{1-\cos(N+1)x}{2\sin^2\frac{x}{2}} \tag{7.1.24}$$

$$= \left(\frac{\sin\frac{(N+1)x}{2}}{\sin\left(\frac{x}{2}\right)}\right)^2 \tag{7.1.25}$$

and the conclusion follows upon dividing by $2\pi(N+1)$. □

Theorem 7.1. *Suppose that $f \in C(\mathbb{T})$. Then $\sigma_N \to f$ in $C(\mathbb{T})$.*

Proof. Since $K_N \geq 0$ and $\int_{\mathbb{T}} K_N(x-y)\,dy = 1$ for any x, we have

$$|\sigma_N(x)-f(x)| = \left|\int_{\mathbb{T}} K_N(x-y)(f(y)-f(x))\,dy\right| \leq \int_{x-\pi}^{x+\pi} K_N(x-y)|f(y)-f(x)|\,dy \tag{7.1.26}$$

If $\epsilon > 0$ is given, then since f must be uniformly continuous on $\mathbb{T}$, there exists $\delta > 0$ such that $|f(x)-f(y)| < \epsilon$ if $|x-y| < \delta$. Thus

$$|\sigma_N(x)-f(x)| \tag{7.1.27}$$

$$\leq \epsilon\int_{|x-y|<\delta} K_N(x-y)\,dy + 2\|f\|_\infty \int_{\delta<|x-y|<\pi} K_N(x-y)\,dy \tag{7.1.28}$$

$$\leq \epsilon + \frac{2\|f\|_\infty}{(N+1)\sin^2\left(\frac{\delta}{2}\right)} \tag{7.1.29}$$

Thus there exists N_0 such that for $N \geq N_0$, $|\sigma_N(x) - f(x)| < 2\epsilon$ for all x, that is to say, $\sigma_N \to f$ uniformly. □

Corollary 7.1. *The functions $\{e_n(x)\}_{n=-\infty}^{\infty}$ form an orthonormal basis of $\mathbf{H} = L^2(-\pi,\pi)$.*

Proof. We have already observed that these functions form an orthonormal set, so it remains only to verify one of the equivalent conditions stated in Theorem 5.4. We will show the closedness property, that is, that set of finite linear combinations of $\{e_n(x)\}_{n=-\infty}^{\infty}$ is dense in $\mathbf{H}$. Given $g \in \mathbf{H}$ and $\epsilon > 0$ we may find $f \in C(\mathbb{T})$ such that $\|f - g\| < \epsilon$, $f \in \mathcal{D}(-\pi,\pi)$ for example. Then choose N such that $\|\sigma_N - f\|_{C(\mathbb{T})} < \epsilon$, which implies $\|\sigma_N - f\| < \sqrt{2\pi}\epsilon$. Thus σ_N is a finite linear combination of the e_n's and

$$\|g - \sigma_N\| < (1 + \sqrt{2\pi})\epsilon \tag{7.1.30}$$

Since ϵ is arbitrary, the conclusion follows. □

Corollary 7.2. *If $f \in \mathbf{H}$ and*

$$s_n(x) = \sum_{k=-n}^{n} c_k e^{ikx} \tag{7.1.31}$$

where

$$c_k = \frac{1}{2\pi}\int_{-\pi}^{\pi} f(x)e^{-ikx}\,dx \tag{7.1.32}$$

then $s_n \to f$ in $\mathbf{H}$.

For $f \in \mathbf{H}$, we will normally write

$$f(x) = \sum_{n=-\infty}^{\infty} c_n e^{inx} \tag{7.1.33}$$

but we emphasize that without further assumptions this only means that the partial sums converge in $L^2(-\pi,\pi)$.

At this point we have looked at the convergence properties of two different sequences of trigonometric polynomials, s_n and σ_N, associated with f. While s_n is simply the nth partial sum of the Fourier series of f, the σ_N's are the so-called *Féjer means* of f. While each Féjer mean is a trigonometric polynomial, the sequence σ_N does not amount to the partial sums of some other Fourier series, since the nth coefficient would also have to depend on N. For $f \in \mathbf{H}$, we have that $s_N \to f$ in $\mathbf{H}$, and so the same is obviously true under the stronger assumption that $f \in C(\mathbb{T})$. On the other hand for $f \in C(\mathbb{T})$ we have shown that $\sigma_N \to f$ uniformly, but it need not be true that $s_N \to f$ uniformly, or even pointwise (example of P. du Bois-Reymond, see Section 1.6.1 of [28]). For $f \in \mathbf{H}$ it can be shown that $\sigma_N \to f$ in $\mathbf{H}$, but on the other hand the best L^2 approximation property of s_N implies that

$$\|s_N - f\| \le \|\sigma_N - f\| \tag{7.1.34}$$

since both s_N and σ_N are in the span of $\{e_k\}_{k=-N}^N$. That is to say, the rate of convergence of s_N to f is faster, in the L^2 sense at least, than that of σ_N. In summary, both s_N and σ_N provide a trigonometric polynomial approximating f, but each has some advantage over the other, depending on what is to be assumed about f.

7.2 ALTERNATIVE FORMS OF FOURIER SERIES

From the basic Fourier series (7.1.1) a number of other closely related and useful expressions can be immediately derived. First suppose that $f \in L^2(-L, L)$ for some $L > 0$. If we let $\tilde{f}(x) = f(Lx/\pi)$ then $\tilde{f} \in L^2(-\pi, \pi)$, so

$$\tilde{f}(x) = \sum_{n=-\infty}^{\infty} c_n e^{inx} \qquad c_n = \frac{1}{2\pi} \int_{-\pi}^{\pi} \tilde{f}(y) e^{-iny}\, dy \tag{7.2.35}$$

Equivalently,

$$f(x) = \sum_{n=-\infty}^{\infty} c_n e^{i\pi nx/L} \qquad c_n = \frac{1}{2L} \int_{-L}^{L} f(y) e^{-i\pi ny/L}\, dy \tag{7.2.36}$$

Likewise Eq. (7.2.36) holds if we just regard f as being $2L$ periodic and square integrable, and in the formula for c_n we could replace $(-L, L)$ by any other interval of length $2L$. The functions $e^{i\pi nx/L}/\sqrt{2L}$ make up an orthonormal basis of $L^2(a, b)$ if $b - a = 2L$.

Next observe that an equivalent form of the first identity in Eq. (7.2.36) is

$$f(x) = \sum_{n=-\infty}^{\infty} c_n \left(\cos \frac{n\pi x}{L} + i \sin \frac{n\pi x}{L} \right) \tag{7.2.37}$$

$$= c_0 + \sum_{n=1}^{\infty} (c_n + c_{-n}) \cos \frac{n\pi x}{L} + i(c_n - c_{-n}) \sin \frac{n\pi x}{L} \tag{7.2.38}$$

If we let

$$a_n = c_n + c_{-n} \quad b_n = i(c_n - c_{-n}) \quad n = 0, 1, 2, \ldots \tag{7.2.39}$$

then we obtain yet another alternative expression,

$$f(x) = \frac{a_0}{2} + \sum_{n=1}^{\infty} a_n \cos \frac{n\pi x}{L} + b_n \sin \frac{n\pi x}{L} \tag{7.2.40}$$

where

$$a_n = \frac{1}{L}\int_{-L}^{L} f(y)\cos\frac{n\pi y}{L}\,dy \quad n = 0,1,\ldots$$
$$b_n = \frac{1}{L}\int_{-L}^{L} f(y)\sin\frac{n\pi y}{L}\,dy \quad n = 1,2,\ldots \tag{7.2.41}$$

We refer to Eqs. (7.2.40), (7.2.41) as the "real form" of the Fourier series, which is natural to use, for example, if f is real valued, since then no complex quantities appear. Again the precise meaning of Eq. (7.2.40) is that $s_n \to f$ in $L^2(-L,L)$ or other interval of length $2L$, where now

$$s_n(x) = \frac{a_0}{2} + \sum_{k=1}^{n} a_k\cos\frac{k\pi x}{L} + b_k\sin\frac{k\pi x}{L} \tag{7.2.42}$$

with results analogous to those mentioned previously for the Féjer means also being valid. It may be verified that the set of functions

$$\left\{\frac{1}{\sqrt{2L}}, \frac{\cos\frac{n\pi x}{L}}{\sqrt{L}}, \frac{\sin\frac{n\pi x}{L}}{\sqrt{L}}\right\}_{n=1}^{\infty} \tag{7.2.43}$$

make up an orthonormal basis of $L^2(-L,L)$.

Another important variant is obtained as follows. If $f \in L^2(0,L)$ then we may define the associated even and odd extensions of f in $L^2(-L,L)$, namely

$$f_e(x) = \begin{cases} f(x) & 0 < x < L \\ f(-x) & -L < x < 0 \end{cases} \qquad f_o(x) = \begin{cases} f(x) & 0 < x < L \\ -f(-x) & -L < x < 0 \end{cases} \tag{7.2.44}$$

If we replace f by f_e in Eqs. (7.2.40), (7.2.41), then we obtain immediately that $b_n = 0$ and the resulting cosine series representation for f,

$$f(x) = \frac{a_0}{2} + \sum_{n=1}^{\infty} a_n\cos\frac{n\pi x}{L} \qquad a_n = \frac{2}{L}\int_0^L f(y)\cos\frac{n\pi y}{L}\,dy \quad n = 0,1,\ldots \tag{7.2.45}$$

Likewise replacing f by f_o gives us the corresponding sine series,

$$f(x) = \sum_{n=1}^{\infty} b_n\sin\frac{n\pi x}{L} \qquad b_n = \frac{2}{L}\int_0^L f(y)\sin\frac{n\pi y}{L}\,dy \quad n = 1,2,\ldots \tag{7.2.46}$$

Note that if f is continuous on $[0,L]$, then the $2L$ periodic extension of f_e is everywhere continuous, but this need not be true in the case of f_o. Thus we might expect that the cosine series of f has typically better convergence properties than that of the sine series.

7.3 MORE ABOUT CONVERGENCE OF FOURIER SERIES

If $f \in L^2(-\pi,\pi)$ it was already observed that since the partial sums s_n converge to f in $L^2(-\pi,\pi)$, some subsequence of the partial sums converges pointwise a.e. In fact it is a famous theorem of Carleson [6] that $s_n \to f$ (i.e., the entire sequence, not just a subsequence) pointwise a.e. This is a complicated proof and even now is not to be found even in advanced textbooks. No better result could be expected since f itself is only defined up to sets of measure zero.

If we were to assume the stronger condition that $f \in C(\mathbb{T})$ then it mighty be natural to conjecture that $s_n \to f$ for every x (recall we know $\sigma_N \to f$ uniformly in this case), but that turns out to be false, as mentioned previously: in fact there exist continuous functions for which $s_n(x)$ is divergent at infinitely many $x \in \mathbb{T}$, see Section 5.11 of [32].

A sufficient condition implying that $s_n(x) \to f(x)$ for every $x \in \mathbb{T}$ is that $f \in C(\mathbb{T})$ be piecewise continuously differentiable on $\mathbb{T}$. In fact the following more precise theorem can be proved.

Theorem 7.2. *Assume that there exist points* $-\pi = x_0 < x_1 < \cdots < x_M = \pi$ *such that* $f \in C^1([x_j, x_{j+1}])$ *for* $j = 0, 1, \ldots, M-1$. *Let*

$$\tilde{f}(x) = \begin{cases} \frac{1}{2}(\lim_{y\to x+} f(y) + \lim_{y\to x-} f(y)) & -\pi < x < \pi \\ \frac{1}{2}(\lim_{y\to -\pi+} f(y) + \lim_{y\to \pi-} f(y)) & x = \pm\pi \end{cases} \tag{7.3.47}$$

Then $\lim_{n\to\infty} s_n(x) = \tilde{f}(x)$ *for* $-\pi \le x \le \pi$.

Under the stated assumptions on f, the theorem states in particular that s_n converges to f at every point of continuity of f (with appropriate modification at the endpoints) and otherwise converges to the average of the left- and right-hand limits. The proof is somewhat similar to that of Theorem 7.1—steps in the derivation are outlined in the exercises.

So far we have discussed the convergence properties of the Fourier series based on assumptions about f, but another point of view we could take is to focus on how convergence properties are influenced by the behavior of the Fourier coefficients c_n. A first simple result of this type is:

Proposition 7.2. *If* $f \in \mathbf{H} = L^2(-\pi, \pi)$ *and its Fourier coefficients satisfy*

$$\sum_{n=-\infty}^{\infty} |c_n| < \infty \tag{7.3.48}$$

then $f \in C(\mathbb{T})$ *and* $s_n \to f$ *uniformly on* $\mathbb{T}$

Proof. By the Weierstrass M-test, the series $\sum_{n=-\infty}^{\infty} c_n e^{inx}$ is uniformly convergent on $\mathbb{R}$ to some limit g, and since each partial sum is continuous, the same must be true of g. Since uniform convergence implies L^2 convergence on any finite interval, we have $s_n \to g$ in $\mathbf{H}$, but also $s_n \to f$ in $\mathbf{H}$ by Corollary 7.2. By uniqueness of the limit $f = g$ and the conclusion follows. □

We say that f has an *absolutely convergent Fourier series* when Eq. (7.3.48) holds. We emphasize here that the conclusion $f = g$ is meant in the sense of L^2, that is, $f(x) = g(x)$ a.e., so by saying that f is continuous, we are really saying that the equivalence class of f contains a continuous function, namely g.

It is not the case that every continuous function has an absolutely convergent Fourier series, according to remarks made earlier in this section. It would therefore be of interest to find other conditions on f which guarantee that Eq. (7.3.48) holds. One such condition follows from the following, which is also of independent interest.

Proposition 7.3. *If $f \in C^m(\mathbb{T})$, then $\lim_{n\to\pm\infty} n^m c_n = 0$.*

Proof. We integrate by parts in Eq. (7.1.2) to get, for $n \neq 0$,

$$c_n = \frac{1}{2\pi}\left[\frac{f(y)e^{-iny}}{-in}\bigg|_{-\pi}^{\pi} + \frac{1}{in}\int_{-\pi}^{\pi} f'(y)e^{-iny}\,dy\right] = \frac{1}{2\pi in}\int_{-\pi}^{\pi} f'(y)e^{-iny}\,dy \tag{7.3.49}$$

if $f \in C^1(\mathbb{T})$. Since $f' \in L^2(\mathbb{T})$, the Riemann-Lebesgue lemma implies that $nc_n \to 0$ as $n \to \pm\infty$. If $f \in C^2(\mathbb{T})$ we could integrate by parts again to get $n^2 c_n \to 0$, etc. □

It is immediate from this result that if $f \in C^2(\mathbb{T})$ then it has an absolutely convergent Fourier series, but in fact even $f \in C^1(\mathbb{T})$ is more than enough, see Exercise 7.6.

One way to regard Proposition 7.3 is that it states that the smoother f is, the more rapidly the Fourier coefficients of f must decay. The next result is a sort of converse statement.

Proposition 7.4. *If $f \in \mathbf{H} = L^2(-\pi,\pi)$ and its Fourier coefficients satisfy*

$$|c_n| \leq \frac{C}{n^{m+\alpha}} \tag{7.3.50}$$

for some C and $\alpha > 1$, then $f \in C^m(\mathbb{T})$.

Proof. When $m = 0$ this is just a special case of Proposition 7.2. When $m = 1$ we see that it is permissible to differentiate the series (7.1.1) term by term, since the differentiated series

$$\sum_{n=-\infty}^{\infty} inc_n e^{inx} \tag{7.3.51}$$

is uniformly convergent, by the assumption (7.3.50). Thus f, f' are both a.e. equal to an absolutely convergent Fourier series, so $f \in C^1(\mathbb{T})$, by Proposition 7.2. The proof for $m = 2, 3, \ldots$ is similar. □

Note that Proposition 7.3 states a necessary condition on the Fourier coefficients for f to be in C^m and Proposition 7.4 states a sufficient condition. The two conditions are not identical, but both point to the general tendency that increased smoothness of f is associated with more rapid decay of the corresponding Fourier coefficients.

7.4 THE FOURIER TRANSFORM ON $\mathbb{R}^N$

We turn now to the notion of Fourier transform. If f is a given function on $\mathbb{R}^N$, we define the Fourier transform of f to be

$$\widehat{f}(y) = \frac{1}{(2\pi)^{N/2}} \int_{\mathbb{R}^N} f(x)e^{-ix\cdot y}\,dx \qquad y \in \mathbb{R}^N \tag{7.4.52}$$

provided that the integral is defined in some sense. This will always be the case, for example, if $f \in L^1(\mathbb{R}^N)$ since then

$$|\widehat{f}(y)| \le \frac{1}{(2\pi)^{N/2}} \int_{\mathbb{R}^N} |f(x)|\,dx < \infty \tag{7.4.53}$$

for any $y \in \mathbb{R}^N$. Thus in fact $\widehat{f} \in L^\infty(\mathbb{R}^N)$ in this case.

There are a number of other commonly used definitions of the Fourier transform, obtained by changing the numerical constant in front of the integral, and/or replacing $-ix \cdot y$ by $ix \cdot y$ and/or including a factor of 2π in the exponent in the integrand. Each convention has some convenient properties in certain situations, but none of them is always the best, hence the lack of a universally agreed upon definition. The differences are nonessential, all having to do with the way certain numerical constants turn up, so the only requirement is that we adopt one specific definition, such as Eq. (7.4.52), and stick with it.

The Fourier transform is a particular kind of linear integral operator, and an alternative operator type notation for it,

$$\mathcal{F}\phi = \widehat{\phi} \tag{7.4.54}$$

is often convenient to use, especially when discussing its mapping properties.

Example 7.1. If $N = 1$ and $f(x) = \chi_{[a,b]}(x)$, the indicator function of the interval $[a, b]$, then the Fourier transform of f is

$$\widehat{f}(y) = \frac{1}{\sqrt{2\pi}} \int_a^b e^{-ixy}\,dy = \frac{e^{-iay} - e^{-iby}}{\sqrt{2\pi}\,iy} \tag{7.4.55}$$

□

Example 7.2. If $N = 1$, $\alpha > 0$ and $f(x) = e^{-\alpha x^2}$ (a *Gaussian* function) then

$$\widehat{f}(y) = \frac{1}{\sqrt{2\pi}} \int_{-\infty}^{\infty} e^{-\alpha x^2} e^{-ixy}\,dx = \frac{e^{-\frac{y^2}{4\alpha}}}{\sqrt{2\pi}} \int_{-\infty}^{\infty} e^{-\alpha(x+\frac{iy}{2\alpha})^2}\,dx \tag{7.4.56}$$

$$= \frac{e^{-\frac{y^2}{4\alpha}}}{\sqrt{2\pi}} \int_{-\infty}^{\infty} e^{-\alpha x^2}\,dx = \frac{e^{-\frac{y^2}{4\alpha}}}{\sqrt{2\pi}} \sqrt{\frac{\pi}{\alpha}} = \frac{1}{\sqrt{2\alpha}} e^{-\frac{y^2}{4\alpha}} \tag{7.4.57}$$

In the above derivation, the key step is the third equality which is justified by contour integration techniques in complex function theory—the integral of

$e^{-\alpha z^2}$ along the real axis is the same as the integral along the parallel line $\operatorname{Im} z = \frac{y}{2\alpha}$ for any y.

Thus the Fourier transform of a Gaussian is another Gaussian, and in particular $\widehat{f} = f$ if $\alpha = \frac{1}{2}$.

It is clear from the Fourier transform definition that if f has the special product form $f(x) = f_1(x_1)f_2(x_2)\dots f_N(x_N)$ then $\widehat{f}(y) = \widehat{f_1}(y_1)\widehat{f_2}(y_2)\dots \widehat{f_N}(y_N)$. The Gaussian in $\mathbb{R}^N$, namely $f(x) = e^{-\alpha|x|^2}$, is of this type, so using Eq. (7.4.57) we immediately obtain

$$\widehat{f}(y) = \frac{e^{-\frac{|y|^2}{4\alpha}}}{(2\alpha)^{N/2}} \tag{7.4.58}$$

□

To state our first theorem about the Fourier transform, let us denote

$$C_0(\mathbb{R}^N) = \left\{ f \in C(\mathbb{R}^N) : \lim_{|x|\to\infty} |f(x)| = 0 \right\} \tag{7.4.59}$$

the space of continuous functions vanishing at ∞. It is a closed subspace of $L^\infty(\mathbb{R}^N)$, hence a Banach space with the L^∞ norm. We emphasize that despite the notation, functions in this space need not be of compact support.

Theorem 7.3. *If $f \in L^1(\mathbb{R}^N)$ then $\widehat{f} \in C_0(\mathbb{R}^N)$ and*

$$\|\widehat{f}\|_{C_0(\mathbb{R}^N)} \le \frac{1}{(2\pi)^{N/2}} \|f\|_{L^1(\mathbb{R}^N)} \tag{7.4.60}$$

Proof. If $y_n \in \mathbb{R}^N$ and $y_n \to y$ then clearly $f(x)e^{-ix\cdot y_n} \to f(x)e^{-ix\cdot y}$ for a.e. $x \in \mathbb{R}^N$. Also, $|f(x)e^{-ix\cdot y_n}| \le |f(x)|$, and since we assume $f \in L^1(\mathbb{R}^N)$ we can immediately apply the dominated convergence theorem to obtain

$$\lim_{n\to\infty} \int_{\mathbb{R}^N} f(x)e^{-ix\cdot y_n}\, dx = \int_{\mathbb{R}^N} f(x)e^{-ix\cdot y}\, dx \tag{7.4.61}$$

that is, $\widehat{f}(y_n) \to \widehat{f}(y)$. Hence $\widehat{f} \in C(\mathbb{R}^N)$.

Next, suppose temporarily that $g \in C^1(\mathbb{R}^N)$ and has compact support. An integration by parts gives us, for $j = 1, 2, \dots, N$ that

$$\widehat{g}(y) = -\frac{1}{(2\pi)^{N/2}} \frac{1}{iy_j} \int_{\mathbb{R}^N} \frac{\partial g}{\partial x_j} e^{-ix\cdot y}\, dx \tag{7.4.62}$$

Thus there exists some C, depending on g, such that

$$|\widehat{g}(y)|^2 \le \frac{C}{y_j^2} \quad j = 1, 2, \dots, N \tag{7.4.63}$$

from which it follows that

$$|\widehat{g}(y)|^2 \le \min_j \left(\frac{C}{y_j^2} \right) \le \frac{CN}{|y|^2} \tag{7.4.64}$$

Thus $\widehat{g}(y) \to 0$ as $|y| \to \infty$ in this case.

Finally, such g's are dense in $L^1(\mathbb{R}^N)$, so given $f \in L^1(\mathbb{R}^N)$ and $\epsilon > 0$, choose g as above such that $\|f - g\|_{L^1(\mathbb{R}^N)} < \epsilon$. We then have, taking into account (7.4.53),

$$|\widehat{f}(y)| \le |\widehat{f}(y) - \widehat{g}(y)| + |\widehat{g}(y)| \le \frac{1}{(2\pi)^{N/2}} \|f - g\|_{L^1(\mathbb{R}^N)} + |\widehat{g}(y)| \quad (7.4.65)$$

and so

$$\limsup_{|y|\to\infty} |\widehat{f}(y)| < \frac{\epsilon}{(2\pi)^{N/2}} \quad (7.4.66)$$

Since $\epsilon > 0$ is arbitrary, the conclusion $\widehat{f} \in C_0(\mathbb{R}^N)$ follows. □

The fact that $\widehat{f}(y) \to 0$ as $|y| \to \infty$ is analogous to the property that the Fourier coefficients $c_n \to 0$ as $n \to \pm\infty$ in the case of Fourier series, and in fact is also called the Riemann-Lebesgue Lemma.

One of the fundamental properties of the Fourier transform is that it is "almost" its own inverse. A first precise version of this is given by the following *Fourier Inversion Theorem.*

Theorem 7.4. *If $f, \widehat{f} \in L^1(\mathbb{R}^N)$ then*

$$f(x) = \frac{1}{(2\pi)^{N/2}} \int_{\mathbb{R}^N} \widehat{f}(y) e^{ix\cdot y}\, dy \quad a.e.\ x \in \mathbb{R}^N \quad (7.4.67)$$

The right-hand side of Eq. (7.4.67) is not precisely the Fourier transform of $\widehat{f}$ because the exponent contains $ix \cdot y$ rather than $-ix \cdot y$, but it does mean that we can think of it as saying that $f(x) = \widehat{\widehat{f}}(-x)$, or

$$\widehat{\widehat{f}} = \check{f}, \quad (7.4.68)$$

where $\check{f}(x) = f(-x)$, is the reflection of f.[1] The requirement in the theorem that both f and $\widehat{f}$ be in L^1 will be weakened later on.

Proof. Since $\widehat{f} \in L^1(\mathbb{R}^N)$ the right-hand side of Eq. (7.4.67) is well defined, and we denote it temporarily by $g(x)$. Define also the family of Gaussian functions,

$$G_\alpha(x) = \frac{e^{-\frac{|x|^2}{4\alpha}}}{(4\pi\alpha)^{N/2}} \quad (7.4.69)$$

We then have

$$g(x) = \lim_{\alpha\to 0+} \frac{1}{(2\pi)^{N/2}} \int_{\mathbb{R}^N} \widehat{f}(y) e^{ix\cdot y} e^{-\alpha|y|^2}\, dy \quad (7.4.70)$$

$$= \lim_{\alpha\to 0+} \frac{1}{(2\pi)^N} \int_{\mathbb{R}^N} \int_{\mathbb{R}^N} f(z) e^{-\alpha|y|^2} e^{-i(z-x)\cdot y}\, dz dy \quad (7.4.71)$$

1. Warning: some authors use the symbol $\check{f}$ to mean the inverse Fourier transform of f.

$$= \lim_{\alpha\to 0+} \frac{1}{(2\pi)^N} \int_{\mathbb{R}^N} f(z) \left(\int_{\mathbb{R}^N} e^{-\alpha|y|^2} e^{-i(z-x)\cdot y}\, dy \right) dz \tag{7.4.72}$$

$$= \lim_{\alpha\to 0+} \int_{\mathbb{R}^N} f(z) \frac{e^{-\frac{|z-x|^2}{4\alpha}}}{(4\pi\alpha)^{N/2}}\, dz \tag{7.4.73}$$

$$= \lim_{\alpha\to 0+} (f * G_\alpha)(x) \tag{7.4.74}$$

Here Eq. (7.4.70) follows from the dominated convergence theorem and Eq. (7.4.72) from Fubini's theorem, which is applicable here because

$$\int_{\mathbb{R}^N} \int_{\mathbb{R}^N} |f(z)e^{-\alpha|y|^2}|\, dzdy < \infty \tag{7.4.75}$$

In Eq. (7.4.73) we have used the explicit calculation (7.4.58) above for the Fourier transform of a Gaussian.

Noting that $\int_{\mathbb{R}^N} G_\alpha(x)\, dx = 1$ for every $\alpha > 0$, we see that the difference $f * G_\alpha(x) - f(x)$ may be written as

$$\int_{\mathbb{R}^N} G_\alpha(y)(f(x-y) - f(x))\, dy \tag{7.4.76}$$

so that

$$\|f * G_\alpha - f\|_{L^1(\mathbb{R}^N)} \le \int_{\mathbb{R}^N} G_\alpha(y)\phi(y)\, dy \tag{7.4.77}$$

where $\phi(y) = \int_{\mathbb{R}^N} |f(x-y) - f(x)|\, dx$. Then ϕ is bounded and continuous at $y = 0$ with $\phi(0) = 0$ (see Exercise 7.11), and we can verify that the hypotheses of Theorem 6.2 are satisfied with f_n replaced by G_{α_n} as long as $\alpha_n \to 0+$. For any sequence $\alpha_n > 0, \alpha_n \to 0$ it follows that $G_{\alpha_n} * f \to f$ in $L^1(\mathbb{R}^N)$, and so there is a subsequence $\alpha_{n_k} \to 0$ such that $(G_{\alpha_{n_k}} * f)(x) \to f(x)$ a.e. We conclude that Eq. (7.4.67) holds. □

7.5 FURTHER PROPERTIES OF THE FOURIER TRANSFORM

Formally speaking we have

$$\frac{\partial}{\partial y_j} \int_{\mathbb{R}^N} f(x)e^{-ix\cdot y}\, dx = \int_{\mathbb{R}^N} -ix_j f(x)e^{-ix\cdot y}\, dx \tag{7.5.78}$$

or in more compact notation

$$\frac{\partial \widehat{f}}{\partial y_j} = (-ix_j f)\widehat{\ } \tag{7.5.79}$$

This is rigorously justified by standard theorems of analysis about differentiation of integrals with respect to parameters provided that $\int_{\mathbb{R}^N} |x_j f(x)|\, dx < \infty$.

A companion property, obtained formally using integration by parts, is that

$$\int_{\mathbb{R}^N} \frac{\partial f}{\partial x_j} e^{-ix\cdot y}\, dx = \int_{\mathbb{R}^N} iy_j f(x) e^{-ix\cdot y}\, dx \tag{7.5.80}$$

or

$$\widehat{\left(\frac{\partial f}{\partial x_j}\right)} = iy_j \widehat{f} \tag{7.5.81}$$

which is rigorously correct provided at least that $f \in C^1(\mathbb{R}^N)$ and $\int_{|x|=R} |f(x)|\, dS \to 0$ as $R \to \infty$. Repeating the above arguments with higher derivatives we obtain

Proposition 7.5. *If α is any multiindex then*

$$D^\alpha \widehat{f}(y) = ((-ix)^\alpha f)\widehat{\ }(y) \tag{7.5.82}$$

provided

$$\int_{\mathbb{R}^N} |x^\alpha f(x)|\, dx < \infty \tag{7.5.83}$$

Also

$$(D^\alpha f)\widehat{\ }(y) = (iy)^\alpha \widehat{f}(y) \tag{7.5.84}$$

provided

$$f \in C^m(\mathbb{R}^n) \quad \int_{|x|=R} |D^\beta f(x)|\, dS \to 0 \text{ as } R \to \infty \quad |\beta| < |\alpha| = m \tag{7.5.85}$$

and $D^\beta f \in L^1(\mathbb{R}^N)$ for $|\beta| \le m$.

We will eventually see that Eqs. (7.5.82), (7.5.84) remain valid, suitably interpreted in a distributional sense, under conditions much more general than Eqs. (7.5.83), (7.5.85). But for now we introduce a new space in which these last two conditions are guaranteed to hold.

Definition 7.1. The *Schwartz space* is defined as

$$\mathcal{S}(\mathbb{R}^N) = \{\phi \in C^\infty(\mathbb{R}^N) : x^\alpha D^\beta \phi \in L^\infty(\mathbb{R}^N) \text{ for all } \alpha, \beta\} \tag{7.5.86}$$

Thus a function is in the Schwartz space if any derivative of it decays more rapidly than the reciprocal of any polynomial. Clearly $\mathcal{S}(\mathbb{R}^N)$ contains the set of test functions $\mathcal{D}(\mathbb{R}^N)$ as well as other kinds of functions such as Gaussians, $e^{-\alpha|x|^2}$ for any $\alpha > 0$.

If $\phi \in \mathcal{S}(\mathbb{R}^N)$ then in particular, for any n

$$|D^\beta \phi(x)| \le \frac{C}{(1+|x|^2)^n} \tag{7.5.87}$$

for some C, and so both Eqs. (7.5.82) and (7.5.84) hold. Thus the two key identities (7.5.82) and (7.5.84) are correct whenever f is in the Schwartz space. It is also immediate from Eq. (7.5.87) that $\mathcal{S}(\mathbb{R}^N) \subset L^1(\mathbb{R}^N) \cap L^\infty(\mathbb{R}^N)$.

Proposition 7.6. *If $\phi \in \mathcal{S}(\mathbb{R}^N)$ then $\widehat{\phi} \in \mathcal{S}(\mathbb{R}^N)$.*

Proof. Note from Eqs. (7.5.82) and (7.5.84) that

$$(iy)^\alpha D^\beta \widehat{\phi}(y) = (iy)^\alpha ((-ix)^\beta \phi)\widehat{\ }(y) = (D^\alpha((-ix)^\beta \phi))\widehat{\ }(y) \tag{7.5.88}$$

holds for $\phi \in \mathcal{S}(\mathbb{R}^N)$. Also, since $\mathcal{S}(\mathbb{R}^N) \subset L^1(\mathbb{R}^N)$ it follows from Eq. (7.4.53) that if $\phi \in \mathcal{S}(\mathbb{R}^N)$ then $\widehat{\phi} \in L^\infty(\mathbb{R}^N)$. Thus we have the following list of implications:

$$\phi \in \mathcal{S}(\mathbb{R}^N) \Longrightarrow (-ix)^\beta \phi \in \mathcal{S}(\mathbb{R}^N) \tag{7.5.89}$$

$$\Longrightarrow D^\alpha((-ix)^\beta \phi) \in \mathcal{S}(\mathbb{R}^N) \tag{7.5.90}$$

$$\Longrightarrow (D^\alpha((-ix)^\beta \phi))\widehat{\ } \in L^\infty(\mathbb{R}^N) \tag{7.5.91}$$

$$\Longrightarrow y^\alpha D^\beta \widehat{\phi} \in L^\infty(\mathbb{R}^N) \tag{7.5.92}$$

$$\Longrightarrow \widehat{\phi} \in \mathcal{S}(\mathbb{R}^N) \tag{7.5.93}$$

□

Corollary 7.3. *The Fourier transform $\mathcal{F} : \mathcal{S}(\mathbb{R}^N) \to \mathcal{S}(\mathbb{R}^N)$ is one to one and onto.*

Proof. The previous theorem says that $\mathcal{F}$ maps $\mathcal{S}(\mathbb{R}^N)$ into $\mathcal{S}(\mathbb{R}^N)$, and if $\mathcal{F}\phi = \widehat{\phi} = 0$ then the inversion theorem (Theorem 7.4) is applicable, since both $\phi, \widehat{\phi}$ are in $L^1(\mathbb{R}^N)$. We conclude $\phi = 0$, that is, $\mathcal{F}$ is one to one. If $\psi \in \mathcal{S}(\mathbb{R}^N)$, let $\phi = \check{\psi}$. Clearly $\phi \in \mathcal{S}(\mathbb{R}^N)$ and one may check directly, again using the inversion theorem, that $\widehat{\phi} = \psi$, so that $\mathcal{F}$ is onto. □

The next result, usually known as the *Parseval identity*, is the key step needed to define the Fourier transform of a function in $L^2(\mathbb{R}^N)$, which turns out to be a very natural setting.

Proposition 7.7. *If $\phi, \psi \in \mathcal{S}(\mathbb{R}^N)$ then*

$$\int_{\mathbb{R}^N} \phi(x)\widehat{\psi}(x)\,dx = \int_{\mathbb{R}^N} \widehat{\phi}(x)\psi(x)\,dx \tag{7.5.94}$$

Proof. The proof is simply an interchange of order in an iterated integral, which is easily justified by Fubini's theorem:

$$\int_{\mathbb{R}^N} \phi(x)\widehat{\psi}(x)\,dx = \frac{1}{(2\pi)^{N/2}} \int_{\mathbb{R}^N} \phi(x) \left(\int_{\mathbb{R}^N} \psi(y) e^{-ix\cdot y}\,dy \right) dx \tag{7.5.95}$$

$$= \frac{1}{(2\pi)^{N/2}} \int_{\mathbb{R}^N} \psi(y) \left(\int_{\mathbb{R}^N} \phi(x) e^{-ix\cdot y}\,dx \right) dy \tag{7.5.96}$$

$$= \int_{\mathbb{R}^N} \widehat{\phi}(y)\psi(y)\,dy \tag{7.5.97}$$

□

There is a slightly different but equivalent formula, which is also sometimes called the Parseval identity, see Exercise 7.12. The content of the following corollary is the *Plancherel identity*.

Corollary 7.4. *For every $\phi \in \mathcal{S}(\mathbb{R}^N)$ we have*

$$\|\phi\|_{L^2(\mathbb{R}^N)} = \|\widehat{\phi}\|_{L^2(\mathbb{R}^N)} \tag{7.5.98}$$

Proof. Given $\phi \in \mathcal{S}(\mathbb{R}^N)$ there exists, by Corollary 7.3, $\psi \in \mathcal{S}(\mathbb{R}^N)$ such that $\widehat{\psi} = \overline{\phi}$. In addition it follows directly from the definition of the Fourier transform and the inversion theorem that $\psi = \overline{\widehat{\phi}}$. Therefore, by Parseval's identity

$$\|\phi\|^2_{L^2(\mathbb{R}^N)} = \int_{\mathbb{R}^N} \phi(x)\widehat{\psi}(x)\,dx = \int_{\mathbb{R}^N} \widehat{\phi}(x)\psi(x) = \int_{\mathbb{R}^N} \widehat{\phi}(x)\overline{\widehat{\phi}}(x)\,dx$$
$$= \|\widehat{\phi}\|^2_{L^2(\mathbb{R}^N)} \tag{7.5.99}$$

□

Recalling that $\mathcal{D}(\mathbb{R}^N)$ is dense in $L^2(\mathbb{R}^N)$ it follows that the same is true of $\mathcal{S}(\mathbb{R}^N)$ and the Plancherel identity therefore implies that the Fourier transform has an extension to all of $L^2(\mathbb{R}^N)$. To be precise, if $f \in L^2(\mathbb{R}^N)$ pick $\phi_n \in \mathcal{S}(\mathbb{R}^N)$ such that $\phi_n \to f$ in $L^2(\mathbb{R}^N)$. Since $\{\phi_n\}$ is Cauchy in $L^2(\mathbb{R}^N)$, Eq. (7.5.98) implies the same for $\{\widehat{\phi}_n\}$, so $g := \lim_{n\to\infty} \widehat{\phi}_n$ exists in the L^2 sense, and we define this limit to be $\widehat{f}$. From elementary considerations this limit is independent of the choice of approximating sequence $\{\phi_n\}$, the extended definition of $\widehat{f}$ agrees with the original definition if $f \in L^1(\mathbb{R}^N) \cap L^2(\mathbb{R}^N)$, and Eq. (7.5.98) continues to hold for all $f \in L^2(\mathbb{R}^N)$. Since $\widehat{\phi}_n \to \widehat{f}$ in $L^2(\mathbb{R}^N)$, it follows by similar reasoning that $\widehat{\widehat{\phi}}_n \to \widehat{\widehat{f}}$. By the inversion theorem we know that $\widehat{\widehat{\phi}}_n = \check{\phi}_n$ which must converge to $\check{f}$, thus $\check{f} = \widehat{\widehat{f}}$, that is, the Fourier inversion theorem continues to hold on $L^2(\mathbb{R}^N)$.

The subset $L^1(\mathbb{R}^N) \cap L^2(\mathbb{R}^N)$ is dense in $L^2(\mathbb{R}^N)$ so we also have that $\widehat{f} = \lim_{n\to\infty} \widehat{f}_n$ if f_n is any sequence in $L^1(\mathbb{R}^N) \cap L^2(\mathbb{R}^N)$ convergent in $L^2(\mathbb{R}^N)$ to f. A natural choice of such a sequence is

$$f_n(x) = \begin{cases} f(x) & |x| < n \\ 0 & |x| > n \end{cases} \tag{7.5.100}$$

leading to the following explicit formula, similar to an improper integral, for the Fourier transform of an L^2 function,

$$\widehat{f}(y) = \lim_{n\to\infty} \frac{1}{(2\pi)^{N/2}} \int_{|x|<n} f(x)e^{-ix\cdot y}\,dx \tag{7.5.101}$$

where again without further assumptions we only know that the limit takes place in the L^2 sense.

Let us summarize.

Theorem 7.5. *For any $f \in L^2(\mathbb{R}^N)$ there exists a unique $\widehat{f} \in L^2(\mathbb{R}^N)$ such that $\widehat{f}$ is given by Eq. (7.4.52) whenever $f \in L^1(\mathbb{R}^N) \cap L^2(\mathbb{R}^N)$ and*

$$\|f\|_{L^2(\mathbb{R}^N)} = \|\widehat{f}\|_{L^2(\mathbb{R}^N)}. \tag{7.5.102}$$

Furthermore, $f, \widehat{f}$ are related by Eq. (7.5.101) and

$$f(x) = \lim_{n\to\infty} \frac{1}{(2\pi)^{N/2}} \int_{|y|<n} \widehat{f}(y) e^{ix\cdot y}\, dy \tag{7.5.103}$$

We conclude this section with one final important property of the Fourier transform.

Proposition 7.8. *If $f, g \in L^1(\mathbb{R}^N)$ then $f * g \in L^1(\mathbb{R}^N)$ and*

$$(f * g)\widehat{\ } = (2\pi)^{N/2} \widehat{f}\widehat{g} \tag{7.5.104}$$

Proof. The fact that $f * g \in L^1(\mathbb{R}^N)$ is immediate from Fubini's theorem, or, alternatively, is a special case of Young's convolution inequality (6.4.78). To prove Eq. (7.5.104) we have

$$(f * g)\widehat{\ }(z) = \frac{1}{(2\pi)^{N/2}} \int_{\mathbb{R}^N} (f * g)(x) e^{-ix\cdot z}\, dx \tag{7.5.105}$$

$$= \frac{1}{(2\pi)^{N/2}} \int_{\mathbb{R}^N} \left(\int_{\mathbb{R}^N} f(x-y) g(y)\, dy \right) e^{-ix\cdot z}\, dx \tag{7.5.106}$$

$$= \frac{1}{(2\pi)^{N/2}} \int_{\mathbb{R}^N} g(y) e^{-iy\cdot z} \left(\int_{\mathbb{R}^N} f(x-y) e^{-i(x-y)\cdot z}\, dx \right) dy \tag{7.5.107}$$

$$= (2\pi)^{N/2} \widehat{f}(z) \widehat{g}(z) \tag{7.5.108}$$

with the exchange of order of integration justified by Fubini's theorem. □

7.6 FOURIER SERIES OF DISTRIBUTIONS

In this and the next section we will see how the theory of Fourier series and Fourier transforms can be extended to a distributional setting. To begin with, let us consider the case of the delta function, viewed as a distribution on $(-\pi, \pi)$. Formally speaking, if $\delta(x) = \sum_{n=-\infty}^{\infty} c_n e^{inx}$, then the coefficients c_n should be given by

$$c_n = \frac{1}{2\pi} \int_{-\pi}^{\pi} \delta(x) e^{-inx}\, dx = \frac{1}{2\pi} \tag{7.6.109}$$

for every n, so that

$$\delta(x) = \frac{1}{2\pi} \sum_{n=-\infty}^{\infty} e^{inx} \tag{7.6.110}$$

Certainly this is not a valid formula in any classical sense, since among other things the terms of the series do not decay to zero. On the other hand, the Nth partial sum of this series is precisely the Dirichlet kernel $D_N(x)$, as in Eq. (7.1.4) or Eq. (7.1.13), and one consequence of Theorem 7.2 is precisely that $D_N \to \delta$ in $\mathcal{D}'(-\pi,\pi)$. Thus we may expect to find Fourier series representations of distributions, provided that we allow for the series to converge in a distributional sense.

Note that since $D_N \to \delta$ we must also have, by Proposition 6.1, that

$$D'_N = \frac{i}{2\pi}\sum_{n=-N}^{N} ne^{inx} \to \delta' \tag{7.6.111}$$

as $N \to \infty$. By repeatedly differentiating, we see that any formal Fourier series $\sum_{n=-\infty}^{\infty} n^m e^{inx}$ is meaningful in the distributional sense, and is simply, up to a constant multiple, some derivative of the delta function. The following proposition shows that we can allow any sequence of Fourier coefficients as long as the rate of growth is at most a power of n.

Proposition 7.9. *Let $\{c_n\}_{n=-\infty}^{\infty}$ be any sequence of constants satisfying*

$$|c_n| \le C|n|^M \tag{7.6.112}$$

for some constant C and positive integer M. Then there exists $T \in \mathcal{D}'(-\pi,\pi)$ such that

$$T = \sum_{n=-\infty}^{\infty} c_n e^{inx} \tag{7.6.113}$$

Proof. Let

$$g(x) = \sum_{n=-\infty}^{\infty} \frac{c_n}{(in)^{M+2}} e^{inx} \tag{7.6.114}$$

which is a uniformly convergent Fourier series, so in particular the partial sums $S_N \to g$ in the sense of distributions on $(-\pi,\pi)$. But then $S_N^{(j)} \to g^{(j)}$ also in the distributional sense, and in particular

$$\sum_{n=-\infty}^{\infty} c_n e^{inx} = g^{(M+2)} =: T \tag{7.6.115}$$

□

It seems clear that any distribution on $\mathbb{R}$ of the form (7.6.113) should be 2π periodic since every partial sum is. To make this precise, define the translate of any distribution $T \in \mathcal{D}'(\mathbb{R}^N)$ by the natural definition $\tau_h T(\phi) = T(\tau_{-h}\phi)$, where as usual $\tau_h\phi(x) = \phi(x-h), h \in \mathbb{R}^N$. We then say that T is h-periodic with period $h \in \mathbb{R}^N$ if $\tau_h T = T$. It is immediate that if T_n is h-periodic and $T_n \to T$ in $\mathcal{D}'(\mathbb{R}^N)$ then T is also h periodic.

Example 7.3. The Fourier series identity (7.6.110) becomes

$$\sum_{n=-\infty}^{\infty} \delta(x-2n\pi) = \frac{1}{2\pi}\sum_{n=-\infty}^{\infty} e^{inx} \tag{7.6.116}$$

when regarded as an identity in $\mathcal{D}'(\mathbb{R})$, since the left side is 2π periodic and coincides with δ on $(-\pi,\pi)$. □

A 2π periodic distribution on $\mathbb{R}$ may also naturally be regarded as an element of the distribution space $\mathcal{D}'(\mathbb{T})$, which is defined as the space of continuous linear functionals on $C^\infty(\mathbb{T})$. Here, convergence in $C^\infty(\mathbb{T})$ means that $\phi_n^{(j)} \to \phi^{(j)}$ uniformly on $\mathbb{T}$ for all $j = 0, 1, 2, \ldots$. Any function $f \in L^1(\mathbb{T})$ gives rise in the usual way to regular distribution T_f defined by $T_f(\phi) = \int_{-\pi}^{\pi} f(x)\phi(x)\,dx$ and if $f \in L^2$ then the nth Fourier coefficient is $c_n = \frac{1}{2\pi}T_f(e^{-inx})$. Since $e^{-inx} \in C^\infty(\mathbb{T})$ it follows that

$$c_n = \frac{1}{2\pi}T(e^{-inx}) \tag{7.6.117}$$

is meaningful for $T \in \mathcal{D}'(\mathbb{T})$, and is defined to be the nth Fourier coefficient of the distribution T. This definition is then consistent with the definition of Fourier coefficient for a regular distribution, and it can be shown (Exercise 7.31) that

$$\sum_{n=-N}^{N} c_n e^{inx} \to T \qquad \text{in } \mathcal{D}'(\mathbb{T}) \tag{7.6.118}$$

Example 7.4. Let us evaluate the distributional Fourier series

$$\sum_{n=0}^{\infty} e^{inx} \tag{7.6.119}$$

The nth partial sum is

$$s_n(x) = \sum_{k=0}^{n} e^{ikx} = \frac{1-e^{i(n+1)x}}{1-e^{ix}} \tag{7.6.120}$$

so that, since $\int_{-\pi}^{\pi} s_n(x)\,dx = 2\pi$, the action of the distribution s_n on a test function ϕ may be written

$$2\pi\phi(0) + \int_{-\pi}^{\pi} \frac{1-e^{i(n+1)x}}{1-e^{ix}}(\phi(x)-\phi(0))\,dx \tag{7.6.121}$$

The function $(\phi(x)-\phi(0))/(1-e^{ix})$ belongs to $L^2(-\pi,\pi)$, hence

$$\int_{-\pi}^{\pi} \frac{e^{i(n+1)x}}{1-e^{ix}}(\phi(x)-\phi(0))\,dx \to 0 \tag{7.6.122}$$

as $n \to \infty$ by the Riemann-Lebesgue lemma.

Next, using obvious trigonometric identities we see that $1/(1-e^{ix}) = \frac{1}{2}(1+i\cot\frac{x}{2})$, and so

$$\int_{-\pi}^{\pi} \frac{\phi(x)-\phi(0)}{1-e^{ix}}\,dx = \lim_{\epsilon\to 0+} \frac{1}{2}\int_{\epsilon<|x|<\pi} (\phi(x)-\phi(0))\left(1+i\cot\frac{x}{2}\right)dx \tag{7.6.123}$$

$$= \frac{1}{2}\int_{-\pi}^{\pi}\phi(x)\,dx - \pi\phi(0) \tag{7.6.124}$$

$$+ \lim_{\epsilon\to 0+}\frac{i}{2}\int_{\epsilon<|x|<\pi}\phi(x)\cot\frac{x}{2}\,dx \tag{7.6.125}$$

The principal value integral in Eq. (7.6.125) is naturally defined to be the action of the distribution $\mathrm{pv}(\cot\frac{x}{2})$, and we obtain the final result, upon letting $n\to\infty$, that

$$\sum_{n=0}^{\infty} e^{inx} = \pi\delta + \frac{1}{2} + \frac{i}{2}\,\mathrm{pv}\left(\cot\frac{x}{2}\right) \tag{7.6.126}$$

By taking the real and imaginary parts of this identity we also find the interesting formulas

$$\sum_{n=0}^{\infty}\cos nx = \pi\delta + \frac{1}{2} \qquad \sum_{n=1}^{\infty}\sin nx = \frac{1}{2}\,\mathrm{pv}\left(\cot\frac{x}{2}\right) \tag{7.6.127}$$

□

7.7 FOURIER TRANSFORMS OF DISTRIBUTIONS

Taking again the example of the delta function, now considered as a distribution on $\mathbb{R}^N$, it appears formally correct that it should have a Fourier transform which is a constant function, namely

$$\widehat{\delta}(y) = \frac{1}{(2\pi)^{N/2}}\int_{\mathbb{R}^N}\delta(x)e^{-ix\cdot y}\,dx = \frac{1}{(2\pi)^{N/2}} \tag{7.7.128}$$

If the inversion theorem remains valid then any constant should also have a Fourier transform, for example, $\widehat{1} = (2\pi)^{N/2}\delta$. On the other hand it will turn out that a rapidly growing function such as e^x does not have a Fourier transform in any reasonable sense.

We will now show that the set of distributions for which the Fourier transform can be defined turns out to be precisely the dual space of the Schwartz space $\mathcal{S}(\mathbb{R}^N)$. This space of linear functionals is also known as the space of *tempered distributions*. To define it we must first have a definition of convergence in $\mathcal{S}(\mathbb{R}^N)$.

Definition 7.2. We say that $\phi_n \to \phi$ in $\mathcal{S}(\mathbb{R}^N)$ if

$$\lim_{n\to\infty}\|x^\alpha D^\beta(\phi_n-\phi)\|_{L^\infty(\mathbb{R}^N)} = 0 \qquad \textit{for any } \alpha,\beta \tag{7.7.129}$$

Proof of the following lemma will be left for the exercises.

Lemma 7.1. *If $\phi_n \to \phi$ in $\mathcal{S}(\mathbb{R}^N)$ then $\widehat{\phi_n} \to \widehat{\phi}$ in $\mathcal{S}(\mathbb{R}^N)$.*

Definition 7.3. The set of tempered distributions on $\mathbb{R}^N$ is the space of continuous linear functionals on $\mathcal{S}(\mathbb{R}^N)$, denoted $\mathcal{S}'(\mathbb{R}^N)$.

It was already observed that $\mathcal{D}(\mathbb{R}^N) \subset \mathcal{S}(\mathbb{R}^N)$ and in addition, it is easy to check that if $\phi_n \to \phi$ in $\mathcal{D}(\mathbb{R}^N)$ then the sequence also converges in $\mathcal{S}(\mathbb{R}^N)$. It therefore follows that

$$\mathcal{S}'(\mathbb{R}^N) \subset \mathcal{D}'(\mathbb{R}^N) \tag{7.7.130}$$

that is, any tempered distribution is also a distribution, as the choice of language suggests. On the other hand, if T_f is the regular distribution corresponding to the L^1_{loc} function $f(x) = e^x$, then $T_f \notin \mathcal{S}'(\mathbb{R}^N)$ since this would require $\int_{-\infty}^{\infty} e^x \phi(x)\, dx$ to be finite for any $\phi \in \mathcal{S}(\mathbb{R}^N)$, which is not true. Thus the inclusion (7.7.130) is strict. We define convergence in $\mathcal{S}'(\mathbb{R}^N)$ in the expected way, analogously to Definition 6.5:

Definition 7.4. If $T, T_n \in \mathcal{S}'(\mathbb{R}^N)$ for $n = 1, 2, \ldots$ then we say $T_n \to T$ in $\mathcal{S}'(\mathbb{R}^N)$ (or in the sense of tempered distributions) if $T_n(\phi) \to T(\phi)$ for every $\phi \in \mathcal{S}(\mathbb{R}^N)$.

A regular distribution T_f will belong to $\mathcal{S}'(\mathbb{R}^N)$ provided it satisfies the condition

$$\lim_{|x|\to\infty} \frac{f(x)}{|x|^m} = 0 \tag{7.7.131}$$

for some m. Such an f is sometimes referred to as a *function of slow growth.* In particular, any polynomial belongs to $\mathcal{S}'(\mathbb{R}^N)$. One may also verify that the delta function belongs to $\mathcal{S}'(\mathbb{R}^N)$ as does any derivative or translate of the delta function.

We can now define the Fourier transform $\widehat{T}$ for any $T \in \mathcal{S}'(\mathbb{R}^N)$. For motivation of the definition, recall the Parseval identity (7.5.94), which amounts to the identity $T_{\widehat{\psi}}(\phi) = T_\psi(\widehat{\phi})$, if we regard ϕ as a function in $\mathcal{S}(\mathbb{R}^N)$ and ψ as a tempered distribution.

Definition 7.5. If $T \in \mathcal{S}'(\mathbb{R}^N)$ then its Fourier transform $\widehat{T}$ is defined by $\widehat{T}(\phi) = T(\widehat{\phi})$ for any $\phi \in \mathcal{S}(\mathbb{R}^N)$.

The action of $\widehat{T}$ on any $\phi \in \mathcal{S}(\mathbb{R}^N)$ is well-defined, since $\widehat{\phi} \in \mathcal{S}(\mathbb{R}^N)$, and linearity of $\widehat{T}$ is immediate. If $\phi_n \to \phi$ in $\mathcal{S}(\mathbb{R}^N)$ then by Lemma 7.1 $\widehat{\phi_n} \to \widehat{\phi}$ in $\mathcal{S}(\mathbb{R}^N)$, so that

$$\widehat{T}(\phi_n) = T(\widehat{\phi_n}) \to T(\widehat{\phi}) = \widehat{T}(\phi) \tag{7.7.132}$$

We have thus verified that $\widehat{T} \in \mathcal{S}'(\mathbb{R}^N)$ whenever $T \in \mathcal{S}'(\mathbb{R}^N)$.

Example 7.5. If $T = \delta$, then from the definition,

$$\widehat{T}(\phi) = T(\widehat{\phi}) = \widehat{\phi}(0) = \frac{1}{(2\pi)^{N/2}} \int_{\mathbb{R}^N} \phi(x)\, dx \tag{7.7.133}$$

Thus, as expected, $\widehat{\delta} = \frac{1}{(2\pi)^{N/2}}$, the constant distribution. □

Example 7.6. If $T = 1$ (the constant distribution) then

$$\widehat{T}(\phi) = T(\widehat{\phi}) = \int_{\mathbb{R}^N} \widehat{\phi}(x)\, dx = (2\pi)^{N/2}\widehat{\widehat{\phi}}(0) = (2\pi)^{N/2}\phi(0) \tag{7.7.134}$$

where the last equality follows from the inversion theorem which is valid for any $\phi \in \mathcal{S}(\mathbb{R}^N)$. Thus again the expected result is obtained,

$$\widehat{1} = (2\pi)^{N/2}\delta \tag{7.7.135}$$

□

The previous two examples verify the validity of one particular instance of the Fourier inversion theorem in the distributional context, but it turns out to be rather easy to prove that it always holds. One more definition is needed first, that of the reflection of a distribution.

Definition 7.6. If $T \in \mathcal{D}'(\mathbb{R}^N)$ then $\check{T}$, the reflection of T, is the distribution defined by $\check{T}(\phi) = T(\check{\phi})$.

We now obtain the Fourier inversion theorem in its most general form, analogous to the statement (7.4.68) first justified when $f, \widehat{f}$ are in $L^1(\mathbb{R}^N)$.

Theorem 7.6. *If $T \in \mathcal{S}'(\mathbb{R}^N)$ then $\widehat{\widehat{T}} = \check{T}$.*

Proof. For any $\phi \in \mathcal{S}(\mathbb{R}^N)$ we have

$$\widehat{\widehat{T}}(\phi) = T(\widehat{\widehat{\phi}}) = T(\check{\phi}) = \check{T}(\phi) \tag{7.7.136}$$

□

The apparent triviality of this proof should not be misconstrued, as it relies on the validity of the inversion theorem in the Schwartz space, and other technical machinery which we have developed.

Here we state several more simple but useful properties. Here and elsewhere, we follow the convention of using x and y as the independent variables before and after Fourier transformation respectively.

Proposition 7.10. *Let $T \in \mathcal{S}'(\mathbb{R}^N)$ and α be a multiindex. Then*

1. $x^\alpha T \in \mathcal{S}'(\mathbb{R}^N)$.
2. $D^\alpha T \in \mathcal{S}'(\mathbb{R}^N)$.
3. $D^\alpha \widehat{T} = ((-ix)^\alpha T)\widehat{\ }$.
4. $(D^\alpha T)\widehat{\ } = (iy)^\alpha \widehat{T}$.
5. *If $T_n \in \mathcal{S}'(\mathbb{R}^N)$ and $T_n \to T$ in $\mathcal{S}'(\mathbb{R}^N)$ then $\widehat{T}_n \to \widehat{T}$ in $\mathcal{S}'(\mathbb{R}^N)$.*

Proof. We give the proof of part 3 only, leaving the rest for the exercises. Just like the inversion theorem, it is more or less a direct consequence of the corresponding identity for functions in $\mathcal{S}(\mathbb{R}^N)$. For any $\phi \in \mathcal{S}(\mathbb{R}^N)$ we have

$$D^\alpha \widehat{T}(\phi) = (-1)^{|\alpha|}\widehat{T}(D^\alpha \phi) \tag{7.7.137}$$
$$= (-1)^{|\alpha|}T((D^\alpha \phi)\widehat{\ }) \tag{7.7.138}$$
$$= (-1)^{|\alpha|}T((iy)^\alpha \widehat{\phi}) \tag{7.7.139}$$
$$= (-ix)^\alpha T(\widehat{\phi}) = ((-ix)^\alpha T)\widehat{\ }(\phi) \tag{7.7.140}$$

as needed, where we used Eq. (7.5.84) to obtain Eq. (7.7.139). □

Example 7.7. If $T = \delta'$ regarded as an element of $\mathcal{S}'(\mathbb{R})$ then

$$\widehat{T} = (\delta')\widehat{\ } = iy\widehat{\delta} = \frac{iy}{\sqrt{2\pi}} \tag{7.7.141}$$

by part 4 of the previous proposition. In other words

$$\widehat{T}(\phi) = \frac{i}{\sqrt{2\pi}}\int_{-\infty}^{\infty} x\phi(x)\,dx \tag{7.7.142}$$

□

Example 7.8. Let $T = H(x)$, the Heaviside function, again regarded as an element of $\mathcal{S}'(\mathbb{R})$. To evaluate the Fourier transform $\widehat{H}$, one possible approach is to use part 4 of Proposition 7.10 along with $H' = \delta$ to first obtain $iy\widehat{H} = 1/\sqrt{2\pi}$. A formal solution is then $\widehat{H} = 1/\sqrt{2\pi}iy$, but it must then be recognized that this distributional equation does not have a unique solution, rather we can add to it any solution of $yT = 0$, for example, $T = C\delta$ for any constant C. It must be verified that there are no other solutions, the constant C must be evaluated, and the meaning of $1/y$ in the distribution sense must be made precise. See Example 8, section 2.4 of [35] for details of how this calculation is completed.

An alternate approach, which yields other useful formulas along the way is as follows. For any $\phi \in \mathcal{S}(\mathbb{R}^N)$ we have

$$\widehat{H}(\phi) = H(\widehat{\phi}) = \int_0^\infty \widehat{\phi}(y)\,dy \tag{7.7.143}$$
$$= \frac{1}{\sqrt{2\pi}}\int_0^\infty \int_{-\infty}^\infty \phi(x)e^{-ixy}\,dxdy \tag{7.7.144}$$
$$= \lim_{R\to\infty} \frac{1}{\sqrt{2\pi}}\int_0^R \int_{-\infty}^\infty \phi(x)e^{-ixy}\,dxdy \tag{7.7.145}$$
$$= \lim_{R\to\infty} \frac{1}{\sqrt{2\pi}}\int_{-\infty}^\infty \phi(x)\left(\int_0^R e^{-ixy}\,dy\right)dx \tag{7.7.146}$$
$$= \lim_{R\to\infty} \frac{1}{\sqrt{2\pi}}\int_{-\infty}^\infty \phi(x)\left(\frac{1-e^{-iRx}}{ix}\right)dx \tag{7.7.147}$$

$$= \lim_{R\to\infty} \frac{1}{\sqrt{2\pi}} \int_{-\infty}^{\infty} \frac{\sin Rx}{x}\phi(x)\,dx + \frac{i}{\sqrt{2\pi}} \int_{-\infty}^{\infty} \frac{\cos Rx - 1}{x}\phi(x)\,dx \tag{7.7.148}$$

It can then be verified that

$$\frac{\sin Rx}{x} \to \pi\delta \qquad \frac{\cos Rx - 1}{x} \to -\,\mathrm{pv}\frac{1}{x} \tag{7.7.149}$$

as $R \to \infty$ in $\mathcal{D}'(\mathbb{R})$. The first limit is just a restatement of the result of part (b) in Exercise 6.7 of Chapter 6, and the second we leave for the exercises. The final result, therefore, is that

$$\widehat{H} = \sqrt{\frac{\pi}{2}}\delta - \frac{i}{\sqrt{2\pi}}\,\mathrm{pv}\frac{1}{x} \tag{7.7.150}$$

□

Example 7.9. Let $T_n = \delta(x-n)$, that is, $T_n(\phi) = \phi(n)$, for $n = 0, \pm 1, \dots$, so that

$$\widehat{T_n}(\phi) = \widehat{\phi}(n) = \frac{1}{\sqrt{2\pi}} \int_{-\infty}^{\infty} \phi(x) e^{-inx}\,dx \tag{7.7.151}$$

Equivalently, $\sqrt{2\pi}\widehat{T_n} = e^{-inx}$. If we now set $T = \sum_{n=-\infty}^{\infty} T_n$ then $T \in \mathcal{S}'(\mathbb{R})$ and

$$\widehat{T} = \frac{1}{\sqrt{2\pi}} \sum_{n=-\infty}^{\infty} e^{-inx} = \frac{1}{\sqrt{2\pi}} \sum_{n=-\infty}^{\infty} e^{inx} = \sqrt{2\pi} \sum_{n=-\infty}^{\infty} \delta(x - 2\pi n) \tag{7.7.152}$$

where the last equality comes from Eq. (7.6.116). The relation $T(\widehat{\phi}) = \widehat{T}(\phi)$, then yields the very interesting identity

$$\sum_{n=-\infty}^{\infty} \widehat{\phi}(n) = \sqrt{2\pi} \sum_{n=-\infty}^{\infty} \phi(2\pi n) \tag{7.7.153}$$

valid at least for $\phi \in \mathcal{S}(\mathbb{R})$, which is known as the *Poisson summation formula.*
□

We conclude this section with some discussion of the Fourier transform and convolution in a distributional setting. Recall we gave a definition of the convolution $T * \phi$ in Definition 6.7, when $T \in \mathcal{D}'(\mathbb{R}^N)$ and $\phi \in \mathcal{D}(\mathbb{R}^N)$. We can use precisely the same definition if $T \in \mathcal{S}'(\mathbb{R}^N)$ and $\phi \in \mathcal{S}(\mathbb{R}^N)$, that is

Definition 7.7. If $T \in \mathcal{S}'(\mathbb{R}^N)$ and $\phi \in \mathcal{S}(\mathbb{R}^N)$ then $(T * \phi)(x) = T(\tau_x \check{\phi})$.

Note that in terms of the action of the distribution T, x is just a parameter, and that we must regard $\check{\phi}$ as a function of some unnamed other variable, say y or $\cdot$. By methods similar to those used in the proof of Theorem 6.3 it can be shown that

$$T * \phi \in C^\infty(\mathbb{R}^N) \cap \mathcal{S}'(\mathbb{R}^N) \tag{7.7.154}$$

and

$$D^\alpha(T * \phi) = D^\alpha T * \phi = T * D^\alpha \phi \tag{7.7.155}$$

In addition we have the following generalization of Proposition 7.8:

Theorem 7.7. *If $T \in \mathcal{S}'(\mathbb{R}^N)$ and $\phi \in \mathcal{S}(\mathbb{R}^N)$ then*

$$(T * \phi)\widehat{\ } = (2\pi)^{N/2}\widehat{T}\widehat{\phi} \tag{7.7.156}$$

Sketch of proof. First observe that from Proposition 7.8 and the inversion theorem we have

$$(\phi\psi)\widehat{\ } = \frac{1}{(2\pi)^{N/2}}(\widehat{\phi} * \widehat{\psi}) \tag{7.7.157}$$

for $\phi, \psi \in \mathcal{S}(\mathbb{R}^N)$. Thus for $\psi \in \mathcal{S}(\mathbb{R}^N)$

$$(\widehat{T}\widehat{\phi})(\psi) = \widehat{T}(\widehat{\phi}\psi) = T((\widehat{\phi}\psi)\widehat{\ }) = \frac{1}{(2\pi)^{N/2}}T(\widehat{\widehat{\phi}} * \widehat{\psi}) = \frac{1}{(2\pi)^{N/2}}T(\check{\phi} * \widehat{\psi}) \tag{7.7.158}$$

On the other hand,

$$(T * \phi)\widehat{\ }(\psi) = (T * \phi)(\widehat{\psi}) \tag{7.7.159}$$

$$= \int_{\mathbb{R}^N} (T * \phi)(x)\widehat{\psi}(x)\,dx \tag{7.7.160}$$

$$= \int_{\mathbb{R}^N} T(\tau_x\check{\phi})\widehat{\psi}(x)\,dx \tag{7.7.161}$$

$$= T\left(\int_{\mathbb{R}^N} \tau_x\check{\phi}(\cdot)\widehat{\psi}(x)\,dx\right) \tag{7.7.162}$$

$$= T\left(\int_{\mathbb{R}^N} \check{\phi}(\cdot - x)\widehat{\psi}(x)\,dx\right) \tag{7.7.163}$$

$$= T(\check{\phi} * \widehat{\psi}) \tag{7.7.164}$$

which completes the proof. □

We have labeled the previous proof a "sketch" because one key step, the equality in Eq. (7.7.162), although formally expected, was not explained adequately. See the conclusion of the proof of Theorem 7.19 in [33] for why it is permissible to move T across the integral in this way.

7.8 EXERCISES

7.1. Find the Fourier series $\sum_{n=-\infty}^{\infty} c_n e^{inx}$ for the function $f(x) = x$ on $(-\pi, \pi)$. Use some sort of computer graphics to plot a few of the partial sums of this series on the interval $[-3\pi, 3\pi]$.

7.2. Use the Fourier series in Problem 7.1 to find the exact value of the series

$$\sum_{n=1}^{\infty} \frac{1}{n^2} \qquad \sum_{n=1}^{\infty} \frac{1}{(2n-1)^2}$$

7.3. Evaluate explicitly the following Fourier series, justifying your steps:

$$\sum_{n=1}^{\infty} \frac{n}{2^n} \cos(nx)$$

(Suggestion: start by evaluating $\sum_{n=1}^{\infty} \frac{e^{inx}}{2^n}$, which is a geometric series.)

7.4. Produce a sketch of the Dirichlet and Féjer kernels D_N and K_N, either by hand or by computer, for some reasonably large value of N. What seems to be the key difference?

7.5. Verify the first identity in Eq. (7.1.19).

7.6. We say that $f \in H^k(\mathbb{T})$ if $f \in \mathcal{D}'(\mathbb{T})$ and its Fourier coefficients c_n satisfy

$$\sum_{n=-\infty}^{\infty} n^{2k} |c_n|^2 < \infty \tag{7.8.165}$$

(a) If $f \in H^1(\mathbb{T})$ show that $\sum_{n=-\infty}^{\infty} |c_n|$ is convergent and so the Fourier series of f is uniformly convergent.

(b) Show that $f \in H^k(\mathbb{T})$ for every k if and only if $f \in C^\infty(\mathbb{T})$.

7.7. Evaluate the Fourier series

$$\sum_{n=1}^{\infty} (-1)^n n \sin(nx)$$

in $\mathcal{D}'(\mathbb{R})$. If possible, plot some partial sums of this series.

7.8. Find the Fourier transform of $H(x)e^{-\alpha x}$ for $\alpha > 0$.

7.9. Prove Theorem 7.2. (Suggestions: For $x \in (-\pi, \pi)$ modify the proof of Theorem 7.1, using the identity

$$s_n(x) - f(x) = \int_{x-\pi}^{x} D_n(x-y)(f(y) - f(x))\,dy + \int_{x}^{x+\pi} D_n(x-y)(f(y) - f(x))\,dy$$

The fact that D_n is not positive is compensated for by the fact that

$$f(y) - f(x+) = O(y - x)$$

as $y \to x+$, and similarly for $y < x$.)

7.10. Let $f \in L^1(\mathbb{R}^N)$.

(a) If $f_\lambda(x) = f(\lambda x)$ for $\lambda > 0$, find a relationship between $\widehat{f_\lambda}$ and $\widehat{f}$.

(b) If $f_h(x) = f(x - h)$ for $h \in \mathbb{R}^N$, find a relationship between $\widehat{f_h}$ and $\widehat{f}$.

7.11. If $f \in L^1(\mathbb{R}^N)$ show that $\tau_h f \to f$ in $L^1(\mathbb{R}^N)$ as $h \to 0$. (Hint: First prove it when f is continuous and of compact support.)

7.12. Show that

$$\int_{\mathbb{R}^N} \phi(x)\overline{\psi(x)}\,dx = \int_{\mathbb{R}^N} \widehat{\phi}(x)\overline{\widehat{\psi}(x)}\,dx \tag{7.8.166}$$

for ϕ and ψ in the Schwartz space. (This is also sometimes called the Parseval identity and leads even more directly to the Plancherel formula.)

7.13. Prove Lemma 7.1.

7.14. In this problem J_n denotes the Bessel function of the first kind and of order n. It may defined in various ways, one of which is

$$J_n(z) = \frac{i^{-n}}{\pi} \int_0^{\pi} e^{iz\cos\theta} \cos(n\theta)\, d\theta \tag{7.8.167}$$

which is the definition you should use in this problem.

(a) Suppose that f is a radially symmetric function in $L^1(\mathbb{R}^2)$, that is, $f(x) = f(r)$ where $r = |x|$. Show that

$$\widehat{f}(y) = \int_0^{\infty} J_0(r|y|) f(r) r\, dr$$

It follows in particular that $\widehat{f}$ is also radially symmetric.

(b) Using the known identity $\frac{d}{dz}(zJ_1(z)) = zJ_0(z)$ compute the Fourier transform of $\chi_{B(0,R)}$ the indicator function of the ball $B(0,R)$ in $\mathbb{R}^2$.

7.15. Prove that $J_0(z)$, defined as in Eq. (7.8.167), is a solution of the zero order Bessel equation

$$u'' + \frac{u'}{z} + u = 0$$

Suggestion: show that

$$zJ_0''(z) + J_0'(z) + zJ_0(z) = \frac{1}{\pi} \int_0^{\pi} \frac{d}{d\theta}(\cos\theta \sin(z\sin\theta))\, d\theta$$

7.16. For $\alpha \in \mathbb{R}$ let $f_\alpha(x) = \cos \alpha x$.

(a) Find the Fourier transform $\widehat{f_\alpha}$.

(b) Find $\lim_{\alpha\to 0} \widehat{f_\alpha}$ and $\lim_{\alpha\to\infty} \widehat{f_\alpha}$ in the sense of distributions.

7.17. Compute the Fourier transform of the Heaviside function $H(x)$ in yet another way by justifying that

$$\widehat{H} = \lim_{n\to\infty} \widehat{H}_n$$

in the sense of distributions, where $H_n(x) = H(x)e^{-\frac{x}{n}}$, and then evaluating this limit.

7.18. Prove the remaining parts of Proposition 7.10.

7.19. Let $f \in C(\mathbb{R})$ be 2π periodic. It then has a Fourier series in the classical sense, but it also has a Fourier transform since f is a tempered distribution. What is the relationship between the Fourier series and the Fourier transform?

7.20. Let $f \in L^2(\mathbb{R}^N)$. Show that f is real valued if and only if $\widehat{f}(-y) = \overline{\widehat{f}(y)}$ for all $y \in \mathbb{R}^N$. What is the analog of this for Fourier series?

7.21. Let f be a continuous 2π periodic function with the usual Fourier coefficients

$$c_n = \frac{1}{2\pi}\int_{-\pi}^{\pi} f(x)e^{-inx}\,dx$$

Show that

$$c_n = -\frac{1}{2\pi}\int_{-\pi}^{\pi} f\left(x+\frac{\pi}{n}\right)e^{-inx}\,dx$$

and therefore

$$c_n = \frac{1}{4\pi}\int_{-\pi}^{\pi}\left(f(x)-f\left(x+\frac{\pi}{n}\right)\right)e^{-inx}\,dx.$$

If f is Lipschitz continuous, use this to show that there exists a constant M such that

$$|c_n| \le \frac{M}{|n|} \qquad n \ne 0$$

7.22. Let $R = (-1,1)\times(-1,1)$ be a square in $\mathbb{R}^2$, let f be the indicator function of R and g be the indicator function of the complement of R.
(a) Compute the Fourier transforms $\widehat{f}$ and $\widehat{g}$.
(b) Is either $\widehat{f}$ or $\widehat{g}$ in $L^2(\mathbb{R}^2)$?

7.23. Verify the second limit in Eq. (7.7.149), that is,

$$\frac{1-\cos Rx}{x} \to \text{pv}\frac{1}{x}$$

in $\mathcal{D}'(\mathbb{R})$ as $R \to \infty$.

7.24. A distribution T on $\mathbb{R}^N$ is even if $\check{T} = T$, and odd if $\check{T} = -T$. Prove that the Fourier transform of an even (resp. odd) tempered distribution is even (resp. odd).

7.25. Let $\phi \in \mathcal{S}(\mathbb{R})$, $\|\phi\|_{L^2(\mathbb{R})} = 1$, and show that

$$\left(\int_{-\infty}^{\infty} x^2|\phi(x)|^2\,dx\right)\left(\int_{-\infty}^{\infty} y^2|\widehat{\phi}(y)|^2\,dy\right) \ge \frac{1}{4} \tag{7.8.168}$$

This is a mathematical statement of the Heisenberg uncertainty principle. (Suggestion: start with the identity

$$1 = \int_{-\infty}^{\infty} |\phi(x)|^2\,dx = -\int_{-\infty}^{\infty} x\frac{d}{dx}|\phi(x)|^2\,dx$$

Make sure to allow ϕ to be complex valued.) Show that equality is achieved in Eq. (7.8.168) if ϕ is a Gaussian.

7.26. Let $\theta(t) = \sum_{n=-\infty}^{\infty} e^{-\pi n^2 t}$. (It is a particular case of a class of special functions known as theta functions.) Use the Poisson summation formula (7.7.153) to show that θ satisfies the functional identity

$$\theta(t) = \sqrt{\frac{1}{t}}\,\theta\left(\frac{1}{t}\right)$$

7.27. Use Eq. (7.7.150) to obtain the Fourier transform of $\mathrm{pv}\frac{1}{x}$,

$$\left(\mathrm{pv}\frac{1}{x}\right)\widehat{\ }(y) = -i\sqrt{\frac{\pi}{2}}\,\mathrm{sgn}\, y \qquad (7.8.169)$$

7.28. The proof of Theorem 7.7 implicitly used the fact that if $\phi, \psi \in \mathcal{S}(\mathbb{R}^N)$ then $\phi * \psi \in \mathcal{S}(\mathbb{R}^N)$. Prove this property.

7.29. Where is the mistake in the following argument? If $u(x) = e^{-x}$ then $u' + u = 0$ so by Fourier transformation

$$iy\widehat{u}(y) + \widehat{u}(y) = (1 + iy)\widehat{u}(y) = 0 \qquad y \in \mathbb{R}$$

Since $1 + iy \neq 0$ for real y, it follows that $\widehat{u}(y) = 0$ for all real y and hence $u(x) = 0$.

7.30. If $f \in L^2(\mathbb{R}^N)$, the autocorrelation function of f is defined to be

$$g(x) = (f * \check{\bar{f}})(x) = \int_{\mathbb{R}^N} f(y)\bar{f}(y - x)\, dy$$

Show that $\widehat{g}(y) = (2\pi)^{N/2}|\widehat{f}(y)|^2$, $\widehat{g} \in L^1(\mathbb{R}^N)$ and that $g \in C_0(\mathbb{R}^N)$. ($\widehat{g}$ is called the power spectrum or spectral density of f.)

7.31. If $T \in \mathcal{D}'(\mathbb{T})$ and $c_n = T(e^{-inx})$, show that $T = \sum_{n=-\infty}^{\infty} c_n e^{inx}$ in $\mathcal{D}'(\mathbb{T})$.

7.32. The ODE $u'' - xu = 0$ is known as Airy's equation, and solutions of it are called Airy functions.

(a) If u is an Airy function which is also a tempered distribution, use the Fourier transform to find a first order ODE for $\widehat{u}(y)$.

(b) Find the general solution of the ODE for $\widehat{u}$.

(c) Obtain the formal solution formula

$$u(x) = C\int_{-\infty}^{\infty} e^{ixy + iy^3/3}\, dy$$

(d) Explain why this formula is not meaningful as an ordinary integral, and how it can be properly interpreted.

(e) Is this the general solution of the Airy equation?

7.33. We say that a function $F \in C(\mathbb{R})$ is *positive definite* if

$$\sum_{j,k=1}^{N} F(x_j - x_k)\xi_j\overline{\xi_k} \geq 0$$

for any N and choice of $x_1, x_2, \ldots, x_N \in \mathbb{R}$ and $\xi_1, \xi_2, \ldots, \xi_N \in \mathbb{C}$. If $f \in L^1(\mathbb{R})$ is nonnegative, show that $\widehat{f}$ is positive definite. (There is an interesting and more difficult converse, known as Bochner's theorem, see Section 60 of [2] or Theorem 3.9.16 of [28].)

Chapter 8

Distributions and Differential Equations

In this chapter we will begin to apply the theory of distributions developed in the previous chapters in a more systematic way to problems in differential equations. The modern theory of partial differential equations, and to a somewhat lesser extent ordinary differential equations, makes extensive use of the so-called *Sobolev spaces*, a class of Banach spaces, which we now proceed to introduce.

8.1 WEAK DERIVATIVES AND SOBOLEV SPACES

If $f \in L^p(\Omega)$ then for any multiindex α we know that $D^\alpha f$ exists as an element of $\mathcal{D}'(\mathbb{R}^N)$, but in general the distributional derivative need not itself be a function. However if there exists $g \in L^q(\Omega)$ such that $D^\alpha f = T_g$ in $\mathcal{D}'(\mathbb{R}^N)$ then we say that f has the *weak α derivative* g in $L^q(\Omega)$. That is to say, the requirement is that

$$\int_\Omega f D^\alpha \phi \, dx = (-1)^{|\alpha|} \int_\Omega g\phi \, dx \quad \forall \phi \in \mathcal{D}(\Omega) \tag{8.1.1}$$

and we write $D^\alpha f \in L^q(\Omega)$. It is important to distinguish the concept of weak derivative and almost everywhere (a.e.) derivative.

Example 8.1. Let $\Omega = (-1, 1)$ and $f(x) = |x|$. Obviously $f \in L^p(\Omega)$ for any $1 \le p \le \infty$, and in the sense of distributions we have $f'(x) = 2H(x) - 1$ (use, e.g., Eq. 6.3.53). Thus $f' \in L^q(\Omega)$ for any $1 \le q \le \infty$. On the other hand $f'' = 2\delta$ which does not coincide with T_g for any g in any L^q space. Thus f has the weak first derivative, but not the weak second derivative, in $L^q(\Omega)$ for any q. The first derivative of f coincides with its a.e. derivative. In the case of the second derivative, $f'' = 2\delta$ in the sense of distributions, and $f'' = 0$ a.e. but this function does not coincide with the weak second derivative, indeed there is no weak second derivative according to the above definition. □

We may now define the spaces $W^{k,p}(\Omega)$, known as Sobolev spaces.

Techniques of Functional Analysis for Differential and Integral Equations
http://dx.doi.org/10.1016/B978-0-12-811426-1.00008-8

Definition 8.1. If $\Omega \subset \mathbb{R}^N$ is an open set, $1 \le p \le \infty$ and $k = 1, 2, \ldots$ then

$$W^{k,p}(\Omega) := \{f \in \mathcal{D}'(\Omega) : D^\alpha f \in L^p(\Omega) \text{ if } |\alpha| \le k\} \tag{8.1.2}$$

We emphasize that the meaning of the condition $D^\alpha f \in L^p(\Omega)$ is that f should have the weak α derivative in $L^p(\Omega)$ as discussed above. Clearly

$$\mathcal{D}(\Omega) \subset W^{k,p}(\Omega) \subset L^p(\Omega) \tag{8.1.3}$$

so that $W^{k,p}(\Omega)$ is always a dense subspace of $L^p(\Omega)$ for $1 \le p < \infty$.

Example 8.2. If $f(x) = |x|$ then referring to the discussion in the previous example we see that $f \in W^{1,p}(-1,1)$ for any $p \in [1,\infty]$, but $f \notin W^{2,p}$ for any p.
□

It may be readily checked that $W^{k,p}(\Omega)$ is a normed linear space with norm

$$\|f\|_{W^{k,p}(\Omega)} = \begin{cases} \left(\sum_{|\alpha| \le k} \|D^\alpha f\|^p_{L^p(\Omega)}\right)^{1/p} & 1 \le p < \infty \\ \max_{|\alpha| \le k} \|D^\alpha f\|_{L^\infty(\Omega)} & p = \infty \end{cases} \tag{8.1.4}$$

Furthermore, the necessary completeness property can be shown (Exercise 8.7, or see Theorem 8.1) so that $W^{k,p}(\Omega)$ is a Banach space. When $p = 2$ the norm may be regarded as arising from the inner product

$$\langle f, g \rangle = \sum_{|\alpha| \le k} \int_\Omega D^\alpha f(x) \overline{D^\alpha g(x)}\, dx \tag{8.1.5}$$

so that it is a Hilbert space. The alternative notation $H^k(\Omega)$ is commonly used in place of $W^{k,2}(\Omega)$.

There is a second natural way to give meaning to the idea of a function $f \in L^p(\Omega)$ having a derivative in an L^q space, which is as follows: if there exists $g \in L^q(\Omega)$ such that there exists $f_n \in C^\infty(\Omega)$ satisfying $f_n \to f$ in $L^p(\Omega)$ and $D^\alpha f_n \to g$ in $L^q(\Omega)$, then we say f has the *strong* α *derivative* g in $L^q(\Omega)$.

It is elementary to see that a strong derivative is also a weak derivative—we simply let $n \to \infty$ in the identity

$$\int_\Omega (D^\alpha f_n)\phi\, dx = (-1)^\alpha \int_\Omega f_n D^\alpha \phi\, dx \tag{8.1.6}$$

for any test function ϕ. Far more interesting is that when $p < \infty$ the converse statement is also true, that is *weak=strong*. This important result, which shall not be proved here, was first established by Friedrichs [13] in some special situations, and then in full generality by Meyers and Serrin [25]. A more thorough discussion may be found, for example, in Chapter 3 of [1]. The key idea is to use convolution, as in Theorem 6.5 to obtain the needed sequence f_n of C^∞ functions. For $f \in W^{k,p}(\Omega)$ the approximating sequence may be supposed to belong to $C^\infty(\Omega) \cap W^{k,p}(\Omega)$, so this space is dense in $W^{k,p}(\Omega)$ and we have

Theorem 8.1. *For any open set $\Omega \subset \mathbb{R}^N$, $1 \le p < \infty$ and $k = 0, 1, 2, \ldots$ the Sobolev space $W^{k,p}(\Omega)$ coincides with the closure of $C^\infty(\Omega) \cap W^{k,p}(\Omega)$ in the $W^{k,p}(\Omega)$ norm.*

We now define another class of Sobolev spaces which will be important for later use.

Definition 8.2. For $\Omega \subset \mathbb{R}^N$, $W_0^{k,p}(\Omega)$ is defined to be the closure of $C_0^\infty(\Omega)$ in the $W^{k,p}(\Omega)$ norm.

Obviously $W_0^{k,p}(\Omega) \subset W^{k,p}(\Omega)$, but it may not be immediately clear whether these are actually the same space. In fact this is certainly true when $k = 0$ since in this case we know $C_0^\infty(\Omega)$ is dense in $L^p(\Omega)$, $1 \le p < \infty$ (Theorem 6.6). It also turns out to be correct for any k, p when $\Omega = \mathbb{R}^N$ (see Corollary 3.19 of [1]). But in general the inclusion is strict, and $f \in W_0^{k,p}(\Omega)$ carries the interpretation that $D^\alpha f = 0$ on $\partial\Omega$ for $|\alpha| \le k - 1$. This topic will be continued in more detail in Chapter 13, see especially Theorem 13.3.

8.2 DIFFERENTIAL EQUATIONS IN $\mathcal{D}'$

If we consider the simplest differential equation

$$u' = f \tag{8.2.7}$$

on an interval $(a, b) \subset \mathbb{R}$, then from elementary calculus we know that if f is continuous on $[a, b]$, then every solution is of the form $u(x) = \int_a^x f(y)\,dy + C$, for some constant C. Furthermore in this case $u \in C^1([a, b])$, and $u'(x) = f(x)$ for every $x \in (a, b)$ and we would refer to u as a classical solution of $u' = f$. If we make the weaker assumption that $f \in L^1(a, b)$ then we can no longer expect u to be C^1 or $u'(x) = f(x)$ to hold at every point, since f itself is only defined up to sets of measure zero. If, however, we let $u(x) = \int_a^x f(y)\,dy + C$ then it is an important result of measure theory that $u'(x) = f(x)$ a.e. on (a, b). The question remains whether *all* solutions of $u' = f$ are of this form, and the answer must now depend on precisely what is meant by "solution". If we were to interpret the differential equation as meaning $u' = f$ a.e. then the answer is no. For example, $u(x) = H(x)$ is a nonconstant function on $(-1, 1)$ with $u'(x) = 0$ for $x \ne 0$. An alternative meaning is that the differential equation should be satisfied in the sense of distributions on (a, b), in which case we have the following theorem.

Theorem 8.2. *Let $f \in L^1(a, b)$.*

1. *If $F(x) = \int_a^x f(y)\,dy$ then $F' = f$ in $\mathcal{D}'(a, b)$.*
2. *If $u' = f$ in $\mathcal{D}'(a, b)$, then there exists a constant C such that*

$$u(x) = \int_a^x f(y)\,dy + C \qquad a < x < b \tag{8.2.8}$$

Proof. If $F(x) = \int_a^x f(y)\,dy$, then $F \in C([a,b])$ and for any $\phi \in C_0^\infty(a,b)$ we have

$$F'(\phi) = -F(\phi') = -\int_a^b F(x)\phi'(x)\,dx \tag{8.2.9}$$

$$= -\int_a^b \left(\int_a^x f(y)\,dy\right)\phi'(x)\,dx \tag{8.2.10}$$

$$= -\int_a^b f(y)\left(\int_y^b \phi'(x)\,dx\right)dy \tag{8.2.11}$$

$$= \int_a^b f(y)\phi(y)\,dy = f(\phi) \tag{8.2.12}$$

Here the interchange of order of integration in the third line is easily justified by Fubini's theorem. This proves part 1.

Now if $u' = f$ in the distributional sense then $T = u - F$ satisfies $T' = 0$ in $\mathcal{D}'(a,b)$, and we will finish by showing that T must be a constant. Choose $\phi_0 \in C_0^\infty(a,b)$ such that $\int_a^b \phi_0(y)\,dy = 1$. If $\phi \in C_0^\infty(a,b)$, set

$$\psi(x) = \phi(x) - \left(\int_a^b \phi(y)\,dy\right)\phi_0(x) \tag{8.2.13}$$

so that $\psi \in C_0^\infty(a,b)$ and $\int_a^b \psi(x)\,dx = 0$. Let

$$\zeta(x) = \int_a^x \psi(y)\,dy \tag{8.2.14}$$

Obviously $\zeta \in C^\infty(a,b)$ since $\zeta' = \psi$, but in fact $\zeta \in C_0^\infty(a,b)$ since $\zeta(a) = \zeta(b) = 0$ and $\zeta' = \psi \equiv 0$ in some neighborhood of a and of b. Finally it follows, since $T' = 0$ that

$$0 = T'(\zeta) = -T(\zeta') = -T(\psi) = \left(\int_a^b \phi(y)\,dy\right)T(\phi_0) - T(\phi) \tag{8.2.15}$$

or equivalently $T(\phi) = \int_a^b C\phi(y)\,dy$ where $C = T(\phi_0)$. Thus T is the distribution corresponding to the constant function C. □

We emphasize that part 2 of this theorem is of interest, and not completely obvious, even when $f = 0$: any distribution whose distributional derivative on some interval is zero must be a constant distribution on that interval. Therefore, any distribution is uniquely determined up to an additive constant by its distributional derivative, which, to repeat, is *not* the case for the a.e. derivative.

Now let $\Omega \subset \mathbb{R}^N$ be an open set and

$$Lu = \sum_{|\alpha| \le m} a_\alpha(x) D^\alpha u \tag{8.2.16}$$

be a differential operator of order m. We assume that $a_\alpha \in C^\infty(\Omega)$ in which case $Lu \in \mathcal{D}'(\Omega)$ is well defined for any $u \in \mathcal{D}'(\Omega)$. We will use the following terminology for the rest of this chapter.

Definition 8.3. If $f \in \mathcal{D}'(\Omega)$ then

- u is a classical solution of $Lu = f$ in Ω if $u \in C^m(\Omega)$ and $Lu(x) = f(x)$ for every $x \in \Omega$.
- u is a weak solution of $Lu = f$ in Ω if $u \in L^1_{loc}(\Omega)$ and $Lu = f$ in $\mathcal{D}'(\Omega)$.
- u is a distributional solution of $Lu = f$ in Ω if $u \in \mathcal{D}'(\Omega)$ and $Lu = f$ in $\mathcal{D}'(\Omega)$.

It is clear that a classical solution is also a weak solution, and a weak solution is a distributional solution. The converse statements are false in general, but may be true in special cases. For example, we have proved above that any distributional solution of $u' = 0$ must be constant, hence in particular any distributional solution of this differential equation is actually a classical solution. On the other hand $u = \delta$ is a distributional solution of $x^2u' = 0$, but is not a classical or weak solution. Of course a classical solution cannot exist if f is not continuous on Ω. A theorem which says that any solution of a certain differential equation must be smoother than what is actually needed for the definition of solution, is called a *regularity* result. Regularity theory is a large and important research topic within the general area of differential equations.

Example 8.3. Let $Lu = u_{xx} - u_{yy}$. If $F, G \in C^2(\mathbb{R})$ and $u(x,y) = F(x+y) + G(x-y)$ then we know u is classical solution of $Lu = 0$. We have also observed, in Example 6.14 that if $F, G \in L^1_{loc}(\mathbb{R})$ then $Lu = 0$ in the sense of distributions, thus u is a weak solution of $Lu = 0$ according to the previous definition. The equation has distributional solutions also, which are not weak solutions, for example, the singular distribution T defined by $T(\phi) = \int_{-\infty}^{\infty} \phi(x,x)\,dx$ in Exercise 6.11 of Chapter 6. □

Example 8.4. If $Lu = u_{xx} + u_{yy}$ then it turns out that all solutions of $Lu = 0$ are classical solutions, in fact, any distributional solution must be in $C^\infty(\Omega)$. This is an example of a particularly important kind of regularity result in PDE theory, and will not be proved here, see, for example, Corollary 2.20 of [12]. The difference between Laplace's equation and the wave equation, that is, that Laplace's equation has only classical solutions, while the wave equation has many nonclassical solutions, is a typical difference between solutions of PDEs of elliptic and hyperbolic types. □

8.3 FUNDAMENTAL SOLUTIONS

Let $\Omega \subset \mathbb{R}^N$, L be a differential operator as in Eq. (8.2.16), and suppose $G(x, y)$ has the following properties[1]:

$$G(\cdot, y) \in \mathcal{D}'(\Omega) \qquad L_x G(x, y) = \delta(x - y) \; \forall y \in \Omega \tag{8.3.17}$$

We then call G a *fundamental solution* of L in Ω. If such a G can be found, then formally if we let

$$u(x) = \int_\Omega G(x, y) f(y)\, dy \tag{8.3.18}$$

we may expect that

$$Lu(x) = \int_\Omega L_x G(x, y) f(y)\, dy = \int_\Omega \delta(x - y) f(y)\, dy = f(x) \tag{8.3.19}$$

That is to say, Eq. (8.3.18) provides a way to obtain solutions of the PDE $Lu = f$, and potentially also a tool to analyze specific properties of solutions. We are of course ignoring here all questions of rigorous justification—whether the formula for u even makes sense if G is only a distribution in x, for what class of f's this might be so, and whether it is permissible to differentiate under the integral to obtain Eq. (8.3.19). A more advanced PDE text such as Hörmander [17] may be consulted for such study. Fundamental solutions are not unique in general, since we could always add to G any function $H(x, y)$ satisfying the homogeneous equation $L_x H = 0$ for fixed y.

We will focus now on the case that $\Omega = \mathbb{R}^N$ and $a_\alpha(x) \equiv a_\alpha$ for every α, that is, L is a constant coefficient operator. In this case, if we can find $\Gamma \in \mathcal{D}'(\mathbb{R}^N)$ for which $L\Gamma = \delta$, then $G(x, y) = \Gamma(x - y)$ is a fundamental solution according to the above definition, and it is normal in this situation to refer to Γ itself as the fundamental solution rather than G.

Formally, the solution formula (8.3.18) becomes

$$u(x) = \int_{\mathbb{R}^N} \Gamma(x - y) f(y)\, dy \tag{8.3.20}$$

an integral operator of convolution type. Again it may not be clear if this makes sense as an ordinary integral, but recall that we have earlier defined (Definition 6.7) the convolution of an arbitrary distribution and test function, namely

$$u(x) = (\Gamma * f)(x) := \Gamma(\tau_x \check{f}) \tag{8.3.21}$$

---

1. The subscript x in L_x is used here to emphasize that the differential operator is acting in the x variable, with y in the role of a parameter.

if $\Gamma \in \mathcal{D}'(\Omega)$ and $f \in C_0^\infty(\mathbb{R}^N)$. Furthermore, using Theorem 6.3, it follows that $u \in C^\infty(\mathbb{R}^N)$ and

$$Lu(x) = (L\Gamma) * f(x) = (\delta * f)(x) = f(x) \tag{8.3.22}$$

We have therefore proved

Proposition 8.1. *If there exists $\Gamma \in \mathcal{D}'(\mathbb{R}^N)$ such that $L\Gamma = \delta$, then for any $f \in C_0^\infty(\mathbb{R}^N)$ the function $u = \Gamma * f$ is a classical solution of $Lu = f$.*

It will essentially always be the case that the solution formula $u = \Gamma * f$ is actually valid for a much larger class of f's than $C_0^\infty(\mathbb{R}^N)$ but this will depend on specific properties of the fundamental solution Γ, which in turn depend on those of the original operator L.

Example 8.5. If $L = \Delta$, the Laplacian operator in $\mathbb{R}^3$, then we have already shown (Example 6.15) that $\Gamma(x) = -1/4\pi|x|$ satisfies $\Delta\Gamma = \delta$ in the sense of distributions on $\mathbb{R}^3$. Thus

$$u(x) = \left(-\frac{1}{4\pi|x|} * f\right)(x) = -\frac{1}{4\pi}\int_{\mathbb{R}^3} \frac{f(y)}{|x-y|}\,dy \tag{8.3.23}$$

provides a solution of $\Delta u = f$ in $\mathbb{R}^3$, at least when $f \in C_0^\infty(\mathbb{R}^3)$. The integral on the right in Eq. (8.3.23) is known as the *Newtonian potential* of f, and can be shown to be a valid solution formula for a much larger class of f's. It is in any case always a "candidate" solution, which can be analyzed directly. A fundamental solution of the Laplacian exists in $\mathbb{R}^N$ for any dimension N, and will be recalled at the end of this section. □

Example 8.6. Consider the wave operator $Lu = u_{tt} - u_{xx}$ in $\mathbb{R}^2$. A fundamental solution for L (see Exercise 8.11) is

$$\Gamma(x,t) = \frac{1}{2}H(t - |x|) \tag{8.3.24}$$

The support of Γ, namely the set $\{(x,t) : |x| \le t\}$ is in this context known as the *forward light cone*, representing the set of points x at which for fixed $t > 0$ a signal emanating from the origin $x = 0$ at time $t = 0$, and traveling with speed one, may have reached.

The resulting solution formula for $Lu = f$ may then be obtained as

$$u(x,t) = \int_{-\infty}^{\infty}\int_{-\infty}^{\infty} \Gamma(x-y, t-s) f(y,s)\,dyds \tag{8.3.25}$$

$$= \frac{1}{2}\int_{-\infty}^{\infty}\int_{-\infty}^{\infty} H(t-s-|x-y|) f(y,s)\,dyds \tag{8.3.26}$$

$$= \frac{1}{2}\int_{-\infty}^{t}\int_{x-t+s}^{x+t-s} f(y,s)\,dyds \tag{8.3.27}$$

In many cases of interest $f(x,t) \equiv 0$ for $t < 0$ in which case we replace the lower limit in the s integral by 0. In any case the region over which f is integrated is the "backward" light cone, with vertex at (x,t). Under this support assumption on f it also follows that $u(x,0) = u_t(x,0) \equiv 0$, so by adding in the corresponding terms in D'Alembert's solution (1.3.81) we find that

$$u(x,t) = \frac{1}{2}\int_0^t \int_{x-t+s}^{x+t-s} f(y,s)\,dyds + \frac{1}{2}(h(x+t) + h(x-t)) + \frac{1}{2}\int_{x-t}^{x+t} g(s)\,ds \tag{8.3.28}$$

is the unique solution of

$$u_{tt} - u_{xx} = f(x,t) \qquad x \in \mathbb{R} \quad t > 0 \tag{8.3.29}$$

$$u(x,0) = h(x) \qquad x \in \mathbb{R} \tag{8.3.30}$$

$$u_t(x,0) = g(x) \qquad x \in \mathbb{R} \tag{8.3.31}$$

It is of interest to note that this solution formula could also be written, formally at least, as

$$u(x,t) = (\Gamma * f)(x,t) + \frac{\partial}{\partial t}(\Gamma \underset{(x)}{*} h)(x,t) + (\Gamma \underset{(x)}{*} g)(x,t) \tag{8.3.32}$$

where the notation $(\Gamma \underset{(x)}{*} h)$ indicates that the convolution takes place in x only, with t as a parameter. Thus the fundamental solution Γ enters into the solution not only of the inhomogeneous equation $Lu = f$ but in solving the Cauchy problem as well. This is not an accidental feature, and we will see other instances of this sort of thing later. □

So far we have seen a couple of examples where an explicit fundamental solution is known, but have given no indication of a general method for finding it, or even determining if a fundamental solution exists. Let us address the second issue first, by stating without proof a remarkable theorem.

Theorem 8.3. *(Malgrange-Ehrenpreis) If $L \neq 0$ is any constant coefficient linear differential operator then there exists a fundamental solution of L.*

The proof of this theorem is well beyond the scope of this book, see, for example, Theorem 8.5 of [33] or Theorem 10.2.1 of [17]. The assumption of constant coefficients is essential here, counterexamples are known otherwise, even in the case of very simple and infinitely differentiable variable coefficients.

8.4 FUNDAMENTAL SOLUTIONS AND THE FOURIER TRANSFORM

If we now consider how it might be possible to compute a fundamental solution for a given operator L, it soon becomes apparent that the Fourier transform is a natural and useful tool. If we start with the distributional PDE

$$L\Gamma = \sum_{|\alpha| \le m} a_\alpha D^\alpha \Gamma = \delta \tag{8.4.33}$$

and take the Fourier transform of both sides, the result is

$$\sum_{|\alpha| \le m} a_\alpha (D^\alpha \Gamma)\widehat{\ } = \sum_{|\alpha| \le m} a_\alpha (iy)^\alpha \widehat{\Gamma} = \frac{1}{(2\pi)^{N/2}} \tag{8.4.34}$$

or

$$P(y)\widehat{\Gamma}(y) = 1 \tag{8.4.35}$$

where $P(y)$, the so-called *symbol* or *characteristic polynomial* of L is defined as

$$P(y) = (2\pi)^{N/2} \sum_{|\alpha| \le m} a_\alpha (iy)^\alpha \tag{8.4.36}$$

Note it was implicitly assumed here that $\widehat{\Gamma}$ exists, which would be the case if Γ were a tempered distribution, but this is not actually guaranteed by Theorem 8.3. This is a rather technical issue which we will not discuss here, but rather take the point of view that we seek a formal solution which, potentially, further analysis may show is a bona fide fundamental solution.

The problem of solving Eq. (8.4.35) for a distribution $\widehat{\Gamma}$ is a special case of the so-called *problem of division*, which is to solve an equation $fS = T$ for a distribution S given a distribution T and smooth function f in a suitable class. Various aspects of this problem may be found in [17].

We have thus obtained $\widehat{\Gamma}(y) = 1/P(y)$, or by the inversion theorem

$$\Gamma(x) = \frac{1}{(2\pi)^{N/2}} \int_{\mathbb{R}^N} \frac{1}{P(y)} e^{ix\cdot y}\, dy \tag{8.4.37}$$

as a candidate for fundamental solution of L. One particular source of difficulty in making sense of the inverse transform of $1/P$ is that in general P has zeros, which might be of arbitrarily high order, making the integrand too singular to have meaning in any ordinary sense. On the other hand, we have seen, at least in one dimension, how well-defined distributions of the "pseudo-function" type may be associated with nonlocally integrable functions such as $1/x^m$. Thus there may be some analogous construction in more than one dimension as well. This is in fact one possible means to proving the Malgrange-Ehrenpreis theorem.

It also suggests that the situation may be somewhat easier to deal with if the zero set of P in $\mathbb{R}^N$ is empty, or at least not very large. As a polynomial, of course, P always has zeros, but some or all of these could be complex, whereas the obstructions to making sense of Eq. (8.4.37) pertain to the real zeros of P only.

If L is a constant coefficient differential operator of order m as above, define

$$P_m(y) = (2\pi)^{N/2} \sum_{|\alpha|=m} a_\alpha (iy)^\alpha \tag{8.4.38}$$

which is known as the *principal symbol* of L.

Definition 8.4. We say that L is *elliptic* if $y \in \mathbb{R}^N$, $P_m(y) = 0$ implies that $y = 0$.

That is to say, the principal symbol has no nonzero real roots. For example, the Laplacian operator $L = \Delta$ is elliptic, as is $\Delta+$ lower order terms, since either way the principal symbol is $P_2(y) = -(2\pi)^{N/2}|y|^2$. On the other hand, the wave operator, written, say, as $Lu = \Delta u - u_{x_{N+1}x_{N+1}}$ is not elliptic, since the principal symbol is $P_2(y) = y_{N+1}^2 - \sum_{j=1}^N y_j^2$ up to a constant.

The following is not so difficult to establish (Exercise 8.18), and may be usefully exploited in working with the representation (8.4.37) in the elliptic case.

Proposition 8.2. *If L is elliptic then*

$$\{y \in \mathbb{R}^N : P(y) = 0\} \tag{8.4.39}$$

the real zero set of P, is compact in $\mathbb{R}^N$, and $\lim_{\substack{|y|\to\infty \\ y\in\mathbb{R}^N}} |P(y)| = \infty$.

We will next derive a fundamental solution for the heat equation by using the Fourier transform, although in a slightly different way from the previous discussion. Consider first the initial value problem for the heat equation

$$u_t - \Delta u = 0 \qquad x \in \mathbb{R}^N \quad t > 0 \tag{8.4.40}$$

$$u(x,0) = h(x) \quad x \in \mathbb{R}^N \tag{8.4.41}$$

with $h \in C_0^\infty(\mathbb{R}^N)$. Assuming a solution exists, define the Fourier transform in the x variables only,

$$\widehat{u}(y,t) = \frac{1}{(2\pi)^{N/2}} \int_{\mathbb{R}^N} u(x,t) e^{-ix\cdot y}\, dx \tag{8.4.42}$$

Taking the partial derivative with respect to t of both sides gives $(\widehat{u})_t = (\widehat{u_t})$ so by the usual Fourier transformation calculation rules,

$$(\widehat{u_t}) = (\widehat{u})_t = -|y|^2 \widehat{u} \tag{8.4.43}$$

and $\widehat{u}(y,0) = \widehat{h}(y)$. We may regard this as an ODE in t satisfied by $\widehat{u}(y,t)$ for fixed y, for which the solution obtained by elementary means is

$$\widehat{u}(y,t) = e^{-|y|^2 t} \widehat{h}(y) \tag{8.4.44}$$

If we let Γ be such that $\widehat{\Gamma}(y,t) = \frac{1}{(2\pi)^{N/2}} e^{-|y|^2 t}$ then by Proposition 7.8 it follows that

$$u(x,t) = (\Gamma \underset{(x)}{*} h)(x,t) \tag{8.4.45}$$

Since $\widehat{\Gamma}$ is a Gaussian in x, the same is true for Γ itself, as long as $t > 0$, and from Eq. (7.4.58) we get

$$\Gamma(x,t) = H(t)\frac{e^{-\frac{|x|^2}{4t}}}{(4\pi t)^{N/2}} \tag{8.4.46}$$

By including the $H(t)$ factor we have for later convenience defined $\Gamma(x,t) = 0$ for $t < 0$. Thus we get an integral representation for the solution of Eqs. (8.4.40)–(8.4.41), namely

$$u(x,t) = \int_{\mathbb{R}^N} \Gamma(x-y,t)h(y)\,dy = \frac{1}{(4\pi t)^{N/2}}\int_{\mathbb{R}^N} e^{-\frac{|x-y|^2}{4t}} h(y)\,dy \tag{8.4.47}$$

valid for $x \in \mathbb{R}^N$ and $t > 0$. As usual, although this was derived for convenience under very restrictive conditions on h, it is actually valid much more generally (see Exercise 8.14).

Now to derive a solution formula for $u_t - \Delta u = f$, let $v = v(x,t;s)$ be the solution of Eqs. (8.4.40)–(8.4.41) with $h(x)$ replaced by $f(x,s)$, regarding s for the moment as a parameter, and define

$$u(x,t) = \int_0^t v(x,t-s;s)\,ds \tag{8.4.48}$$

Assuming that f is sufficiently regular, it follows that

$$u_t(x,t) = v(x,0,t) + \int_0^t v_t(x,t-s,s)\,ds \tag{8.4.49}$$

$$= f(x,t) + \int_0^t \Delta v(x,t-s,s)\,ds \tag{8.4.50}$$

$$= f(x,t) + \Delta u(x,t) \tag{8.4.51}$$

Inserting the formula (8.4.47) with h replaced by $f(\cdot,s)$ gives

$$u(x,t) = \int_0^t \int_{\mathbb{R}^N} \Gamma(x-y,t-s)f(y,s)\,dyds = (\Gamma * f)(x,t) \tag{8.4.52}$$

with Γ given again by Eq. (8.4.46). Strictly speaking, we should assume that $f(x,t) \equiv 0$ for $t < 0$ in order that the integral on the right in Eq. (8.4.52) coincide with the convolution in $\mathbb{R}^{N+1}$, but this is without loss of generality, since we only seek to solve the PDE for $t > 0$. The procedure used above for obtaining the solution of the inhomogeneous PDE starting with the solution of a corresponding initial value problem is known as *Duhamel's method*, and is generally applicable, with suitable modifications, for time dependent PDEs in which the coefficients are independent of time.

Since $u(x,t)$ in Eq. (8.4.48) evidently satisfies $u(x,0) \equiv 0$, it follows (compare to Eq. 8.3.32) that

$$u(x,t) = (\Gamma \underset{(x)}{*} h)(x,t) + (\Gamma * f)(x,t) \tag{8.4.53}$$

is a solution[2] of

$$u_t - \Delta u = f(x,t) \qquad x \in \mathbb{R}^N \quad t > 0 \tag{8.4.54}$$

$$u(x,0) = h(x) \qquad x \in \mathbb{R}^N \tag{8.4.55}$$

Let us also observe here that if

$$F(x) = \frac{1}{(2\pi)^{N/2}} e^{-\frac{|x|^2}{4}} \tag{8.4.56}$$

then $F \geq 0$, $\int_{\mathbb{R}^N} F(x)\,dx = 1$, and

$$\Gamma(x,t) = \left(\frac{1}{\sqrt{t}}\right)^N F\left(\frac{x}{\sqrt{t}}\right) \tag{8.4.57}$$

for $t > 0$. From Theorem 6.2, and the observation that a sequence of the form (6.3.37) satisfies the assumptions of that theorem, it follows that $n^N F(nx) \to \delta$ in $\mathcal{D}(\mathbb{R}^N)$ as $n \to \infty$. Choosing $n = \frac{1}{\sqrt{t}}$ we conclude that

$$\lim_{t \to 0+} \Gamma(\cdot,t) = \delta \qquad \text{in } \mathcal{D}'(\mathbb{R}^N) \tag{8.4.58}$$

In particular $\lim_{t \to 0+} (\Gamma \underset{(x)}{*} h)(x,t) = h(x)$ for all $x \in \mathbb{R}^N$, at least when $h \in C_0^\infty(\mathbb{R}^N)$.

8.5 FUNDAMENTAL SOLUTIONS FOR SOME IMPORTANT PDES

We conclude this chapter by collecting all in one place a number of important fundamental solutions. Some of these have been discussed already, some will be left for the exercises, and in several other cases we will be content with a reference.

Laplace Operator

For $L = \Delta$ in $\mathbb{R}^N$ there exist the following fundamental solutions[3]:

$$\Gamma(x) = \begin{cases} \frac{|x|}{2} & N = 1 \\ \frac{1}{2\pi} \log |x| & N = 2 \\ \frac{C_N}{|x|^{N-2}} & N \geq 3 \end{cases} \tag{8.5.59}$$

2. Note we do not say "the solution" here, in fact the solution is not unique without further restrictions.

3. Some texts will use consistently the fundamental solution of $-\Delta$ rather than Δ, in which case all of the signs will be reversed.

where

$$C_N = \frac{1}{(2-N)\Omega_{N-1}} \qquad \Omega_{N-1} = \int_{|x|=1} dS(x) \tag{8.5.60}$$

Thus C_N is a geometric constant, related to the area of the unit sphere in $\mathbb{R}^N$—an equivalent formula in terms of the volume of the unit ball in $\mathbb{R}^N$ is also commonly used. Of the various cases, $N = 1$ is elementary to check, $N = 2$ is requested in Exercise 6.20 of Chapter 6, and we have done the $N \geq 3$ case in Example 6.15.

Heat Operator

For the heat operator $L = \frac{\partial}{\partial t} - \Delta$ in $\mathbb{R}^{N+1}$, we have derived earlier in this section the fundamental solution

$$\Gamma(x,t) = H(t)\frac{e^{-\frac{|x|^2}{4t}}}{(4\pi t)^{N/2}} \tag{8.5.61}$$

for all N.

Wave Operator

For the wave operator $L = \frac{\partial^2}{\partial t^2} - \Delta$ in $\mathbb{R}^{N+1}$, the fundamental solution is significantly dependent on N. The cases of $N = 1, 2, 3$ are as follows:

$$\Gamma(x,t) = \begin{cases} \frac{1}{2}H(t-|x|) & N=1 \\ \frac{1}{2\pi}\frac{H(t-|x|)}{\sqrt{t^2-|x|^2}} & N=2 \\ \frac{\delta(t-|x|)}{4\pi|x|} & N=3 \end{cases} \tag{8.5.62}$$

We have discussed the $N = 1$ case earlier in this section, and refer to [11] or [19] for the cases $N = 2, 3$. As a distribution, the meaning of the fundamental solution in the $N = 3$ case is just what one expects from the formal expression, namely

$$\Gamma(\phi) = \int_{\mathbb{R}^3}\int_{-\infty}^{\infty} \frac{\delta(t-|x|)}{4\pi|x|}\phi(x,t)\,dtdx = \int_{\mathbb{R}^3} \frac{\phi(x,|x|)}{4\pi|x|}\,dx \tag{8.5.63}$$

for any test function ϕ. Note the tendency for the fundamental solution to become more and more singular, as N increases. This pattern persists in higher dimensions, as the fundamental solution starts to contain expressions involving δ' and even higher derivatives of the δ function.

Schrödinger Operator

The Schrödinger operator is defined as $L = \frac{\partial}{\partial t} - i\Delta$ in $\mathbb{R}^{N+1}$. The derivation of a fundamental solution here is nearly the same as for the heat equation, the result being

$$\Gamma(x,t) = H(t)\frac{e^{-\frac{|x|^2}{4it}}}{(4i\pi t)^{N/2}} \tag{8.5.64}$$

In quantum mechanics Γ is frequently referred to as the "propagator". See [29] for much material about the Schrödinger equation.

Helmholtz Operator

The Helmholtz operator is defined by $Lu = \Delta u - \lambda u$. For $\lambda > 0$ and dimensions $N = 1, 2$ and 3 fundamental solutions are

$$\Gamma(x) = \begin{cases} -\frac{e^{-(\sqrt{\lambda}|x|)}}{2\sqrt{\lambda}} & N = 1 \\ \frac{i}{4}H_0^{(1)}(\sqrt{\lambda}|x|) & N = 2 \\ -\frac{e^{-\sqrt{\lambda}|x|}}{4\pi|x|} & N = 3 \end{cases} \tag{8.5.65}$$

where $H_0^{(1)}$ is a so-called Hankel function[4] of order 0. See Chapter 6 of [3] for derivations of these formulas when $N = 2, 3$, while the $N = 1$ case is left for the exercises. This is a case where it may be convenient to use the Fourier transform method directly, since the symbol of L, $P(y) = -|y|^2 - \lambda$ has no real zeros and its reciprocal decays sufficiently fast at ∞.

Klein-Gordon Operator

The Klein-Gordon operator is defined by $Lu = \frac{\partial^2 u}{\partial t^2} - \Delta u - \lambda u$ in $\mathbb{R}^{N+1}$. We mention only the case $N = 1$, $\lambda > 0$, in which case a fundamental solution is

$$\Gamma(x,t) = \frac{1}{2}H(t - |x|)J_0(\sqrt{\lambda(t^2 - x^2)}) \quad N = 1 \tag{8.5.66}$$

where J_0 is the Bessel function of the first kind and order zero (see Exercise 7.14 of Chapter 7). This may be derived, for example, by the method presented in Problem 2, Section 5.1 of [19], and choosing $\psi = \delta$.

Biharmonic Operator

The biharmonic operator is $L = \Delta^2$, that is, $Lu = \Delta(\Delta u)$. It arises especially in connection with the theory of plates and shells, so that $N = 2$ is the most interesting case. A fundamental solution is

$$\Gamma(x) = \frac{1}{8\pi}|x|^2 \log|x| \quad N = 2 \tag{8.5.67}$$

for which a derivation is outlined in Exercise 8.12.

---

4. It is also sometimes called a Bessel function of the third kind.

8.6 EXERCISES

8.1. Show that an equivalent definition of $W^{s,2}(\mathbb{R}^N) = H^s(\mathbb{R}^N)$ for $s = 0, 1, 2, \dots$ is

$$H^s(\mathbb{R}^N) = \left\{ f \in \mathcal{S}'(\mathbb{R}^N) : \int_{\mathbb{R}^n} |\widehat{f}(y)|^2 (1 + |y|^2)^s \, dy < \infty \right\} \qquad (8.6.68)$$

The second definition makes sense even if s isn't a positive integer and leads to one way to define fractional and negative order differentiability. Implicitly it requires that $\widehat{f}$ (but not f itself) must be a function.

8.2. Using the definition (8.6.68), show that $H^s(\mathbb{R}^N) \subset C_0(\mathbb{R}^N)$ if $s > \frac{N}{2}$. Show that $\delta \in H^s(\mathbb{R}^N)$ if $s < -\frac{N}{2}$.

8.3. If $s_1 < s < s_2, f \in H^{s_1}(\mathbb{R}^N) \cap H^{s_2}(\mathbb{R}^N)$ and $\epsilon > 0$ show that there exists C independent of f such that

$$\|f\|_{H^s(\mathbb{R}^N)} \le \epsilon \|f\|_{H^{s_2}(\mathbb{R}^N)} + C\|f\|_{H^{s_1}(\mathbb{R}^N)}$$

8.4. If Ω is a bounded open set in $\mathbb{R}^3$, and $u(x) = \frac{1}{|x|}$, show that $u \in W^{1,p}(\Omega)$ for $1 \le p < \frac{3}{2}$. Along the way, you should show carefully that the distributional first derivative $\frac{\partial u}{\partial x_i}$ agrees with the corresponding pointwise derivative.

8.5. Prove that if $f \in W^{1,p}(a, b)$ for $p > 1$ then

$$|f(x) - f(y)| \le \|f\|_{W^{1,p}(a,b)} |x - y|^{1-\frac{1}{p}} \qquad (8.6.69)$$

so in particular $W^{1,p}(a, b) \subset C([a, b])$. (Caution: You would like to use the fundamental theorem of calculus here, but it isn't quite obvious whether it is valid assuming only that $f \in W^{1,p}(a, b)$.)

8.6. Suppose that $\Omega_1 \subset \Omega_2$ are open sets in $\mathbb{R}^N$, $u \in H_0^1(\Omega_1)$, and

$$v(x) = \begin{cases} u(x) & x \in \Omega_1 \\ 0 & x \in \Omega_2 \backslash \Omega_1 \end{cases}$$

Show that $v \in H_0^1(\Omega_2)$. Show by example that this need not be true if $H_0^1(\Omega)$ is replaced by $H^1(\Omega)$.

8.7. Prove directly that $W^{k,p}(\Omega)$ is complete (relying of course on the fact that $L^p(\Omega)$ is complete).

8.8. Show that Theorem 8.1 is false for $p = \infty$.

8.9. If f is a nonzero constant function on $[0, 1]$, show that $f \notin W_0^{1,p}(0, 1)$ for $1 \le p \le \infty$.

8.10. Let $Lu = u'' - u$ and $E(x) = H(x) \sinh x$, $x \in \mathbb{R}$.

(a) Show that E is a fundamental solution of L.

(b) What is the corresponding solution formula for $Lu = f$?

(c) The fundamental solution E is not the same as the one given in Eq. (8.5.65) for $\lambda = 1$. Does this call for any explanation?

8.11. Show that $E(x,t) = \frac{1}{2}H(t - |x|)$ is a fundamental solution for the wave operator $Lu = u_{tt} - u_{xx}$.

8.12. The fourth order operator $Lu = u_{xxxx} + 2u_{xxyy} + u_{yyyy}$ in $\mathbb{R}^2$ is the *biharmonic operator* which arises in the theory of deformation of elastic plates.

(a) Show that $L = \Delta^2$, that is, $Lu = \Delta(\Delta u)$ where Δ is the Laplacian.

(b) Find a fundamental solution of L. (Suggestions: To solve $LE = \delta$, first solve $\Delta F = \delta$ and then $\Delta E = F$. Since F will depend on $r = \sqrt{x^2 + y^2}$ only, you can look for a solution $E = E(r)$ also.)

8.13. Let $Lu = u'' + \alpha u'$ where $\alpha > 0$ is a constant.

(a) Find a fundamental solution of L which is a tempered distribution.

(b) Find a fundamental solution of L which is not a tempered distribution.

8.14. Show directly that $u(x,t)$ defined by Eq. (8.4.47) is a classical solution of the heat equation for $t > 0$, under the assumption that h is bounded and continuous on $\mathbb{R}^N$.

8.15. Assuming that Eq. (8.4.47) is valid and $h \in L^p(\mathbb{R}^N)$, derive the decay property

$$\|u(\cdot,t)\|_{L^\infty(\mathbb{R}^N)} \le \frac{\|h\|_{L^p(\mathbb{R}^N)}}{t^{N/2p}} \tag{8.6.70}$$

for $1 \le p \le \infty$.

8.16. If

$$G(x,y) = \begin{cases} y(x-1) & 0 < y < x < 1 \\ x(y-1) & 0 < x < y < 1 \end{cases}$$

show that G is a fundamental solution of $Lu = u''$ in $(0,1)$.

8.17. Is the heat operator $L = \frac{\partial}{\partial t} - \Delta$ elliptic?

8.18. Prove Proposition 8.2.

Chapter 9

Linear Operators

9.1 LINEAR MAPPINGS BETWEEN BANACH SPACES

Let $\mathbf{X}, \mathbf{Y}$ be Banach spaces. We say that

$$T : D(T) \subset \mathbf{X} \to \mathbf{Y} \tag{9.1.1}$$

is linear if

$$T(c_1x_1 + c_2x_2) = c_1T(x_1) + c_2T(x_2) \qquad \forall x_1, x_2 \in D(T) \quad \forall c_1, c_2 \in \mathbb{C} \tag{9.1.2}$$

Here $D(T)$ is the domain of T which we *do not* assume is all of $\mathbf{X}$. Note, however, that it must be a subspace of $\mathbf{X}$ according to this definition. Likewise $R(T)$, the range of T, must be a subspace of $\mathbf{Y}$. If $\overline{D(T)} = \mathbf{X}$ we say T is *densely defined* . As before it is common to write Tx instead of $T(x)$ when T is linear, and we will often use this notation.

The definition of operator norm given earlier in Eq. (4.3.9) for the case when $D(T) = \mathbf{X}$ may be modified for the present case.

Definition 9.1. The *norm* of the operator T is

$$\|T\|_{\mathbf{X},\mathbf{Y}} = \sup_{\substack{x \in D(T) \\ x \neq 0}} \frac{\|Tx\|_{\mathbf{Y}}}{\|x\|_{\mathbf{X}}} \tag{9.1.3}$$

In general, $\|T\|_{\mathbf{X},\mathbf{Y}} = \infty$ may occur.

Definition 9.2. If $\|T\|_{\mathbf{X},\mathbf{Y}} < \infty$ we will say that T is bounded on its domain. If in addition $D(T) = \mathbf{X}$ we say T is bounded on $\mathbf{X}$, or more simply that T is bounded, if there is no possibility of confusion.

If it is clear from context what $\mathbf{X}, \mathbf{Y}$ are, we may write $\|x\|$ instead of $\|x\|_{\mathbf{X}}$, etc. We point out, however, that many linear operators of interest may be defined for many different choices of $\mathbf{X}, \mathbf{Y}$, and it will be important to be able to specify precisely which spaces we have in mind.

It is immediate from the definition that

- $\|T\| = \sup_{\substack{x \in D(T) \\ \|x\|=1}} \|Tx\|$.
- $\|Tx\| \le \|T\| \, \|x\|$ for all $x \in D(T)$.

Techniques of Functional Analysis for Differential and Integral Equations
http://dx.doi.org/10.1016/B978-0-12-811426-1.00009-X

Regarding T as a linear mapping from the normed linear space $D(T)$ into $\mathbf{Y}$, Proposition 4.4 can be restated as follows.

Theorem 9.1. *Let $T : D(T) \subset \mathbf{X} \to \mathbf{Y}$ be linear. Then the following are equivalent:*

(a) *T is bounded on its domain.*
(b) *T is continuous at every point of $D(T)$.*
(c) *T is continuous at some point of $D(T)$.*
(d) *T is continuous at 0.*

We also have (see Exercise 9.3)

Proposition 9.1. *If T is bounded on its domain then it has a unique norm preserving extension to $\overline{D(T)}$. That is to say, there exists a unique linear operator $S : \overline{D(T)} \subset \mathbf{X} \to \mathbf{Y}$ such that $Sx = Tx$ for $x \in D(T)$ and $\|S\| = \|T\|$.*

It follows that if T is densely defined and bounded on its domain, then it automatically has a unique bounded extension to all of $\mathbf{X}$. In such a case we will always assume that T has been replaced by this extension, unless otherwise stated.

Recall the notations introduced previously,

$$\mathcal{B}(\mathbf{X}, \mathbf{Y}) = \{T : \mathbf{X} \to \mathbf{Y} : T \text{ is linear and } \|T\|_{\mathbf{X},\mathbf{Y}} < \infty\} \tag{9.1.4}$$

$$\mathcal{B}(\mathbf{X}) = \mathcal{B}(\mathbf{X}, \mathbf{X}) \qquad \mathbf{X}^* = \mathcal{B}(\mathbf{X}, \mathbb{C}) \tag{9.1.5}$$

9.2 EXAMPLES OF LINEAR OPERATORS

We next discuss a number of examples of linear operators.

Example 9.1. Let $\mathbf{X} = \mathbb{C}^N, \mathbf{Y} = \mathbb{C}^M$ and $Tx = Ax$ for some $M \times N$ complex matrix A, that is,

$$(Tx)_k = \sum_{j=1}^{N} a_{kj} x_j \quad k = 1, \ldots, M \tag{9.2.6}$$

if a_{jk} is the (j, k) entry of A. Clearly T is linear and in Exercise 9.6 you are asked to verify that T is bounded for any choice of the norms on $\mathbf{X}, \mathbf{Y}$. The exact value of the operator norm of T, however, will depend on exactly which norms are used in $\mathbf{X}, \mathbf{Y}$.

Suppose we use the usual Euclidean norm $\| \cdot \|_2$ in both spaces. Then using the Schwarz inequality we may obtain

$$\|Tx\|^2 = \sum_{k=1}^{M} \left| \sum_{j=1}^{N} a_{kj} x_j \right|^2 \le \sum_{k=1}^{M} \left(\sum_{j=1}^{N} |a_{kj}|^2 \right) \left(\sum_{j=1}^{N} |x_j|^2 \right) \tag{9.2.7}$$

$$= \left(\sum_{k=1}^{M} \sum_{j=1}^{N} |a_{kj}|^2 \right) \|x\|^2 \tag{9.2.8}$$

from which we conclude that

$$\|T\| \le \left(\sum_{k=1}^{M} \sum_{j=1}^{N} |a_{kj}|^2 \right)^{1/2} \tag{9.2.9}$$

The right-hand side of Eq. (9.2.9) is known as the Frobenius norm of the matrix A, and it is easy to check that it satisfies all of the axioms of a norm on the vector space of $M \times N$ matrices. Note however that Eq. (9.2.9) is only an inequality and it is known to be strict in general, as will be clarified below.

If $p, q \in [1, \infty]$ let us temporarily use the notation $\|T\|_{p,q}$ for the norm of T when we use the p-norm in $\mathbf{X}$ and the q-norm in $\mathbf{Y}$, or $\|T\|_p$ in the case $q = p$. The problem of computing $\|T\|_{p,q}$ in a more or less explicit way from the entries of A is difficult in general, but several special cases are well known.

- If $p = q = 1$ then $\|T\|_1 = \max_j \sum_{k=1}^{M} |a_{kj}|$, the maximum absolute column sum of A.
- If $p = q = \infty$ then $\|T\|_\infty = \max_k \sum_{j=1}^{N} |a_{kj}|$, the maximum absolute row sum of A.
- If $p = q = 2$ then $\|T\|_2$ is the largest singular value of A, or equivalently $\|T\|_2^2$ is the largest eigenvalue of the square Hermitian matrix A^*A.

The notation $\|A\|_{p,q}$, etc. may also be used when we take the point of view that the norm is a property of the matrix itself rather than the linear operator defined by the matrix. Details about these points may be found in most textbooks on linear algebra or numerical analysis, see, for example, Chapter 2 of [15] or Chapter 7 of [23]. □

Example 9.2. Let $\mathbf{X} = \mathbf{Y} = L^p(\mathbb{R}^N)$ and T be the *translation operator* defined on $D(T) = \mathbf{X}$ by

$$Tu(x) = \tau_h u(x) = u(x - h) \tag{9.2.10}$$

for some fixed $h \in \mathbb{R}^N$. Then T is linear and

$$\|Tu\| = \|u\| \tag{9.2.11}$$

for any u so that $\|T\| = 1$. □

Example 9.3. Let $\Omega \subset \mathbb{R}^N$, $\mathbf{X} = \mathbf{Y} = L^p(\Omega)$, $a \in L^\infty(\Omega)$ and define the *multiplication operator* T on $D(T) = \mathbf{X}$ by

$$Tu(x) = a(x)u(x) \tag{9.2.12}$$

The obvious inequality

$$\|Tu\|_{L^p} \le \|a\|_{L^\infty} \|u\|_{L^p} \tag{9.2.13}$$

implies that $\|T\| \le \|a\|_{L^\infty}$. We claim that actually equality holds. The case $a \equiv 0$ is trivial, otherwise in the case $1 \le p < \infty$ we can see it as follows. For any $0 < \epsilon < \|a\|_{L^\infty}$ there must exist a measurable set $\Sigma \subset \Omega$ of measure $m(\Sigma) = \eta > 0$ such that $|a(x)| \ge \|a\|_{L^\infty} - \epsilon$ for $x \in \Sigma$. If we now choose $u = \chi_\Sigma$, the characteristic function of Σ, then $\|u\|_{L^p} = \eta^{1/p}$ and

$$\|Tu\|_{L^p}^p = \int_\Sigma |a(x)|^p \, dx \ge \eta(\|a\|_{L^\infty} - \epsilon)^p \tag{9.2.14}$$

Thus

$$\frac{\|Tu\|_{L^p}}{\|u\|_{L^p}} \ge \|a\|_{L^\infty} - \epsilon \tag{9.2.15}$$

which immediately implies that $\|T\| \ge \|a\|_{L^\infty}$ as needed. The case $p = \infty$ is left as an exercise. □

Example 9.4. One of the most important classes of operators we will be concerned with in this book is integral operators. Let $\Omega \subset \mathbb{R}^N$, $\mathbf{X} = \mathbf{Y} = L^2(\Omega)$, $K \in L^2(\Omega \times \Omega)$ and define the operator T by

$$Tu(x) = \int_\Omega K(x,y)u(y) \, dy \tag{9.2.16}$$

It may not be immediately clear how we should define $D(T)$, but note by the Schwarz inequality that

$$\|Tu\|_{L^2}^2 = \int_\Omega \left| \int_\Omega K(x,y)u(y) \, dy \right|^2 dx \tag{9.2.17}$$

$$\le \int_\Omega \left(\int_\Omega |K(x,y)|^2 \, dy \right) \left(\int_\Omega |u(y)|^2 \, dy \right) dx \tag{9.2.18}$$

$$= \left(\int_\Omega \int_\Omega |K(x,y)|^2 \, dy \, dx \right) \left(\int_\Omega |u(y)|^2 \, dy \right) \tag{9.2.19}$$

This shows simultaneously that $Tu \in L^2$ whenever $u \in L^2$, so that we may take $D(T) = L^2(\Omega)$, and that

$$\|T\| \le \|K\|_{L^2(\Omega\times\Omega)} \tag{9.2.20}$$

We refer to K as the kernel[1] of the operator T. Note the formal similarity between this calculation and that of Example 9.1. Just as in that case, the inequality for $\|T\|$ is strict, in general. □

Example 9.5. Let $h \in L^1_{loc}(\mathbb{R}^N)$ and define the convolution operator

$$Tu(x) = (h * u)(x) = \int_{\mathbb{R}^N} h(x-y)u(y)\,dy \tag{9.2.21}$$

This is obviously an operator of the type (9.2.16) with $\Omega = \mathbb{R}^N$ but for which $K(x,y) = h(x-y)$ does not satisfy the L^2 condition in the previous example, except in trivial cases. Thus it is again not immediately apparent how we should define $D(T)$. Recall, however, Young's convolution inequality (6.4.78) which implies immediately that

$$\|Tu\|_{L^r} \le \|h\|_{L^p}\|u\|_{L^q} \tag{9.2.22}$$

if

$$p,q,r \in [1,\infty] \qquad \frac{1}{p} + \frac{1}{q} = 1 + \frac{1}{r} \tag{9.2.23}$$

Thus if $h \in L^p(\mathbb{R}^N)$ we may take $D(T) = \mathbf{X} = L^q(\mathbb{R}^N)$ and $\mathbf{Y} = L^r(\mathbb{R}^N)$ with p,q,r are related as above, in which case $\|T\| \le \|h\|_{L^p}$. □

Example 9.6. If we let

$$Tu(x) = \frac{1}{(2\pi)^{N/2}}\int_{\mathbb{R}^N} u(y)e^{-ix\cdot y}\,dy \tag{9.2.24}$$

then $Tu(x) = \widehat{u}(x)$, is the Fourier transform of u studied in Chapter 7. It is again a special case of Eq. (9.2.16) but with kernel K not satisfying the L^2 integrability condition. From the earlier discussion of properties of the Fourier transform we have the following:

1. (see Theorem 7.3) T is a bounded linear operator from $\mathbf{X} = L^1(\mathbb{R}^N)$ into $\mathbf{Y} = C_0(\mathbb{R}^N)$ with norm

$$\|T\| \le \frac{1}{(2\pi)^{N/2}} \tag{9.2.25}$$

In fact it is easy to see that equality holds here, see Exercise 9.17.

1. which is not to be confused with the null space of T!

2. T is a bounded linear operator from $\mathbf{X} = L^2(\mathbb{R}^N)$ onto $\mathbf{Y} = L^2(\mathbb{R}^N)$ with norm $\|T\| = 1$. Indeed $\|Tu\| = \|u\|$ for all $u \in L^2(\mathbb{R}^N)$ by the Plancherel identity (7.5.102).

It can also be shown, although this is more difficult (see Chapter I, section 2 of [37]), that T is a bounded linear operator from $\mathbf{X} = L^p(\mathbb{R}^N)$ into $\mathbf{Y} = L^q(\mathbb{R}^N)$ if

$$1 < p \le 2 \le q < \infty \qquad \frac{1}{p} + \frac{1}{q} = 1 \tag{9.2.26}$$

If $u \in L^p(\mathbb{R}^N)$ for $p > 2$ then it is a tempered distribution, so $\widehat{u}$ always exists in a distributional sense, but it may not be a function, see Chapter I, section 4.13 of [37] □

Example 9.7. Let $a \in L^\infty(\mathbb{R}^N)$ and define the linear operator T, known as a Fourier multiplication operator, by

$$\widehat{Tu}(y) = a(y)\widehat{u}(y) \tag{9.2.27}$$

where as usual $\widehat{u}$ denotes the Fourier transform. If we use $\mathcal{F}$ as an alternative special notation for the Fourier transform, and let S denote the multiplication operator defined in Example 9.3, then it is equivalent to defining $T = \mathcal{F}^{-1}S\mathcal{F}$. If we take $\mathbf{X} = \mathbf{Y} = L^2(\mathbb{R}^N)$ then from the known properties of $\mathcal{F}, S$ we get immediately from the Plancherel identity that

$$\|Tu\|_{L^2} = \|\widehat{Tu}\|_{L^2} = \|a\widehat{u}\|_{L^2} \le \|a\|_{L^\infty}\|\widehat{u}\|_{L^2} = \|a\|_{L^\infty}\|u\|_{L^2} \tag{9.2.28}$$

implying that $\|T\| \le \|a\|_{L^\infty}$. As in the case of the ordinary multiplication operator one can show that equality must hold.

Note that formally we have

$$Tu(x) = \frac{1}{(2\pi)^N}\int_{\mathbb{R}^N} e^{ix\cdot y}a(y)\left(\int_{\mathbb{R}^N} e^{-iz\cdot y}u(z)\,dz\right)dy \tag{9.2.29}$$

$$= \frac{1}{(2\pi)^N}\int_{\mathbb{R}^N} u(z)\left(\int_{\mathbb{R}^N} a(y)e^{i(x-z)\cdot y}\,dy\right)dz \tag{9.2.30}$$

$$= \int_{\mathbb{R}^N} u(z)h(x-z)\,dz \tag{9.2.31}$$

provided that $\widehat{h}(y) = \frac{a(y)}{(2\pi)^{N/2}}$. Thus the Fourier multiplication operators appears to be just a special kind of convolution operator. However $a \in L^\infty(\mathbb{R}^N)$ could happen even if $h \notin L^p(\mathbb{R}^N)$ for any p, in which case the previous discussion about convolution operators is not applicable. The simplest example of this is when $a(y) \equiv 1$, corresponding to T being the identity mapping and h being the delta function.

A more significant example is obtained by taking $N = 1$ and $a(y) = -i\,\mathrm{sgn}\,(y)$. By Eq. (7.8.169) we see that $a(y) = \sqrt{2\pi}\,\widehat{h}(y)$ if $h(x) = \frac{1}{\pi}\,\mathrm{pv}\frac{1}{x}$, where here the Fourier transform is meant in the sense of distributions. Thus, we have at least formally that

$$Tu(x) = \left(\frac{1}{\pi}\,\mathrm{pv}\frac{1}{x} * u\right)(x) = \frac{1}{\pi}\,\mathrm{pv}\int_{-\infty}^{\infty} \frac{u(y)}{x-y}\,dy \tag{9.2.32}$$

This operator is known as the *Hilbert transform*, and will be from now on denoted by $\mathcal{H}$. Since we have not rigorously established the validity of the formulas (9.2.32), or even explained why the principal value integral in Eq. (9.2.32) should exist in general for $u \in L^2(\mathbb{R})$, we will always use the above, completely unambiguous definition of $\mathcal{H}$ as a Fourier multiplication operator when anything needs to be proved. For example, since $|a(y)| \equiv 1$, we get $|\widehat{\mathcal{H}u}(y)| \equiv |\widehat{u}(y)|$ and then

$$\|\mathcal{H}u\|_{L^2} = \|\widehat{\mathcal{H}u}\|_{L^2} = \|\widehat{u}\|_{L^2} = \|u\|_{L^2} \tag{9.2.33}$$

In particular $\|\mathcal{H}\| = 1$ as an operator on $L^2(\mathbb{R})$. The Hilbert transform is the archetypical example of a singular integral operator, see, for example, Chapter II of [36].

A Fourier multiplication operator is often referred to as a *filter*, especially in the electrical engineering and signals processing literature. The idea here is that if $u = u(t), t \in \mathbb{R}$ represents a signal, then $\widehat{u}(k)$ corresponds to the signal in the "frequency domain," in the sense that the Fourier inversion formula

$$u(t) = \frac{1}{\sqrt{2\pi}} \int_{-\infty}^{\infty} e^{ikt}\widehat{u}(k)\,dk \tag{9.2.34}$$

represents the signal as a superposition of fixed frequency signals e^{ikt}, with $\widehat{u}(k)$ then being the weight given to the component of frequency k. The effect of a filter is thus to modify the frequency component $\widehat{u}(k)$ by multiplying it by $a(k)$. The operator T coming from the choice

$$a(k) = \begin{cases} 1 & |k| < k_0 \\ 0 & |k| \ge k_0 \end{cases} \tag{9.2.35}$$

leaves low frequencies ($|k| < k_0$) unchanged and removes all of the high frequency components, and is for this reason sometimes called an ideal low-pass filter. Likewise $1 - a(k)$ gives an ideal high-pass filter. A band-pass filter would be one which for which $a(k) = 1$ on some interval of frequencies $[k_1, k_2]$ and is zero otherwise. □

Example 9.8. If **H** is a Hilbert space and $M \subset \mathbf{H}$ is a closed subspace, we have seen in Chapter 5 that the orthogonal projection P_M is a linear operator defined on all of **H**. It is immediate from the relation (5.4.32) that $\|P_M x\| \le \|x\|$ for all $x \in \mathbf{H}$. Aside from the trivial case $P_M = 0$ there must exist $x \in \mathbf{H}$, $x \neq 0$ such that $P_M x = x$, from which it follows that $\|P_M\| = 1$. □

Example 9.9. Let $\mathbf{X} = \mathbf{Y} = \ell^2$ (the sequence space defined in Example 5.3). If $x = \{x_1, x_2, \dots\} \in \ell^2$ set

$$S_+ x = \{0, x_1, x_2, \dots\} \tag{9.2.36}$$

$$S_- x = \{x_2, x_3, \dots\} \tag{9.2.37}$$

which are called respectively the right- and left-shift operators on ℓ^2. Clearly $\|S_+ x\| = \|x\|$ for any x, and $\|S_- x\| \le \|x\|$ with equality if $x_1 = 0$. Thus, $\|S_+\| = \|S_-\| = 1$. Note that $S_- S_+ = I$ (the identity map), while $S_+ S_- = P_M$ where M is the closed subspace $M = \{x \in \ell^2 : x_1 = 0\}$. □

Example 9.10. Let Ω be an open set in $\mathbb{R}^N$, m a positive integer and

$$Tu(x) = \sum_{|\alpha| \le m} a_\alpha(x) D^\alpha u \tag{9.2.38}$$

where the coefficients $a_\alpha \in C(\overline{\Omega})$. If $\mathbf{X} = \mathbf{Y} = L^p(\Omega)$, $1 \le p < \infty$ then we can let $D(T) = C^m(\overline{\Omega})$ which is a dense subset of $\mathbf{X}$ (since it contains $C_0^\infty(\Omega)$, for example). Thus T is a densely defined linear operator, but it is not bounded in general. For example, take $\mathbf{X} = \mathbf{Y} = L^2(0,1)$, $Tu = u'$ and $u_n(x) = \sin n\pi x$. Then by explicit calculation we find $\|u_n\| = 1/\sqrt{2}$ and $\|Tu_n\| = n\pi/\sqrt{2}$, so that $\|Tu_n\|/\|u_n\| \to \infty$ as $n \to \infty$.

Note that in the constant coefficient case with $\Omega = \mathbb{R}^N$ we have $\widehat{Tu}(y) = P(y)\widehat{u}(y)$, provided u is a tempered distribution, where P is the characteristic polynomial of T as discussed earlier in Section 8.3. Thus T is formally a Fourier multiplication operator but with a multiplier $a(y) = P(y)$ which is not in L^∞. □

Example 9.11. A pseudodifferential operator (ΨDO) is an operator of the form

$$Tu(x) = \int_{\mathbb{R}^N} a(x,y) e^{ix\cdot y} \widehat{u}(y)\, dy \tag{9.2.39}$$

for some function a, known as the symbol of T. If $a(x,y) = a(y)$ then T is a Fourier multiplication operator, bounded if $a \in L^\infty(\mathbb{R}^N)$, while if $a = a(x)$ it is an ordinary multiplication operator. □

9.3 LINEAR OPERATOR EQUATIONS

Given a linear operator $T : D(T) \subset \mathbf{X} \to \mathbf{Y}$, we wish to study the operator equation

$$Tu = f \tag{9.3.40}$$

where f is a given member of $\mathbf{Y}$. In the usual way, if T is one-to-one, that is, if $N(T) = \{0\}$, then we may define the corresponding inverse operator $T^{-1} : R(T) \to D(T)$. It is easy to check that T^{-1} is also linear when it exists, but it need not be bounded even if T is, or it may be bounded even if T is not. Some key questions which always arise in connection with Eq. (9.3.40) are:

- For what f's does there exist a solution u, that is, what is the range $R(T)$?
- If a solution exists, is it unique? If not, how can we describe the set of all solutions? Since any two solutions differ by a solution of $Tu = 0$ this amounts to characterizing the null space $N(T)$.

The investigation of these questions will clearly require us to be precise about what the spaces $\mathbf{X}, \mathbf{Y}$ are. For reasons which will become more apparent below, we will mostly focus on the case that $\mathbf{X} = \mathbf{Y} = \mathbf{H}$, a Hilbert space, but the study of more general situations can be found in more advanced texts.

Let us first consider the case when $\mathbf{X} = \mathbb{C}^N, \mathbf{Y} = \mathbb{C}^M$ so $Tu = Au$ for some $M \times N$ matrix $A = [a_{kj}]$. Then

- $R(T)$ is the column space of A, that is, the set of all linear combinations of the columns of A. This is immediate from the definition of matrix multiplication.
- $R(T) = N(T^*)^{\perp}$, where T^* is the matrix multiplication operator with matrix A^*, the conjugate transpose (or Hermitian conjugate, or adjoint matrix) of A. See, for example, Theorem 2′, Chapter 3 of [23] or almost any other linear algebra textbook.

The second item provides a complete characterization of when $Tu = f$ is solvable, namely, a solution exists if and only if $f \perp v$ for every $v \in N(T^*)$. If the subspace $N(T^*)$ has the basis $\{v_1, \ldots, v_p\}$ then it is equivalent to requiring $\langle f, v_k \rangle = 0, k = 1, \ldots, p$. This amounts to p solvability, or consistency, conditions on f, which are necessary and sufficient for the existence of a solution of $Tu = f$. Eventually we will prove a version of this statement in a Hilbert space setting, for certain types of operator T. The main point, at present, is that the operator T^* plays a key role in understanding the solvability of $Tu = f$, and so something similar can be expected in the infinite dimensional case. The operator T^* is the so-called *adjoint operator* of T, and in the next section we show how it can always be defined at least in the case that T is bounded. The case of unbounded T is more subtle, and will be taken up in the following chapter.

9.4 THE ADJOINT OPERATOR

In the finite dimensional example of the previous section, note that T^* has the property

$$\langle Tu, v \rangle = \langle u, T^* v \rangle \qquad \forall u \in \mathbb{C}^N \quad v \in \mathbb{C}^M \tag{9.4.41}$$

since either side is equal to $\sum_{k=1}^{M} \sum_{j=1}^{N} a_{kj} u_j \overline{v_k}$.

Now suppose $\mathbf{X} = \mathbf{Y} = \mathbf{H}$, a Hilbert space and T is a bounded linear operator on $\mathbf{H}$. With the above motivation we seek another bounded linear operator T^* with the property that

$$\langle Tu, v \rangle = \langle u, T^* v \rangle \qquad \forall u, v \in \mathbf{H} \tag{9.4.42}$$

If such a T^* can be found, observe that if there exists any solution u of $Tu = f$ then we must have

$$\langle f, v\rangle = \langle Tu, v\rangle = \langle u, T^* v\rangle = \langle u, 0\rangle = 0 \tag{9.4.43}$$

for any $v \in N(T^*)$, so that $f \perp v$ must hold for all such v. We have thus shown already that $R(T) \perp N(T^*)$, or equivalently

$$R(T) \subset N(T^*)^\perp \qquad N(T^*) \subset R(T)^\perp \tag{9.4.44}$$

In particular $f \perp N(T^*)$ is a necessary condition for the solvability of $Tu = f$. The sufficiency of this condition need not be true in general as we will see by examples, but it does hold for some important classes of operator T.

Theorem 9.2. *If* $\mathbf{H}$ *is a Hilbert space and* $T \in \mathcal{B}(\mathbf{H})$ *then there exists a unique* $T^* \in \mathcal{B}(\mathbf{H})$, *the adjoint of* T, *such that Eq. (9.4.42) holds. In addition,* $(T^*)^* = T$ *and* $\|T^*\| = \|T\|$.

Proof. Fix $v \in \mathbf{H}$ and let $\ell(u) = \langle Tu, v\rangle$. Clearly ℓ is linear on $\mathbf{H}$ and

$$|\ell(u)| = |\langle Tu, v\rangle| \le \|Tu\|\,\|v\| \le \|T\|\,\|u\|\,\|v\| \tag{9.4.45}$$

therefore $\ell \in \mathbf{H}^*$ with $\|\ell\| \le \|T\|\,\|v\|$. By the Riesz representation theorem (Theorem 5.6) there exists a unique $v^* \in \mathbf{H}$ such that

$$\ell(u) = \langle u, v^*\rangle \qquad \forall u \in \mathbf{H} \tag{9.4.46}$$

We define $T^* v = v^*$ so that $T^* : \mathbf{H} \to \mathbf{H}$ and Eq. (9.4.42) is true. We claim next that T^* is linear. To see this, note that for any $v_1, v_2 \in \mathbf{H}$, $u \in \mathbf{H}$ and scalars c_1, c_2

$$\langle u, T^*(c_1 v_1 + c_2 v_2)\rangle = \langle Tu, c_1 v_1 + c_2 v_2\rangle \tag{9.4.47}$$
$$= \overline{c}_1 \langle Tu, v_1\rangle + \overline{c}_2 \langle Tu, v_2\rangle \tag{9.4.48}$$
$$= \overline{c}_1 \langle u, T^* v_1\rangle + \overline{c}_2 \langle u, T^* v_2\rangle \tag{9.4.49}$$
$$= \langle u, c_1 T^* v_1 + c_2 T^* v_2\rangle \tag{9.4.50}$$

Since u is arbitrary we must have $T^*(c_1 v_1 + c_2 v_2) = c_1 T^* v_1 + c_2 T^* v_2$ as needed.

Next we claim that T^* is bounded. To verify this, note that $\|T^* v\| = \|v^*\| = \|\ell\| \le \|T\|\,\|v\|$ implying that

$$\|T^*\| \le \|T\| \tag{9.4.51}$$

To check the uniqueness property suppose that there exists some other bounded linear operator S such that $\langle Tu, v\rangle = \langle u, Sv\rangle$ for all $u, v \in \mathbf{H}$. It would then follow that $\langle u, T^* v - Sv\rangle = 0$ for all u, implying $T^* v = Sv$ for all v, in other words $S = T^*$ must hold.

Finally, since $T^* \in \mathcal{B}(\mathbf{H})$, it also has an adjoint $T^{**} := (T^*)^*$ satisfying $\langle T^* u, v\rangle = \langle u, T^{**} v\rangle$ for all u, v. But we also have

$$\langle T^* u, v\rangle = \overline{\langle v, T^* u\rangle} = \overline{\langle Tv, u\rangle} = \langle u, Tv\rangle \tag{9.4.52}$$

so by uniqueness of the adjoint we must have $T^{**} = T$. From Eq. (9.4.51) with T replaced by T^* it follows that $\|T\| = \|T^{**}\| \le \|T^*\|$ and consequently we obtain $\|T\| = \|T^*\|$. □

Certain special classes of operator are defined, depending on the relationship between T and T^*.

Definition 9.3. If $T \in \mathcal{B}(\mathbf{H})$ then

- If $T^* = T$ we say T is *self-adjoint.*
- If $T^* = -T$ we say T is *skew-adjoint.*
- If $T^* = T^{-1}$ we say T is *unitary.*

Proposition 9.2. *If $S, T \in \mathcal{B}(\mathbf{H})$ then $ST \in \mathcal{B}(\mathbf{H})$ and*

$$(ST)^* = T^* S^* \tag{9.4.53}$$

If $T^{-1} \in \mathcal{B}(\mathbf{H})$ then $(T^*)^{-1} \in \mathcal{B}(\mathbf{H})$ and

$$(T^{-1})^* = (T^*)^{-1} \tag{9.4.54}$$

The proofs of these two properties will be left for the exercises.

9.5 EXAMPLES OF ADJOINTS

We now revisit several of the examples from Section 9.2, with focus on computing the corresponding adjoint operators. We remark that the uniqueness assertion of Theorem 9.2 is a relatively elementary thing, but note how it gets used repeatedly below to establish what the adjoint of a given operator T is.

Example 9.12. In the case $\mathbf{H} = \mathbb{C}^N$ with $Tu = Au$, A an $N \times N$ matrix, we already know that

$$\langle Tu, v\rangle = \langle Au, v\rangle = \langle u, A^* v\rangle \tag{9.5.55}$$

where A^* is the conjugate transpose matrix of A. Thus by uniqueness $T^* v = A^* v$, as expected. T is then obviously self-adjoint if $A^* = A$, consistent with the usual definition from linear algebra. A is also said to be a Hermitian matrix in this case, or symmetric if A is real. Likewise the meaning of a skew-adjoint operator or unitary operator coincides with the way the terms are normally used in linear algebra. □

Note that we haven't considered here the case of an $M \times N$ matrix with $M \ne N$ since then the domain and range spaces would be different, requiring a somewhat more general way of defining the adjoint.

Example 9.13. Consider the multiplication operator $Tu(x) = a(x)u(x)$ on $L^2(\Omega)$, where $a \in L^\infty(\Omega)$. Then

$$\langle Tu, v\rangle = \int_\Omega a(x)u(x)\overline{v(x)}\,dx = \int_\Omega u(x)\overline{\left(\overline{a(x)}v(x)\right)}\,dx \tag{9.5.56}$$

from which it follows that $T^* v(x) = \overline{a(x)} v(x)$. T is self-adjoint if a is real valued, skew-adjoint if a is purely imaginary and unitary if $|a(x)| \equiv 1$. □

Example 9.14. Next we look at the integral operator (9.2.16) on $L^2(\Omega)$, with $K \in L^2(\Omega \times \Omega)$ so that T is bounded. Assuming that the use of Fubini's theorem below can be justified, we get

$$\langle Tu, v \rangle = \int_\Omega \left(\int_\Omega K(x,y) u(y)\, dy \right) \overline{v(x)}\, dx \tag{9.5.57}$$

$$= \int_\Omega u(y) \left(\overline{\int_\Omega \overline{K(x,y)} v(x)\, dx} \right) dy \tag{9.5.58}$$

which is the same as $\langle u, T^* v \rangle$ if and only if

$$T^* v(y) = \int_\Omega \overline{K(x,y)} v(x)\, dx \tag{9.5.59}$$

or equivalently

$$T^* v(x) = \int_\Omega \overline{K(y,x)} v(y)\, dy \tag{9.5.60}$$

Thus T^* is the integral operator with kernel $\overline{K(y,x)}$, and note again the formal analogy to the case of the matrix multiplication operator. The use of Fubini's theorem to exchange the order of integrals above can be justified by observing that $K(x,y)u(y)v(x) \in L^1(\Omega \times \Omega)$ under our assumptions (Exercise 9.9). T will be self-adjoint, for example, if K is real valued and symmetric in x, y. □

Example 9.15. Consider next $T = \mathcal{F}$, the Fourier transform on $L^2(\mathbb{R}^N)$. Based on the previous example we may expect that

$$T^* v(x) = \frac{1}{(2\pi)^{N/2}} \int_{\mathbb{R}^N} e^{ix\cdot y} v(y)\, dy \tag{9.5.61}$$

since the kernel here is the conjugate transpose of that for T. This is correct, but can't be proven as above since the use of Fubini's theorem can't be directly justified. Instead we proceed by first recalling the Parseval identity (7.5.94)

$$\int_{\mathbb{R}^N} \widehat{u}(x) v(x)\, dx = \int_{\mathbb{R}^N} u(x) \widehat{v}(x)\, dx \tag{9.5.62}$$

Thus

$$\langle Tu, v \rangle = \int_{\mathbb{R}^N} \widehat{u}(x) \overline{v(x)}\, dx = \int_{\mathbb{R}^N} u(x) \widehat{\overline{v(x)}}\, dx \tag{9.5.63}$$

so that $T^* v = \overline{\widehat{\overline{v}}}$. One can now check, by unwinding the definitions, that this is the same as $(T^* v)(x) = (Tv)(-x)$, which amounts to Eq. (9.5.61). Furthermore, we now recognize from the Fourier inversion theorem that Eq. (9.5.61) may be restated as

$$T^*v = T^{-1}v \tag{9.5.64}$$

so in particular the Fourier transform is a unitary operator on $L^2(\mathbb{R}^N)$. □

Example 9.16. If T is the Fourier multiplication operator $T = \mathcal{F}^{-1}S\mathcal{F}$ on $L^2(\mathbb{R}^N)$, where S is the multiplication operator with L^∞ multiplier a, then we can obtain using Eq. (9.4.53) that $T^* = \mathcal{F}^{-1}S^*\mathcal{F}$, that is, T^* is the Fourier multiplication operator with multiplier $\overline{a}$. In particular, the Hilbert transform is skew-adjoint, $\mathcal{H}^* = -\mathcal{H}$, since $\overline{a(y)} = -a(y)$ in this case. □

9.6 CONDITIONS FOR SOLVABILITY OF LINEAR OPERATOR EQUATIONS

Let us return now to the general study of operator equations $Tu = f$, when T is a bounded linear operator on a Hilbert space $\mathbf{H}$.

Proposition 9.3. *If $T \in \mathcal{B}(\mathbf{H})$ then $N(T^*) = R(T)^\perp$. In particular, $\overline{R(T)} = N(T^*)^\perp$.*

Proof. By Eq. (9.4.44) we have $N(T^*) \subset R(T)^\perp$. Conversely, if $v \in R(T)^\perp$ then $\langle u, T^*v\rangle = \langle Tu, v\rangle = 0$ for all $u \in \mathbf{H}$. Thus $T^*v = 0$ must hold so $v \in N(T^*)$. By Proposition 5.2 $M^{\perp\perp} = \overline{M}$ for any subspace M, so the second conclusion follows. □

Corollary 9.1. *If $T \in \mathcal{B}(\mathbf{H})$ and T has closed range then $R(T) = N(T^*)^\perp$, that is to say, $Tu = f$ has a solution if and only if $f \perp N(T^*)$.*

Definition 9.4. If T is any linear operator, we define $\operatorname{rank}(T) = \dim R(T)$, and say that T is a finite rank operator whenever $\operatorname{rank}(T) < \infty$.

Recall that any finite dimensional subspace is closed, by Theorem 4.1, thus we have also established the following:

Corollary 9.2. *If $T \in \mathcal{B}(\mathbf{H})$ is a finite rank operator then $R(T) = N(T^*)^\perp$.*

Aside from the completely finite dimensional situation, there are other finite rank operators which will be of interest to us.

Example 9.17. Let $\mathbf{H} = L^2(0,1)$ and $Tu(x) = \int_0^1 xyu(y)\,dy$. Then $R(T) = \operatorname{span}(e)$ where $e(x) = x$, so $\operatorname{rank}(T) = 1$. Here T is self-adjoint so $N(T^*) = N(T) = \{e\}^\perp$ so the conclusion of the corollary is obvious.

More generally, let $\mathbf{H} = L^2(\Omega)$ for some open set $\Omega \subset \mathbb{R}^N$ and let T be an integral operator as in Eq. (9.2.16) with kernel

$$K(x,y) = \sum_{j=1}^M \phi_j(x)\psi_j(y) \tag{9.6.65}$$

for some $\phi_j, \psi_j \in L^2(\Omega)$. We may always assume that the ϕ_j's and ψ_j's are linearly independent. Such a kernel K is sometimes said to be *degenerate*. In this case we have $R(T) = \text{span}(\phi_1, \dots, \phi_M)$ so that rank $(T) = M$. The condition $f \perp N(T^*)$ amounts to requiring infinitely many consistency conditions, since $N(T^*)$ has infinite dimension. □

Since $N(T^*)^\perp$ is always a closed subspace, the identity $R(T) = N(T^*)^\perp$ can only hold if T has closed range. This property of the range is not true in general, as is shown by the following example.

Example 9.18. Let $\mathbf{H} = L^2(0,1)$ and $Tu(x) = \int_0^x u(y)\,dy$. We may think of this operator as the special case of Eq. (9.2.16) in which

$$K(x,y) = \begin{cases} 1 & y < x \\ 0 & y > x \end{cases} \tag{9.6.66}$$

This kernel is clearly in $L^2((0,1) \times (0,1))$ so that $T \in \mathcal{B}(\mathbf{H})$. Let f_n be any sequence of continuously differentiable functions such that $f_n(0) = 0$ for all n, and f_n converges in $\mathbf{H}$ to $f(x) = H(x - 1/2)$. Each f_n is in the range of T since $f_n = Tu_n$ if $u_n = f_n'$. But $f \notin R(T)$ since the range of T contains only continuous functions. Thus T does not have closed range. □

9.7 FREDHOLM OPERATORS AND THE FREDHOLM ALTERNATIVE

The following is a very useful concept.

Definition 9.5. $T \in \mathcal{B}(\mathbf{H})$ is of Fredholm type (or more informally, a Fredholm operator) if

- $N(T), N(T^*)$ are both finite dimensional,
- $R(T)$ is closed.

For such an operator T we define ind (T), the *index of* T, as

$$\text{ind}\,(T) = \dim(N(T)) - \dim(N(T^*)) \tag{9.7.67}$$

For our purposes the case of Fredholm operators of index 0 will be the most important one. If we can show somehow that an operator T belongs to this class then we obtain immediately the conclusion that "uniqueness is equivalent to existence". That is to say, the property that $Tu = f$ has at most one solution for any $f \in \mathbf{H}$ is equivalent to the property that $Tu = f$ has at least one solution for any $f \in \mathbf{H}$. The following elaboration of this is known as the Fredholm Alternative Theorem. The proof comes directly by consideration of the two cases when the common dimension of $N(T), N(T^*)$ is zero or nonzero, and using Proposition 9.3.

Theorem 9.3. *Let $T \in \mathcal{B}(\mathbf{H})$ be a Fredholm operator of index 0. Then either*

1. $N(T) = N(T^*) = \{0\}$ *and the equation $Tu = f$ has a unique solution for every $f \in \mathbf{H}$, or*

2. $\dim(N(T)) = \dim(N(T^*)) = M > 0$, *the equation* $Tu = f$ *has a solution* u^* *if and only if* f *satisfies the* M *compatibility conditions* $f \perp N(T^*)$, *and the general solution of* $Tu = f$ *can be written as* $\{u = u^* + v : v \in N(T)\}$.

Example 9.19. Every linear operator on $\mathbb{C}^N$ is of Fredholm type and index 0, since by a well known fact from matrix theory, a matrix and its transpose have null spaces of the same dimension. □

In the infinite dimensional situation it is easy to find examples of nonzero index—the simplest example is a shift operator.

Example 9.20. If we define S_+, S_- as in Eqs. (9.2.36), (9.2.37) then by Exercise 9.10 $S_+^* = S_-, S_-^* = S_+$, and it is then easy to see that $\text{ind}\,(S_+) = -1$ and $\text{ind}\,(S_-) = 1$. Clearly by shifting to the left or right by more than one entry, we can create an example of a Fredholm operator with any integer as its index. □

We will see in Chapter 12 that the operator $\lambda I + T$, where T is an integral operator of the form (9.2.16), with $K \in L^2(\Omega \times \Omega)$ and $\lambda \neq 0$, is always a Fredholm operator of index 0. Combining this fact with Theorem 9.3 will yield a great deal of information about the solvability of second kind integral equations.

9.8 CONVERGENCE OF OPERATORS

Recall that if $\mathbf{X}, \mathbf{Y}$ are Banach spaces we have defined a norm on $\mathcal{B}(\mathbf{X}, \mathbf{Y})$ for which all of the norm axioms are satisfied, so that $\mathcal{B}(\mathbf{X}, \mathbf{Y})$ is a normed linear space, and in fact is itself a Banach space (see Exercise 4.3 of Chapter 4).

Definition 9.6. We say $T_n \to T$ *uniformly* if $\|T_n - T\| \to 0$, that is, $T_n \to T$ in the topology of $\mathcal{B}(\mathbf{X}, \mathbf{Y})$. We say $T_n \to T$ *strongly* if $T_n x \to Tx$ for every $x \in \mathbf{X}$.

It is immediate from the definitions that uniform convergence implies strong convergence, but the converse is false (see Exercise 9.18). As usual we can define an infinite series of operators as the limit of the partial sums, and speak of uniform or strong convergence of the series. The series $\sum_{n=1}^{\infty} T_n$ will converge uniformly to some limit $T \in \mathcal{B}(\mathbf{X}, \mathbf{Y})$ if

$$\sum_{n=1}^{\infty} \|T_n\| < \infty \tag{9.8.68}$$

and in this case $\|T\| \leq \sum_{n=1}^{\infty} \|T_n\|$ (see Exercise 9.19). An important special case is given by the following.

Theorem 9.4. *If* $T \in \mathcal{B}(\mathbf{X})$, $\lambda \in \mathbb{C}$ *and* $\|T\| < |\lambda|$ *then* $(\lambda I - T)^{-1} \in \mathcal{B}(\mathbf{X})$,

$$(\lambda I - T)^{-1} = \sum_{n=0}^{\infty} \frac{T^n}{\lambda^{n+1}} \tag{9.8.69}$$

where the series is uniformly convergent, and

$$\|(\lambda I - T)^{-1}\| \leq \frac{1}{|\lambda| - \|T\|} \tag{9.8.70}$$

Proof. If T_n is replaced by T^n/λ^{n+1} then (9.8.68) holds for the series on the right-hand side of (9.8.69), so it is uniformly convergent to some $S \in \mathcal{B}(\mathbf{X})$. If S_N denotes the Nth partial sum then

$$S_N(\lambda I - T) = I - \frac{T^{N+1}}{\lambda^{N+1}} \tag{9.8.71}$$

Since $\|T^{N+1}/\lambda^{N+1}\| < (\|T\|/|\lambda|)^{N+1} \to 0$ we obtain $S(\lambda I - T) = I$ in the limit as $N \to \infty$. Likewise $(\lambda I - T)S = I$, so that (9.8.69), and subsequently (9.8.70) holds. □

The formula (9.8.69) is easily remembered as the "geometric series" for $(\lambda I - T)^{-1}$.

9.9 EXERCISES

In these exercises assume that $\mathbf{X}$ is a Banach space and $\mathbf{H}$ is a Hilbert space.

9.1. If $T_1, T_2 \in \mathcal{B}(\mathbf{X})$ show that $\|T_1 + T_2\| \leq \|T_1\| + \|T_2\|$, $\|T_1 T_2\| \leq \|T_1\| \, \|T_2\|$, and $\|T^n\| \leq \|T\|^n$.

9.2. If $A = \begin{bmatrix} 1 & -2 \\ 3 & 4 \end{bmatrix}$ compute the Frobenius norm of A and $\|A\|_p$ for $p = 1, 2$, and ∞.

9.3. Prove Proposition 9.1.

9.4. Define the averaging operator

$$Tu(x) = \frac{1}{x}\int_0^x u(y)\, dy$$

Show that T is bounded on $L^p(0, \infty)$ for $1 < p < \infty$. (Suggestions: Assume first that $u \geq 0$ and is a continuous function of compact support. If $v = Tu$ show that

$$\int_0^\infty v^p(x)\, dx = -p\int_0^\infty v^{p-1}(x) x v'(x)\, dx$$

Note that $xv' = u - v$ and apply Hölder's inequality. Then derive the general case. The resulting inequality is known as Hardy's inequality.)

9.5. Let T be the Fourier multiplication operator on $L^2(\mathbb{R})$ with multiplier $a(y) = H(y)$ (the Heaviside function), and define

$$M_+ = \{u \in L^2(\mathbb{R}) : \widehat{u}(y) = 0 \ \ \forall y < 0\}$$
$$M_- = \{u \in L^2(\mathbb{R}) : \widehat{u}(y) = 0 \ \ \forall y > 0\}$$

(a) Show that $T = \frac{1}{2}(I + i\mathbb{H})$, where $\mathbb{H}$ is the Hilbert transform.
(b) Show that if u is real valued, then u is uniquely determined by either the real or imaginary part of Tu.
(c) Show that $L^2(\mathbb{R}) = M_+ \oplus M_-$. (See Problem 5.19 for this notation.)
(d) Show that $T = P_{M_+}$.
(e) If $u \in M_+$ show that $u = i\mathbb{H}u$. In particular, if $u(x) = \alpha(x) + i\beta(x)$ then

$$\beta = \mathbb{H}\alpha \qquad \alpha = -\mathbb{H}\beta$$

(Comments: Tu is sometimes called the *analytic signal* of u. This terminology comes from the fact that Tu can be shown to always have an extension as an analytic function to the upper half of the complex plane. It is often convenient to work with Tu instead of u, because it avoids ambiguities due to k and $-k$ really being the same frequency—the analytic signal has only positive frequency components. By (b), u and Tu are in one-to-one correspondence, at least for real signals. The relationships between α and β in (e) are sometimes called the *Kramers-Kronig* relations. Note that it means that M_+ contains no purely real valued functions except for $u = 0$, and likewise for M_-.)

9.6. Show that a linear operator $T : \mathbb{C}^N \to \mathbb{C}^M$ is always bounded for any choice of norms on $\mathbb{C}^N$ and $\mathbb{C}^M$.

9.7. If $T, T^{-1} \in \mathcal{B}(\mathbf{H})$ show that $(T^*)^{-1} \in \mathcal{B}(\mathbf{H})$ and $(T^{-1})^* = (T^*)^{-1}$.

9.8. If $S, T \in \mathcal{B}(\mathbf{H})$, show that
(i) $(S+T)^* = S^* + T^*$
(ii) $(ST)^* = T^*S^*$
(These properties, together with (iii) $(\lambda T)^* = \bar{\lambda}T^*$ for scalars λ and (iv) $T^{**} = T$, which we have already proved, are the axioms for an *involution* on $\mathcal{B}(\mathbf{H})$, that is to say the mapping $T \to T^*$ is an involution. The term involution is also used more generally to refer to any mapping which is its own inverse.)

9.9. Give a careful justification of how (9.5.58) follows from (9.5.57) with reference to an appropriate version of Fubini's theorem.

9.10. Let S_+, S_- be the left- and right-shift operators on ℓ^2. Show that $S_+ = S_-^*$ and $S_- = S_+^*$.

9.11. Let T be the Volterra integral operator $Tu(x) = \int_0^x u(y)\,dy$, considered as an operator on $L^2(0,1)$. Find T^* and $N(T^*)$.

9.12. Suppose $T \in \mathcal{B}(\mathbf{H})$ is self-adjoint and there exists a constant $c > 0$ such that $\|Tu\| \geq c\|u\|$ for all $u \in \mathbf{H}$. Show that there exists a solution of $Tu = f$ for all $f \in \mathbf{H}$. Show by example that the conclusion may be false if the assumption of self-adjointness is removed.

9.13. Let M be the multiplication operator $Mu(x) = xu(x)$ in $L^2(0,1)$. Show that $R(M)$ is dense but not closed.

9.14. If $T \in \mathcal{B}(\mathbf{H})$ show that T^* restricted to $R(T)$ is one-to-one.

9.15. An operator $T \in \mathcal{B}(\mathbf{H})$ is said to be *normal* if it commutes with its adjoint, that is, $T^*T = TT^*$. Thus, for example, any self-adjoint, skew-adjoint, or unitary operator is normal. For a normal operator T show that

(a) $\|Tu\| = \|T^*u\|$ for every $u \in \mathbf{H}$.

(b) T is one to one if and only if it has dense range.

(c) Show that any multiplication operator (Example 9.3) or Fourier multiplication operator (Example 9.7) is normal in L^2.

(d) Show that the shift operators S_+, S_- are not normal in ℓ^2.

9.16. If $\mathcal{U}(\mathbf{H})$ denotes the set of unitary operators on $\mathbf{H}$, show that $\mathcal{U}(\mathbf{H})$ is a group under composition. Is $\mathcal{U}(\mathbf{H})$ a subspace of $\mathcal{B}(\mathbf{H})$?

9.17. Prove that if T is the Fourier transform regarded as a linear operator from $L^1(\mathbb{R}^N)$ into $C_0(\mathbb{R}^N)$ then $\|T\| = \frac{1}{(2\pi)^{N/2}}$.

9.18. Give an example of a sequence $T_n \in \mathcal{B}(\mathbf{H})$ which is strongly convergent but not uniformly convergent.

9.19. If $T_n \in \mathcal{B}(\mathbf{X})$ and $\sum_{n=1}^{\infty} \|T_n\| < \infty$, show that the series $\sum_{n=1}^{\infty} T_n$ is uniformly convergent.

9.20. If $\mathbf{X}, \mathbf{Y}$ are vector spaces, and $T : \mathbf{X} \to \mathbf{Y}$ is linear, then S is a *left inverse* for T if $STx = x$ for every $x \in \mathbf{X}$ and is a *right inverse* if $TSy = y$ for every $y \in \mathbf{Y}$. Show that a linear operator with a left inverse is one-to-one and an operator with a right inverse is onto. If $\mathbf{X} = \mathbf{Y}$ is finite dimensional then it is known from linear algebra that a left inverse must also be a right inverse. Show by means of examples that this is false if $\mathbf{X} \neq \mathbf{Y}$ or if $\mathbf{X} = \mathbf{Y}$ is infinite dimensional.

9.21. If $T \in \mathcal{B}(\mathbf{H})$, the *numerical range* of T is the set

$$\left\{\lambda \in \mathbb{C} : \lambda = \frac{\langle Tx, x\rangle}{\langle x, x\rangle} \text{ for some } x \in \mathbf{H}\right\}$$

If T is self-adjoint show that the numerical range of T is contained in the interval $[-\|T\|, \|T\|]$ of the real axis. What is the corresponding statement for a skew-adjoint operator?

9.22. Find the explicit expression for an ideal low pass filter (recall the definition in Example 9.7) in the form of a convolution operator.

Chapter 10

Unbounded Operators

10.1 GENERAL ASPECTS OF UNBOUNDED LINEAR OPERATORS

Let us return to the general definition of linear operator given at the beginning of the previous chapter, without any assumption about continuity of the operator. For simplicity we will assume a Hilbert space setting, although much of what is stated below remains true for mappings between Banach spaces. We have the following important definition.

Definition 10.1. If $\mathbf{H}$ is a Hilbert space and T: $D(T) \subset \mathbf{H} \to \mathbf{H}$ is a linear operator then we say T is *closed* if whenever $u_n \in D(T)$, $u_n \to u$ and $Tu_n \to v$ then $u \in D(T)$ and $Tu = v$.

We emphasize that this definition is strictly weaker than continuity of T, since for a closed operator it is quite possible that $u_n \to u$ but the image sequence $\{Tu_n\}$ is divergent. This could not happen for a bounded linear operator. It is immediate that any $T \in \mathcal{B}(\mathbf{H})$ must be closed.

A common alternate way to define a closed operator employs the concept of the graph of T.

Definition 10.2. If T: $D(T) \subset \mathbf{H} \to \mathbf{H}$ is a linear operator then we define the graph of T to be

$$G(T) = \{(u, v) \in \mathbf{H} \times \mathbf{H}: v = Tu\} \tag{10.1.1}$$

The definition of $G(T)$ (and for that matter the definition of closedness) makes sense even if T is not linear, but we will only use it in the linear case. It is easy to check that $\mathbf{H} \times \mathbf{H}$ is a Hilbert space with the inner product

$$\langle (u_1, v_1), (u_2, v_2) \rangle = \langle u_1, u_2 \rangle + \langle v_1, v_2 \rangle \tag{10.1.2}$$

In particular, $(u_n, v_n) \to (u, v)$ in $\mathbf{H} \times \mathbf{H}$ if and only if $u_n \to u$ and $v_n \to v$ in $\mathbf{H}$. One may now verify (Exercise 10.2).

Techniques of Functional Analysis for Differential and Integral Equations
http://dx.doi.org/10.1016/B978-0-12-811426-1.00010-6

Proposition 10.1. *T: $D(T) \subset \mathbf{H} \to \mathbf{H}$ is a closed linear operator if and only if $G(T)$ is a closed subspace of $\mathbf{H} \times \mathbf{H}$.*

We emphasize that closedness of T does not mean that $D(T)$ is closed in $\mathbf{H}$—this is false in general. In fact we have the so-called Closed Graph Theorem.

Theorem 10.1. *If T is a closed linear operator and $D(T)$ is a closed subspace of $\mathbf{H}$, then T must be continuous on $D(T)$.*

We refer to Theorem 2.15 of [33] or Theorem 2.9 of [5] for a proof. In particular if T is closed and unbounded then $D(T)$ cannot be all of $\mathbf{H}$.

By far the most common types of unbounded operator which we will be interested in are differential operators.

For use in the next example, let us recall that a function f defined on a closed interval $[a,b]$ is *absolutely continuous* on $[a,b]$ ($f \in AC([a,b])$) if for any $\epsilon > 0$ there exists $\delta > 0$ such that if $\{(a_k, b_k)\}_{k=1}^n$ is a disjoint collection of intervals in $[a,b]$, and $\sum_{k=1}^n |b_k - a_k| < \delta$ then $\sum_{k=1}^n |f(b_k) - f(a_k)| < \epsilon$. Clearly an absolutely continuous function is continuous.

Theorem 10.2. *The following are equivalent.*

1. *f is absolutely continuous on $[a,b]$.*
2. *f is differentiable a.e. on $[a,b]$, $f' \in L^1(a,b)$ and*

$$f(x) = f(a) + \int_a^x f'(y)\,dy \quad \forall x \in [a,b] \tag{10.1.3}$$

3. *$f \in W^{1,1}(a,b)$*

Furthermore, when f satisfies these equivalent conditions, the distributional derivative of f coincides with its pointwise a.e. derivative.

Here, the equivalence of 1 and 2 is an important theorem of analysis, see for example, Theorem 11, Section 6.5 of [30], Theorem 7.29 of [40], or Theorem 7.20 of [32], while the equivalence of 2 and 3, follows from Theorem 8.2 and the definition of the Sobolev space $W^{1,1}$. The final statement is also a direct consequence of Theorem 8.2.

Example 10.1. Let $\mathbf{H} = L^2(0,1)$ and $Tu = u'$ on the domain

$$D(T) = \{u \in H^1(0,1): u(0) = 0\} \tag{10.1.4}$$

Here T is unbounded, as in Example 9.10, $D(T)$ is a dense subspace of $\mathbf{H}$, since for example, it contains $\mathcal{D}(0,1)$, but it is not all of $\mathbf{H}$. We claim that T is closed. To see this, suppose $u_n \in D(T)$, $u_n \to u$ in $\mathbf{H}$ and $v_n = u_n' \to v$ in $\mathbf{H}$. By our assumptions, Eq. (10.1.3) is valid, so

$$u_n(x) = \int_0^x v_n(y)\,dy \tag{10.1.5}$$

We can then find a subsequence $n_k \to \infty$ and a subset $\Sigma \subset (0,1)$ such that $u_{n_k}(x) \to u(x)$ for $x \in \Sigma$ and the complement of Σ has measure zero. For any x we also have that $v_{n_k} \to v$ in $L^2(0,x)$, so that passing to the limit in Eq. (10.1.5) through the subsequence n_k we obtain

$$u(x) = \int_0^x v(s)\,ds \quad \forall x \in \Sigma \tag{10.1.6}$$

If we denote the right-hand side by w then from Theorem 10.2 we get that $w \in D(T)$, with $w' = v$ in the sense of distributions. Since $u = w$ a.e., u and w coincide as elements of $L^2(0,1)$ and so we get the necessary conclusion that $u \in D(T)$ with $u' = v$. □

The proper definition of $D(T)$ was essential in this example. If we had defined instead $D(T) = \{u \in C^1([0,1]) : u(0) = 0\}$ then we would not have been able to reach the conclusion that $u \in D(T)$.

An operator which is not closed may still be *closeable*, meaning that it has a closed extension. Let us define this concept carefully.

Definition 10.3. If S, T are linear operators on $\mathbf{H}$, we say that S is an extension of T if $D(T) \subset D(S)$ and $Tu = Su$ for $u \in D(T)$. In this case we write $T \subset S$. T is closeable if it has a closed extension.

If T is not closed, then its graph $G(T)$ is not closed, but it always has a closure $\overline{G(T)}$ in the topology of $\mathbf{H} \times \mathbf{H}$, which is then a natural candidate for the graph of a closed operator which extends T. This procedure may fail, however, because it may happen that $(u, v_1), (u, v_2) \in \overline{G(T)}$ with $v_1 \neq v_2$ so that $\overline{G(T)}$ would not correspond to a single-valued operator. If we know somehow that this cannot happen, then $\overline{G(T)}$ will be the graph of some linear operator S (you should check that $\overline{G(T)}$ is a subspace of $\mathbf{H} \times \mathbf{H}$) which is obviously closed and extends T, thus T will be closeable.

It is useful to have a clearer criterion for the closability of a linear operator T. Note that if $(u, v_1), (u, v_2)$ are both in $\overline{G(T)}$, with $v_1 \neq v_2$, then $(0, v) \in \overline{G(T)}$ for $v = v_1 - v_2 \neq 0$. This means that there must exist $u_n \to 0$, $u_n \in D(T)$ such that $v_n = Tu_n \to v \neq 0$. If we can show that no such sequence u_n can exist, then evidently no such pair of points can exist in $\overline{G(T)}$, so that T will be closeable. The converse statement is also valid and is easy to check. Thus we have established the following.

Proposition 10.2. *A linear operator on* $\mathbf{H}$ *is closeable if and only if* $u_n \in D(T)$, $u_n \to 0$, $Tu_n \to v$ *implies* $v = 0$.

Example 10.2. Let $Tu = u'$ on $L^2(0,1)$ with domain $D(T) = \{u \in C^1([0,1] : u(0) = 0\}$. We have previously observed that T is not closed, but we can check that the above criterion holds, so that T is closeable. Let $u_n \in D(T)$ and $u_n \to 0, u_n' \to v$ in $L^2(0,1)$. As before,

$$u_n(x) = \int_0^x u_n'(s)\,ds \tag{10.1.7}$$

Picking a subsequence $n_k \to \infty$ for which $u_{n_k} \to 0$ a.e., we get

$$\int_0^x v(s)\,ds = 0 \quad \text{a.e.} \tag{10.1.8}$$

The left-hand side is absolutely continuous so equality must hold for every $x \in [0,1]$ and by Theorem 10.2 we conclude that $v = 0$ a.e. □

An operator which is closeable may in general have many different closed extensions. However, there always exists a minimal extension in this case, denoted $\overline{T}$, the closure of T, defined by $G(\overline{T}) = \overline{G(T)}$. It can be alternatively characterized as follows: $\overline{T}$ is the unique linear operator on **H** with the properties that (i) $T \subset \overline{T}$ and (ii) if $T \subset S$ and S is closed then $\overline{T} \subset S$.

If T: $D(T) \subset \mathbf{H} \to \mathbf{H}$ and S: $D(S) \subset \mathbf{H} \to \mathbf{H}$ are closed linear operators then the sum $S + T$ is defined and linear on $D(S + T) = D(S) \cap D(T)$, but need not be closed, in general. Choose, for example, any closed and densely defined linear operator T with $D(T) \neq \mathbf{H}$ and $S = -T$. Then the sum $S + T$ is the zero operator, on the dense domain $D(S \cap T) = D(T) \neq \mathbf{H}$, which is not a closed operator. In this example $S + T$ is closeable, but even that need not be true, see Exercise 10.13. One can show, however, that if T is closed and S is bounded, then $S + T$ is closed. Likewise the product ST is defined on $D(ST) = \{x \in D(T) : Tx \in D(S)\}$ and need not be closed even if S, T are. If $S \in \mathcal{B}(\mathbf{H})$ and T is closed then TS will be closed, but ST need not be (see Exercise 10.11).

Finally consider the inverse operator T^{-1}: $R(T) \to D(T)$, which is well defined if T is one-to-one.

Proposition 10.3. *If T is one-to-one and closed then T^{-1} is also closed.*

Proof. Let $u_n \in D(T^{-1})$, $u_n \to u$, and $T^{-1}u_n \to v$. Then if $v_n = T^{-1}u_n$ we have $v_n \in D(T)$, $v_n \to v$, and $Tv_n = u_n \to u$. Since T is closed it follows that $v \in D(T)$ and $Tv = u$, or equivalently $u \in R(T) = D(T^{-1})$ and $T^{-1}u = v$ as needed. □

10.2 THE ADJOINT OF AN UNBOUNDED LINEAR OPERATOR

To some extent it is possible to define an adjoint operator, even in the unbounded case, and obtain some results about the solvability of the operator equation $Tu = f$ analogous to those proved earlier in the case of bounded T.

For the rest of this section we assume that T: $D(T) \subset \mathbf{H} \to \mathbf{H}$ is linear and densely defined. We will say that (v, v^*) is an *admissible pair for* T^* if

$$\langle Tu, v\rangle = \langle u, v^*\rangle \quad \forall u \in D(T) \tag{10.2.9}$$

We then define

$$D(T^*) = \{v \in \mathbf{H}: \text{there exists } v^* \in \mathbf{H} \text{ such that } (v, v^*) \text{ is an admissible pair for } T^*\} \tag{10.2.10}$$

and

$$T^* v = v^* \quad v \in D(T^*) \tag{10.2.11}$$

For this to be an appropriate definition, we should check that for any v there is at most one v^* for which (v, v^*) is admissible. Indeed if there were two such elements, then the difference $v_1^* - v_2^*$ would satisfy $\langle u, v_1^* - v_2^* \rangle = 0$ for all $u \in D(T)$. Since we assume $D(T)$ is dense, it follows that $v_1^* = v_2^*$.

Note that for $v \in D(T^*)$, if we define $\phi_v(u) = \langle Tu, v \rangle$ for $u \in D(T)$, then ϕ_v is bounded on $D(T)$, since

$$|\phi_v(u)| = |\langle u, v^* \rangle| = |\langle u, T^* v \rangle| \le \|u\| \, \|T^* v\| \tag{10.2.12}$$

The converse statement is also true (see Exercise 10.5) so that it is equivalent to define $D(T^*)$ as the set of all $v \in \mathbf{H}$ such that $u \to \langle Tu, v \rangle$ is a bounded linear functional on $D(T)$.

The domain $D(T^*)$ always contains at least the zero element, since $(0, 0)$ is always an admissible pair. There are known examples for which $D(T^*)$ contains no other points (see Exercise 10.4).

Here is a useful characterization of T^* in terms of its graph $G(T^*) \subset \mathbf{H} \times \mathbf{H}$.

Proposition 10.4. *If T is a densely defined linear operator on* $\mathbf{H}$ *then*

$$G(T^*) = (V(G(T)))^{\perp} \tag{10.2.13}$$

where V is the unitary operator on $\mathbf{H} \times \mathbf{H}$ *defined by*

$$V(x, y) = (-y, x) \quad x, y \in \boldsymbol{H} \tag{10.2.14}$$

We leave the proof as an exercise.

Proposition 10.5. *If T is a densely defined linear operator on* $\mathbf{H}$ *then T^* is a closed linear operator on* $\mathbf{H}$.

We emphasize that it is *not* assumed here that T is closed. The conclusion that T^* must be closed also follows directly from Eq. (10.2.13), but we give a more direct proof below.

Proof. First we verify the linearity of T^*. If $v_1, v_2 \in D(T^*)$ and c_1, c_2 are scalars, then there exist unique elements v_1^*, v_2^* such that

$$\langle Tu, v_1 \rangle = \langle u, v_1^* \rangle \quad \langle Tu, v_2 \rangle = \langle u, v_2^* \rangle \qquad \text{for all } u \in D(T) \tag{10.2.15}$$

Then

$$\begin{aligned} \langle Tu, c_1 v_1 + c_2 v_2 \rangle &= \overline{c}_1 \langle Tu, v_1 \rangle + \overline{c}_2 \langle Tu, v_2 \rangle = \overline{c}_1 \langle u, v_1^* \rangle + \overline{c}_2 \langle u, v_2^* \rangle \\ &= \langle u, c_1 v_1^* + c_2 v_2^* \rangle \end{aligned} \tag{10.2.16}$$

for all $u \in D(T)$, and so $(c_1 v_1 + c_2 v_2, c_1 v_1^* + c_2 v_2^*)$ is an admissible pair for T^*. In particular $c_1 v_1 + c_2 v_2 \in D(T^*)$ and

$$T^*(c_1v_1 + c_2v_2) = c_1v_1^* + c_2v_2^* = c_1T^*v_1 + c_2T^*v_2 \tag{10.2.17}$$

To see that T^* is closed, let $v_n \in D(T^*)$, $v_n \to v$, and $T^*v_n \to w$. If $u \in D(T)$ then we must have

$$\langle Tu, v_n\rangle = \langle u, T^*v_n\rangle \tag{10.2.18}$$

Letting $n \to \infty$ yields $\langle Tu, v\rangle = \langle u, w\rangle$. Thus (v, w) is an admissible pair for T^* implying that $v \in D(T^*)$ and $T^*v = w$, as needed. □

With a small modification of the proof, we obtain that Proposition 9.3 remains valid.

Theorem 10.3. *If $T\colon D(T) \subset \mathbf{H} \to \mathbf{H}$ is a densely defined linear operator then $N(T^*) = R(T)^\perp$.*

Proof. Let $f \in R(T)$ and $v \in N(T^*)$. We have $f = Tu$ for some $u \in D(T)$ and

$$\langle f, v\rangle = \langle Tu, v\rangle = \langle u, T^*v\rangle = 0 \tag{10.2.19}$$

so $N(T^*) \subset R(T)^\perp$. To get the reverse inclusion, let $v \in R(T)^\perp$, so that $\langle Tu, v\rangle = 0 = \langle u, 0\rangle$ for any $u \in D(T)$. This means that $(v, 0)$ is an admissible pair for T^*, so $v \in D(T^*)$ and $T^*v = 0$. Thus $R(T)^\perp \subset N(T^*)$ as needed. □

Example 10.3. Let us revisit the densely defined differential operator in Example 10.1. We seek here to find the adjoint operator T^*, and emphasize that one must determine $D(T^*)$ as part of the answer. It is typical in computing adjoints of unbounded operators that precisely identifying the domain of the adjoint is more difficult than finding a formula for the adjoint.

Let $v \in D(T^*)$ and $T^*v = g$, so that $\langle Tu, v\rangle = \langle u, g\rangle$ for all $u \in D(T)$. That is to say,

$$\int_0^1 u'(x)\overline{v(x)}\,dx = \int_0^1 u(x)\overline{g(x)}\,dx \qquad \forall u \in D(T) \tag{10.2.20}$$

Let

$$G(x) = -\int_x^1 g(y)\,dy \tag{10.2.21}$$

so that $G(1) = 0$ and $G'(x) = g(x)$ a.e., since g is integrable. Integration by parts then gives

$$\int_0^1 u(x)\overline{g(x)}\,dx = \int_0^1 u(x)\overline{G'(x)}\,dx = -\int_0^1 u'(x)\overline{G(x)}\,dx \tag{10.2.22}$$

since the boundary term vanishes. Thus we have

$$\int_0^1 u'(x)\overline{(v(x) + G(x))}\,dx = 0 \tag{10.2.23}$$

Now in Eq. (10.2.23) choose $u(x) = \int_0^x v(y)+G(y)\,dy$, which is legitimate since $u \in D(T)$. The result is that

$$\int_0^1 |v(x) + G(x)|^2 \, dx = 0 \tag{10.2.24}$$

which can only occur if $v(x) = -G(x) = \int_x^1 g(y)\,dy$ a.e., implying that $T^*v = g = -v'$. The above representation for v also shows that $v' \in L^2(0,1)$ and $v(1) = 0$, that is

$$D(T^*) \subset \{v \in L^2(0,1) \colon v' \in L^2(0,1), v(1) = 0\} \tag{10.2.25}$$

We claim that the reverse inclusion is also correct: If v belongs to the set on the right and $u \in D(T)$ then

$$\langle Tu, v\rangle = \int_0^1 u'(x)\overline{v(x)}\,dx = -\int_0^1 u(x)\overline{v'(x)}\,dx = \langle u, -v'\rangle \tag{10.2.26}$$

Thus $(v, -v')$ is an admissible pair for T^*, from which we conclude that $v \in D(T^*)$ and $T^*v = -v'$ as needed. We clearly have $N(T^*) = \{0\}$ and $R(T) = \mathbf{H}$.

In summary we have established that $T^*v = -v'$ with domain

$$D(T^*) = \{v \in H^1(0,1) \colon v(1) = 0\} \tag{10.2.27}$$

□

We remark that if we had originally defined T on the smaller domain $\{u \in C^1([0,1]) \colon u(0) = 0\}$ we would have obtained exactly the same result for T^* as mentioned above. This is a special case of the general fact that $T^* = \overline{T}^*$ (see Exercise 10.14). For this unclosed version of T we still have $N(T^*) = \{0\}$ but concerning the range can only state that $\overline{R(T)} = \mathbf{H}$.

Definition 10.4. If $T = T^*$ we say T is self-adjoint.

In this definition it is crucial that equality of the operators T and T^* must include the fact that their domains are identical.

Example 10.4. If in the previous example we defined $Tu = iu'$ on the same domain we would find that $T^*v = iv'$ on the domain (10.2.27). Even though the expressions for T, T^* are the same, T is not self-adjoint since the two domains are different. □

A closely related property is that of *symmetry*.

Definition 10.5. We say that T is symmetric if $\langle Tu, v\rangle = \langle u, Tv\rangle$ for all $u, v \in D(T)$.

Example 10.5. Let $Tu = iu'$ be the unbounded operator on $\mathbf{H} = L^2(0,1)$ with domain

$$D(T) = \{u \in H^1(0,1) \colon u(0) = u(1) = 0\} \tag{10.2.28}$$

One sees immediately that T is symmetric; however, it is still not self-adjoint since $D(T^*) \neq D(T)$ again, see Exercise 10.6. □

If T is symmetric and $u \in D(T)$, then (v, Tv) is an admissible pair for T^*, thus $D(T) \subset D(T^*)$ and $T^*v = Tv$ for $v \in D(T)$. In other words, T^* is always an extension of T whenever T is symmetric. We see, in particular, that any self-adjoint operator is closed and any symmetric operator is closeable.

Proposition 10.6. *If T is densely defined and one-to-one, and if also $R(T)$ is dense, then T^* is also one-to-one and $(T^*)^{-1} = (T^{-1})^*$.*

Proof. By our assumptions, $S = (T^{-1})^*$ exists. We are done if we show $ST^*u = u$ for all $u \in D(T^*)$ and $T^*Sv = v$ for all $v \in D(S)$.

First let $u \in D(T^*)$ and $v \in D(T^{-1})$. Then

$$\langle v, u\rangle = \langle TT^{-1}v, u\rangle = \langle T^{-1}v, T^*u\rangle \tag{10.2.29}$$

This means that (T^*u, u) is an admissible pair for $(T^{-1})^*$ and $ST^*u = (T^{-1})^*T^*u = u$ as needed.

Next, if $u \in D(T)$ and $v \in D(S)$ then

$$\langle u, v\rangle = \langle T^{-1}Tu, v\rangle = \langle Tu, Sv\rangle \tag{10.2.30}$$

Therefore (Sv, v) is admissible for T^*, so that $Sv \in D(T^*)$ and $T^*Sv = v$. □

Theorem 10.4. *If T, T^* are both densely defined then $T \subset T^{**}$, and in particular T is closeable.*

Proof. If we assume that T^* is densely defined, then T^{**} exists and is closed. If $u \in D(T)$ and $v \in D(T^*)$ then $\langle T^*v, u\rangle = \langle v, Tu\rangle$, which is to say that (u, Tu) is an admissible pair for T^{**}. Thus $u \in D(T^{**})$ and $T^{**}u = Tu$, or equivalently $T \subset T^{**}$. Thus T has a closed extension, namely T^{**}. □

There is an interesting converse statement, which we state but will not prove here, see [2] section 46, or [33] Theorem 13.12.

Theorem 10.5. *If T is densely defined and closeable then T^* must be densely defined, and $\overline{T} = T^{**}$. In particular if T is closed and densely defined then $T = T^{**}$.*

10.3 EXTENSIONS OF SYMMETRIC OPERATORS

It has been observed previously that if T is a densely defined symmetric operator then the adjoint T^* is always an extension of T. It is an interesting question whether such a T always possesses a self-adjoint extension—the extension would necessarily be different from T^* at least if T is closed, since then if T^* is self-adjoint so is T, by Theorem 10.5.

We say that a linear operator T is positive if $\langle Tu, u\rangle \geq 0$ for all $u \in D(T)$.

Theorem 10.6. *If T is a densely defined, positive, symmetric operator on a Hilbert space* $\mathbf{H}$ *then T has a positive self-adjoint extension.*

Proof. Define

$$\langle u, v\rangle_e = \langle u, v\rangle + \langle Tu, v\rangle \quad u, v \in D(T) \tag{10.3.31}$$

with corresponding norm denoted by $\|u\|_e$. It may be easily verified that all of the inner product axioms are satisfied by $\langle \cdot, \cdot\rangle_e$ on $D(T)$, and $\|u\| \le \|u\|_e$. Let $\mathbf{H}^e$ be the dense closed subspace of $\mathbf{H}$ obtained as the closure of $D(T)$ in the $\|\cdot\|_e$ norm, and regard it as equipped with the $\langle\cdot,\cdot\rangle_e$ inner product. For any $z \in \mathbf{H}$ the functional $\psi_z(u) = \langle u, z\rangle$ belongs to the dual space of $\mathbf{H}^e$ since $|\psi_z(u)| \le \|u\|\,\|z\| \le \|u\|_e\|z\|$, in particular $\|\psi_z\|_e \le \|z\|$ as a linear functional on $\mathbf{H}^e$. Thus by the Riesz Representation Theorem there exists a unique element $\Lambda z \in \mathbf{H}^e \subset \mathbf{H}$ such that

$$\psi_z(u) = \langle u, \Lambda z\rangle_e \qquad u \in \mathbf{H}^e \tag{10.3.32}$$

with $\|\Lambda z\| \le \|\Lambda z\|_e \le \|z\|$.

It may be checked that $\Lambda\colon \mathbf{H} \to \mathbf{H}$ is linear, and regarded as an operator on $\mathbf{H}$ we claim it is also self-adjoint. To see this observe that for any $u, z \in \mathbf{H}$ we have

$$\langle \Lambda u, z\rangle = \psi_z(\Lambda u) = \langle \Lambda u, \Lambda z\rangle_e = \overline{\langle \Lambda z, \Lambda u\rangle_e} = \overline{\psi_u(\Lambda z)} = \overline{\langle \Lambda z, u\rangle} = \langle u, \Lambda z\rangle \tag{10.3.33}$$

Choosing $u = z$ we also see that Λ is positive, namely

$$\langle \Lambda z, z\rangle = \langle \Lambda z, \Lambda z\rangle_e \ge 0 \tag{10.3.34}$$

Next Λ is one-to-one, since if $\Lambda z = 0$ and $u \in \mathbf{H}^e$ it follows that

$$0 = \langle u, \Lambda z\rangle_e = \langle u, z\rangle \quad \forall u \in \mathbf{H}^e \tag{10.3.35}$$

and since $\mathbf{H}^e$ is dense in $\mathbf{H}$ the conclusion follows. The range of Λ is also dense in $\mathbf{H}^e$, hence in $\mathbf{H}$, because otherwise there must exist $u \in \mathbf{H}^e$ such that $0 = \langle u, \Lambda z\rangle_e = \langle u, z\rangle$ for all $z \in \mathbf{H}$. From the above considerations and Proposition 10.6 we conclude that $S = \Lambda^{-1}$ exists and is a densely defined self-adjoint operator on $\mathbf{H}$.

We will complete the proof by showing that the self-adjoint operator $S - I$ is a positive extension of T. For $z, w \in D(T)$ we have

$$\langle z, w\rangle_e = \langle (I+T)z, w\rangle = \overline{\psi_{(I+T)z}(w)} = \overline{\langle w, \Lambda(I+T)z\rangle_e} = \langle \Lambda(I+T)z, w\rangle_e \tag{10.3.36}$$

and so

$$\Lambda(I+T)z = z \quad \forall z \in D(T) \tag{10.3.37}$$

by the assumed density of $D(T)$. In particular $D(T) \subset R(\Lambda) = D(S)$ and $(I+T)$ $z = \Lambda^{-1}z = Sz$ for $z \in D(T)$, as needed.

Finally we note that $\langle \Lambda z, z\rangle = \langle \Lambda z, \Lambda z\rangle_e \geq ||\Lambda z||^2$ for any $z \in \mathbf{H}$. Thus for $x = \Lambda z \in R(\Lambda) = D(S)$ if follows that $\langle x, Sx\rangle \geq ||x||^2$, which amounts to the positivity of $S - I$. □

A positive symmetric operator may have more than one self-adjoint extension, but the specific one constructed in the above proof is usually known as the *Friedrichs extension*. To clarify what all of the objects in the proof are, it may be helpful to think of the case that $Tu = -\Delta u$ on the domain $D(T) = C^2(\Omega) \cap C_0(\overline{\Omega})$. In this case $\|u\|_e = \|u\|_{H^1(\Omega)}$, $H^e = H_0^1(\Omega)$ (except endowed with the usual H^1 norm) and the Friedrichs extension will turn out to be the Dirichlet Laplacian discussed in detail in Section 13.4.

The condition of positivity for T may be weakened, see Exercise 10.16.

10.4 EXERCISES

10.1. Let T, S be densely defined linear operators on a Hilbert space. If $T \subset S$, show that $S^* \subset T^*$.

10.2. Verify that $\mathbf{H} \times \mathbf{H}$ is a Hilbert space with the inner product given by Eq. (10.1.2), and prove Proposition 10.1.

10.3. Prove the null space of a closed operator is closed.

10.4. Let $\phi \in \mathbf{H} = L^2(\mathbb{R})$ be any nonzero function and define the linear operator

$$Tu = \left(\int_{-\infty}^{\infty} u(x)\, dx \right) \phi$$

on the domain $D(T) = L^1(\mathbb{R}) \cap L^2(\mathbb{R})$.

(a) Show that T is unbounded and densely defined.

(b) Show that T^* is not densely defined, more specifically show that T^* is the zero operator with domain $\{\phi\}^\perp$. (Since $D(T^*)$ is not dense, it then follows from Theorem 10.5 that T is not closeable.)

10.5. If $T: D(T) \subset \mathbf{H} \to \mathbf{H}$ is a densely defined linear operator, $v \in \mathbf{H}$ and the map $u \to \langle Tu, v\rangle$ is bounded on $D(T)$, show that there exists $v^* \in \mathbf{H}$ such that (v, v^*) is an admissible pair for T^*.

10.6. Let $\mathbf{H} = L^2(0, 1)$ and $T_1 u = T_2 u = iu'$ with domains

$$D(T_1) = \{u \in H^1(0, 1)\colon u(0) = u(1)\}$$

$$D(T_2) = \{u \in H^1(0, 1)\colon u(0) = u(1) = 0\}$$

Show that T_1 is self-adjoint, and that T_2 is closed and symmetric but not self-adjoint. What is T_2^*?

10.7. If T is symmetric and $R(T) = \mathbf{H}$ show that T is self-adjoint. (Suggestion: It is enough to show that $D(T^*) \subset D(T)$.)

10.8. Show that if T is self-adjoint and one-to-one then T^{-1} is also self-adjoint. (Hint: All you really need to do is show that T^{-1} is densely defined.)

10.9. If T is self-adjoint, S is symmetric and $T \subset S$, show that $T = S$. (Thus a self-adjoint operator has no proper symmetric extension).

10.10. Let T, S be densely defined linear operators on $\mathbf{H}$ and assume that $D(T+S) = D(T) \cap D(S)$ is also dense. Show that $T^* + S^* \subset (T+S)^*$. Give an example showing that $T^* + S^*$ and $(T+S)^*$ may be unequal.

10.11. Assume that T is closed and S is bounded

(a) Show that $S + T$ is closed.

(b) Show that TS is closed, but that ST is not closed, in general.

10.12. Prove Proposition 10.4.

10.13. Let $\mathbf{H} = \ell^2$ and define

$$Sx = \left\{ \sum_{n=1}^{\infty} nx_n, 4x_2, 9x_3, \ldots \right\} \tag{10.4.38}$$

$$Tx = \{0, -4x_2, -9x_3, \ldots\} \tag{10.4.39}$$

on $D(S) = D(T) = \{x \in \ell^2 \colon \sum_{n=1}^{\infty} n^4 |x_n|^2 < \infty\}$. Show that S, T are closed, but $S + T$ is not closable. (Hint: For example, $e_n/n \to 0$ but $(S+T)e_n/n \to e_1$.)

10.14. If T is closable, show that T and $\overline{T}$ have the same adjoint.

10.15. Suppose that T is densely defined and symmetric with dense range. Prove that $N(T) = \{0\}$.

10.16. We say that a linear operator on a Hilbert space $\mathbf{H}$ is bounded below, if there exists a constant $c_0 > 0$ such that

$$\langle Tu, u \rangle \geq -c_0 \|u\|^2 \quad \forall u \in D(T)$$

Show that Theorem 10.6 remains valid if the condition that T be positive is replaced by the assumption that T is bounded below, with the corresponding conclusion being that T has a self-adjoint extension which is bounded below. (Hint: $T + c_o I$ is positive.)

Chapter 11

Spectrum of an Operator

11.1 RESOLVENT AND SPECTRUM OF A LINEAR OPERATOR

Let T be a closed, densely defined linear operator on a Hilbert space $\mathbf{H}$. As usual, we use I to denote the identity operator on $\mathbf{H}$.

Definition 11.1. We say that $\lambda \in \mathbb{C}$ is a *regular point* for T if $\lambda I - T$ is one-to-one and onto. We then define $\rho(T)$, the *resolvent set* of T and $\sigma(T)$, the *spectrum* of T by

$$\rho(T) = \{\lambda \in \mathbb{C}\colon \lambda \text{ is a regular point for } T\} \qquad \sigma(T) = \mathbb{C}\backslash\rho(T) \tag{11.1.1}$$

We note that since T is assumed to be closed, $(\lambda I - T)^{-1}$ is also a closed operator for any λ for which it is defined, as a consequence of Proposition 10.3. If $\lambda \in \rho(T)$ it follows that $(\lambda I - T)^{-1}$ is a closed operator on a closed set, and so $(\lambda I - T)^{-1} \in \mathcal{B}(\mathbf{H})$ by the Closed Graph Theorem, Theorem 10.1.

Example 11.1. Let $\mathbf{H} = \mathbb{C}^N$, $Tu = Au$ for some $N \times N$ matrix A. From linear algebra we know that $\lambda I - T$ is one-to-one and onto precisely if $\lambda I - A$ is a nonsingular matrix. Equivalently, λ is in the resolvent set if and only if λ is not an eigenvalue of A, where the eigenvalues are the roots of the Nth degree polynomial $\det(\lambda I - A)$. Thus $\sigma(T)$ consists of a finite number of points $\lambda_1, \dots, \lambda_M$, where $1 \le M \le N$, and all other points of the complex plane make up the resolvent set $\rho(T)$. □

In the case of a finite dimensional Hilbert space there is thus only one kind of point in the spectrum, where $(\lambda I - T)$ is neither one-to-one nor onto. But in general there are more possibilities. The following definition presents a traditional division of the spectrum into three parts.

Definition 11.2. Let $\lambda \in \sigma(T)$. Then

1. If $\lambda I - T$ is not one-to-one then we say $\lambda \in \sigma_p(T)$, the point spectrum of T.
2. If $\lambda I - T$ is one-to-one, $\overline{R(\lambda I - T)} = \mathbf{H}$, but $R(\lambda I - T) \ne \mathbf{H}$, then we say $\lambda \in \sigma_c(T)$, the continuous spectrum of T.
3. If $\lambda I - T$ is one-to-one but $\overline{R(\lambda I - T)} \ne \mathbf{H}$ then we say $\lambda \in \sigma_r(T)$, the residual spectrum of T.

Techniques of Functional Analysis for Differential and Integral Equations
http://dx.doi.org/10.1016/B978-0-12-811426-1.00011-8

Thus $\sigma(T)$ is the disjoint union of $\sigma_p(T), \sigma_c(T)$, and $\sigma_r(T)$. The point spectrum is also sometimes called the discrete spectrum. In the case of $\mathbf{H} = \mathbb{C}^N$, $\sigma(T) = \sigma_p(T)$ by the above discussion, but in general all three parts of the spectrum may be nonempty, as we will see from examples. There are further subclassifications of the spectrum, which are sometimes useful, see the exercises.

In the case that $\lambda \in \sigma_p(T)$ there must exist $u \neq 0$ such that $Tu = \lambda u$, and we then say that λ is an eigenvalue of T and u is a corresponding eigenvector. In the case that $\mathbf{H}$ is a space of functions we will often refer to u an eigenfunction instead. Obviously any nonzero scalar multiple of an eigenvector is also an eigenvector, and the set of all eigenvectors for a given λ, together with the zero element, make up $N(T - \lambda I)$, the null space of $T - \lambda I$, which will also be called the eigenspace of the eigenvalue λ. The dimension of $N(T - \lambda I)$ is the multiplicity of λ and may be infinity.[1] It is easy to check that if T is a closed operator then any eigenspace of T is closed.

If $\lambda \in \sigma_c(T)$ then $(\lambda I - T)^{-1}$ is defined on the dense domain $R(\lambda I - T)$, and must be unbounded. Indeed otherwise $(\lambda I - T)^{-1}$ would have a bounded extension to all of $\mathbf{H}$ and so could not be itself closed. In the case $\lambda \in \sigma_r(T)$ the operator $(\lambda I - T)^{-1}$ is no longer densely defined. Its domain may or may not be closed and it may or may not be bounded on its domain.

The concepts of resolvent set and spectrum, and the division of the spectrum just introduced, are closely connected with what is meant by a well-posed or ill-posed problem, as discussed in Section 1.4, and which we can restate in somewhat more precise terms here. If T: $D(T) \subset \mathbf{X} \to \mathbf{Y}$ is an operator between Banach spaces $\mathbf{X}, \mathbf{Y}$ (T may even be nonlinear here) then the problem of solving the operator equation $T(u) = f$ is said to be *well posed* with respect to $\mathbf{X}, \mathbf{Y}$ if

1. A solution u exists for every $f \in \mathbf{Y}$.
2. The solution is unique in $\mathbf{X}$.
3. The solution depends continuously on f in the sense that if $T(u_n) = f_n$ and $f_n \to f$ in $\mathbf{Y}$ then $u_n \to u$ in $\mathbf{X}$, where u is the unique solution of $T(u) = f$.

If the problem is not well-posed then it is said to be ill-posed. Now observe that if T is a linear operator on $\mathbf{H}$ and $\lambda \in \rho(T)$ then the problem of solving $\lambda u - Tu = f$ is well posed with respect to $\mathbf{H}$. Existence holds since $\lambda I - T$ is onto, uniqueness since it is one-to-one, and the continuous dependence property follows from the fact noted above that $(\lambda I - T)^{-1}$ must be bounded. On the other hand the three subsets of $\sigma(T)$ correspond more or less to the failure of one of the three conditions as mentioned above: $\lambda \in \sigma_p(T)$ means that uniqueness fails, $\lambda \in \sigma_c(T)$ means that the inverse map is defined on a dense subspace on which it is discontinuous, and $\lambda \in \sigma_r(T)$ implies that existence fails in a more dramatic way, namely the closure of the range of $\lambda I - T$ is a proper subspace of $\mathbf{H}$.

---

1. This agrees with the *geometric* multiplicity concept in linear algebra. In general there is no meaning for algebraic multiplicity in this setting.

Because the operator $(\lambda I - T)^{-1}$ arises so frequently, we introduce the notation

$$R_\lambda = (\lambda I - T)^{-1} \tag{11.1.2}$$

which is called the *resolvent operator of* T. Thus $\lambda \in \rho(T)$ if and only if $R_\lambda \in \mathcal{B}(\mathbf{H})$. It may be checked that the resolvent identity

$$R_\lambda - R_\mu = (\mu - \lambda) R_\lambda R_\mu \tag{11.1.3}$$

is valid (see Exercise 11.3).

Below we will look at a number of examples of operators and their spectra, but first we will establish a few general results. Among the most fundamental of these is that the resolvent set of any linear operator is open, so that the spectrum is closed. More generally, the property of being in the resolvent set is preserved under any sufficiently small bounded perturbation.

Proposition 11.1. *Let T, S be closed linear operators on* $\mathbf{H}$ *such that* $0 \in \rho(T)$ *and* $S \in \mathcal{B}(\mathbf{H})$ *with* $\|S\|\,\|T^{-1}\| < 1$. *Then* $0 \in \rho(T+S)$.

Proof. Since $\|T^{-1}S\| \le \|T^{-1}\|\,\|S\| < 1$ it follows from Theorem 9.4 that $(I + T^{-1}S)^{-1} \in \mathcal{B}(\mathbf{H})$. If we now set $A = (I + T^{-1}S)^{-1}T^{-1}$ then $A \in \mathcal{B}(\mathbf{H})$ also, and

$$A(T+S) = (I+T^{-1}S)^{-1}T^{-1}(T+S) = (I+T^{-1}S)^{-1}(I+T^{-1}S) = I \tag{11.1.4}$$

Similarly $(T+S)A = I$, so $(T+S)$ has a bounded inverse, as needed. □

We may now immediately obtain the properties of resolvent set and spectrum mentioned above.

Theorem 11.1. *If T is a closed linear operator on* $\mathbf{H}$ *then $\rho(T)$ is open and $\sigma(T)$ is closed in* $\mathbb{C}$. *In addition if* $T \in \mathcal{B}(\mathbf{H})$ *and* $\lambda \in \sigma(T)$ *then* $|\lambda| \le \|T\|$, *so that $\sigma(T)$ is compact.*

Proof. Let $\lambda \in \rho(T)$ so $(\lambda I - T)^{-1} \in \mathcal{B}(\mathbf{H})$. If $|\epsilon| < 1/\|(\lambda I - T)^{-1}\|$ we can apply Proposition 11.1 with T replaced by $\lambda I - T$ and $S = \epsilon I$ to get that $0 \in \rho((\lambda + \epsilon)I - T)$, or equivalently $\lambda + \epsilon \in \rho(T)$ for all sufficiently small $|\epsilon|$. When $T \in \mathcal{B}(\mathbf{H})$, the conclusion that $\sigma(T)$ is contained in the closed disk centered at the origin of radius $\|T\|$ is part of the statement of Theorem 9.4. □

Definition 11.3. The spectral radius of T is

$$r(T) = \sup\{|\lambda| \colon \lambda \in \sigma(T)\} \tag{11.1.5}$$

That is to say, $r(T)$ is the radius of the smallest disk in the complex plane centered at the origin and containing the spectrum of T. By the previous theorem we have always $r(T) \le \|T\|$. This inequality can be strict, even in the case that $\mathbf{H} = \mathbb{C}^2$, as may be seen in the example

$$Tu = \begin{bmatrix} 0 & 1 \\ 0 & 0 \end{bmatrix} \tag{11.1.6}$$

for which $r(T) = 0$ but $\|T\| = 1$. We do, however, have the following theorem, generalizing the well-known spectral radius formula from matrix theory.

Theorem 11.2. *If* $T \in \mathcal{B}(\mathbf{H})$ *then* $r(T) = \lim_{n\to\infty} \|T^n\|^{1/n}$.

We will not prove this here, but see for example Proposition 9.7 of [18] or Theorem 10.13 of [33].

It is a natural question to ask whether it is possible that either of $\rho(T), \sigma(T)$ can be empty. In fact both can happen, see Exercise 11.8. However, in the case of a bounded operator T we already know that the resolvent contains all λ for which $|\lambda| > \|T\|$, and it turns out that the spectrum is nonempty in this case also.

Theorem 11.3. *If* $T \in \mathcal{B}(\mathbf{H})$ *then* $\sigma(T) \neq \emptyset$.

Proof. Let $x, y \in \mathbf{H}$ and define

$$f(\lambda) = \langle x, R_\lambda y \rangle \tag{11.1.7}$$

If $\sigma(T) = \emptyset$ then f is defined for all $\lambda \in \mathbb{C}$, and is differentiable with respect to the complex variable λ, so that f is an entire function. On the other hand, for $|\lambda| > \|T\|$ we have by Eq. (9.8.70) that

$$\|R_\lambda\| \leq \frac{1}{|\lambda| - \|T\|} \to 0 \text{ as } |\lambda| \to \infty \tag{11.1.8}$$

Thus by Liouville's theorem $f(\lambda) \equiv 0$. Since x is arbitrary we must have $R_\lambda y = 0$ for any $y \in \mathbf{H}$ which is clearly false. □

11.2 EXAMPLES OF OPERATORS AND THEIR SPECTRA

The purpose of introducing the concepts of resolvent and spectrum is to provide a systematic way of analyzing the solvability properties for operator equations of the form $\lambda u - Tu = f$. Even if we are actually only interested in the case when $\lambda = 0$ (or some other fixed value) it is somehow still revealing to study the whole family of problems, as λ varies over $\mathbb{C}$. In this section we will look in detail at some examples.

Example 11.2. If $\mathbf{H} = \mathbb{C}^N$ and $Tu = Au$ for some $N \times N$ matrix A, then by previous discussion we have

$$\sigma(T) = \sigma_p(T) = \{\lambda_1, \ldots, \lambda_m\} \tag{11.2.9}$$

for some $1 \leq m \leq N$, where $\lambda_1, \ldots, \lambda_m$ are the distinct eigenvalues of A. Each eigenspace $N(\lambda_j I - T)$ has dimension equal to the geometric multiplicity of λ_j and the sum of these dimensions is also some integer between 1 and N. □

Example 11.3. Let $\Omega \subset \mathbb{R}^N$ be a bounded open set, $\mathbf{H} = L^2(\Omega)$ and let T be the multiplication operator $Tu(x) = a(x)u(x)$ for some $a \in C(\overline{\Omega})$. If we begin by looking for eigenvalues of T then we seek nontrivial solutions of $Tu = \lambda u$, that is to say

$$(\lambda - a(x))u(x) = 0 \tag{11.2.10}$$

If $a(x) \neq \lambda$ a.e. then $u \equiv 0$ is the only solution, so $\lambda \notin \sigma_p(T)$.

It is useful here to introduce a notation for the level sets of m,

$$E_\lambda = \{x \in \Omega: a(x) = \lambda\} \tag{11.2.11}$$

If E_λ has positive measure for some λ, and $\Sigma \subset E_\lambda$ is of positive, finite measure, then the characteristic function $u(x) = \chi_\Sigma(x)$ is an eigenfunction for the eigenvalue λ. In fact so is any other L^2 function whose support lies within E_λ, and thus the corresponding eigenspace is infinite dimensional. We therefore have

$$\sigma_p(T) = \{\lambda \in \mathbb{C}: m(E_\lambda) > 0\} \tag{11.2.12}$$

Note that $\sigma_p(T)$ is at most countably infinite, since for example $A_n = \{\lambda \in \mathbb{C}: m(E_\lambda) > \frac{1}{n}\}$ is at most countable for every n and $\sigma_p(T) = \cup_{n=1}^\infty A_n$.

Now let us consider the other parts of the spectrum. Consider the equation $\lambda u - Tu = f$ whose only possible solution is $u(x) = f(x)/(\lambda - a(x))$. For $\lambda \notin \sigma_p(T)$, $u(x)$ is well defined a.e., but it does not necessarily follow that $u \in L^2(\Omega)$ even if f is. If $\lambda \notin \overline{R(a)}$ (here $R(a)$ is the ordinary range of the function a) then there exists $\delta > 0$ such that $|a(x) - \lambda| \geq \delta$ for all $x \in \Omega$, from which it follows that $u = (\lambda I - T)^{-1}f$ exists in $L^2(\Omega)$ and satisfies $|u(x)| \leq \delta^{-1}|f(x)|$. Thus $\|(\lambda I - T)^{-1}\| \leq \delta^{-1}$ and so $\lambda \in \rho(T)$.

If, on the other hand $\lambda \in \overline{R(a)}$ it is always possible to find $f \in L^2(\Omega)$ such that $u(x) = f(x)/(\lambda - a(x))$ is not in $L^2(\Omega)$. This means in particular that $\lambda I - T$ is not onto, that is, λ is either in the continuous or residual spectrum. In fact it is not hard to verify that the range of $\lambda I - T$ must be dense in this case. To see this, suppose $\lambda \in \sigma(T)\backslash\sigma_p(T)$ so that $m(E_\lambda) = 0$. Then for any n there must exist an open set $\mathcal{O}_n$ containing E_λ such that $m(\mathcal{O}_n) < \frac{1}{n}$. For any function $f \in L^2(\Omega)$ let $\mathcal{U}_n = \Omega\backslash\mathcal{O}_n$ and $f_n = f\chi_{\mathcal{U}_n}$. Then $f_n \in R(\lambda I - T)$ since $\lambda - a(x)$ will be bounded away from zero on $\mathcal{U}_n$, and $f_n \to f$ in $L^2(\Omega)$ as needed.

To summarize, we have the following conclusions about the spectrum of T:

- $\rho(T) = \{\lambda \in \mathbb{C}: \lambda \notin \overline{R(a)}\}$
- $\sigma_p(T) = \{\lambda \in \overline{R(a)}: m(E_\lambda) > 0\}$
- $\sigma_c(T) = \{\lambda \in \overline{R(a)}: m(E_\lambda) = 0\}$
- $\sigma_r(T) = \emptyset$

□

Example 11.4. Next we consider the Volterra type integral operator $Tu(x) = \int_0^x u(s)\, ds$ on $\mathbf{H} = L^2(0, 1)$. We first observe that any $\lambda \neq 0$ is in the resolvent set of T. To see this, consider the problem of solving $(\lambda I - T)u = f$, that is

$$\lambda u(x) - \int_0^x u(s)\, ds = f(x) \quad 0 < x < 1 \tag{11.2.13}$$

with $f \in L^2(0,1)$. This is precisely Eq. (1.2.28) whose unique solution is given in Eq. (1.2.31) and which is well defined for any $f \in L^2(0,1)$. Thus any nonzero λ is in $\rho(T)$. By Theorem 11.3 we can immediately conclude that 0 must be in $\sigma(T)$. It is clear that $\lambda = 0$ cannot be an eigenvalue, since $\int_0^x u(s)\,ds = 0$ implies $u(x) = 0$ a.e., by the Fundamental Theorem of Calculus. On the other hand $R(T)$ is dense, since for example it contains $\mathcal{D}(0,1)$. One could also verify directly that T^{-1} is unbounded. We conclude then that

$$\sigma(T) = \sigma_c(T) = \{0\} \tag{11.2.14}$$

□

In the next example we see a typical way that residual spectrum appears.

Example 11.5. Let $\mathbf{H} = \ell^2$ and $T = S_+$ the right shift operator introduced in Eq. (9.2.36). As usual we first look for eigenvalues. The equation $Tx = \lambda x$ is satisfied precisely if $\lambda x_1 = 0$ and $\lambda x_{n+1} = x_n$ for $n = 1, 2, \ldots$. Thus if $\lambda \neq 0$ we immediately conclude that $x = 0$. If $Tx = 0$ we also see by direct inspection that $x = 0$, thus the point spectrum is empty. Since T is a bounded operator of norm 1, we also know that if $|\lambda| > 1$ then $\lambda \in \rho(T)$. Since $R(T) \subset \{x \in \ell^2\colon x_1 = 0\}$ it follows that $R(T)$ is not dense in ℓ^2, and since we already know 0 is not an eigenvalue it must be that $0 \in \sigma_r(T)$. See Exercise 11.6 for classification of the remaining λ values. □

Finally we consider an example of an unbounded operator.

Example 11.6. Let $\mathbf{H} = L^2(0,1)$ and $Tu = -u''$ on the domain

$$D(T) = \{u \in H^2(0,1)\colon u(0) = u(1) = 0\} \tag{11.2.15}$$

The equation $\lambda u - Tu = 0$ is equivalent to the ODE boundary value problem

$$u'' + \lambda u = 0 \quad 0 < x < 1 \qquad u(0) = u(1) = 0 \tag{11.2.16}$$

which was already discussed in Chapter 1, see Eq. (1.3.88). We found that a nontrivial solution $u_n(x) = \sin n\pi x$ exists for $\lambda = \lambda_n = (n\pi)^2$ and there are no other eigenvalues. Notice that the spectrum is unbounded here, as typically happens for unbounded operators.

We claim that all other $\lambda \in \mathbb{C}$ are in the resolvent set of T. To see this, we begin by representing the general solution of $u'' + \lambda u = f$ for $f \in L^2(0,1)$ as

$$u(x) = C_1 \sin\sqrt{\lambda}x + C_2 \cos\sqrt{\lambda}x + \frac{1}{\sqrt{\lambda}} \int_0^x \sin\sqrt{\lambda}(x-y) f(y)\,dy \tag{11.2.17}$$

which may be derived from the usual variation of parameters method.[2] To satisfy the boundary conditions $u(0) = u(1) = 0$ we must have $C_2 = 0$ and

2. It is correct for all complex $\lambda \neq 0$, taking $\sqrt{\lambda}$ to denote the principal branch of the square root function. We leave the remaining case $\lambda = 0$ as an exercise.

$$C_1 \sin \sqrt{\lambda} + \frac{1}{\sqrt{\lambda}} \int_0^1 \sin \sqrt{\lambda}(1-y) f(y)\, dy = 0 \tag{11.2.18}$$

which uniquely determines C_1 as long as $\lambda \neq (n\pi)^2$. Using this expression for C_1 we obtain a formula for $u = (\lambda I - T)^{-1} f$ of the form

$$u(x) = \int_0^1 G_\lambda(x,y) f(y)\, dy \tag{11.2.19}$$

with a bounded kernel $G_\lambda(x,y)$. By previous discussion we know that such an integral operator is bounded on $L^2(0,1)$ and so $\lambda \in \rho(T)$. □

11.3 PROPERTIES OF SPECTRA

In this section we will note some relationships that exist between the spectrum of an operator and the spectrum of its adjoint, and also observe that if an operator T belongs to some special class, then its spectrum will often have some corresponding special properties.

Theorem 11.4. *Let T be a closed, densely defined operator.*

1. *If $\lambda \in \rho(T)$ then $\overline{\lambda} \in \rho(T^*)$.*
2. *If $\lambda \in \sigma_r(T)$ then $\overline{\lambda} \in \sigma_p(T^*)$.*
3. *If $\lambda \in \sigma_p(T)$ then $\overline{\lambda} \in \sigma_r(T^*) \cup \sigma_p(T^*)$.*

Proof. If $\lambda \in \rho(T)$ then

$$N(\overline{\lambda} I - T^*) = N((\lambda I - T)^*) = R(\lambda I - T)^\perp = \{0\} \tag{11.3.20}$$

where Theorem 10.3 is used for the second equality. In particular $\overline{\lambda} I - T^*$ is invertible. Also

$$\overline{R(\overline{\lambda} I - T^*)} = N(\lambda I - T)^\perp = \{0\}^\perp = \mathbf{H} \tag{11.3.21}$$

so that $(\overline{\lambda} I - T^*)^{-1}$ is densely defined. Proposition 10.6 is then applicable so that

$$(\overline{\lambda} I - T^*)^{-1} = ((\lambda I - T)^*)^{-1} = ((\lambda I - T)^{-1})^* \in \mathcal{B}(\mathbf{H}) \tag{11.3.22}$$

Therefore $\overline{\lambda} \in \rho(T^*)$.

Next, if $\lambda \in \sigma_r(T)$ then $R(\lambda I - T) = M$ for some subspace M whose closure is not all of $\mathbf{H}$. Thus

$$N(\overline{\lambda} I - T^*) = R(\lambda I - T)^\perp = M^\perp = \overline{M}^\perp \neq \{0\} \tag{11.3.23}$$

and so $\overline{\lambda} \in \sigma_p(T^*)$.

Finally, if $\lambda \in \sigma_p(T)$ then

$$\overline{R(\overline{\lambda} I - T^*)} = N(\lambda I - T)^\perp \neq \mathbf{H} \tag{11.3.24}$$

so $\overline{\lambda} \notin \sigma_c(T^*)$, as needed. □

Next we turn to some special properties of self-adjoint and unitary operators.

Theorem 11.5. *Suppose that T is a densely defined operator with $T^* = T$. We then have*

1. $\sigma(T) \subset \mathbb{R}$.
2. $\sigma_r(T) = \emptyset$.
3. *If* $\lambda_1, \lambda_2 \in \sigma_p(T)$, $\lambda_1 \neq \lambda_2$ *then* $N(\lambda_1 I - T) \perp N(\lambda_2 I - T)$.

Proof. To prove the first statement, let $\lambda = \xi + i\eta$ with $\eta \neq 0$. Then

$$\|\lambda u - Tu\|^2 = \langle \xi u + i\eta u - Tu, \xi u + i\eta u - Tu\rangle = \|\xi u - Tu\|^2 + |\eta|^2\|u\|^2 \quad (11.3.25)$$

since $\langle \xi u - Tu, i\eta u\rangle + \langle i\eta u, \xi u - Tu\rangle = 0$. In particular

$$\|\lambda u - Tu\| \geq |\eta|\|u\| \quad (11.3.26)$$

so $\lambda I - T$ is one to one, that is, $\lambda \notin \sigma_p(T)$. Likewise $\lambda \notin \sigma_r(T)$ since otherwise, by Theorem 11.4 we would have $\bar{\lambda} \in \sigma_p(T^*) = \sigma_p(T)$ which is impossible by the same argument. Thus if $\lambda \in \sigma(T)$ then it can only be in the continuous spectrum so $R(\lambda I - T)$ is dense in $\mathbf{H}$. But Eq. (11.3.26) with $\eta \neq 0$ also implies that $R(\lambda I - T)$ is closed and that $\|(\lambda I - T)^{-1}\| \leq 1/|\eta|$. Thus $\lambda \in \rho(T)$.

Next, if $\lambda \in \sigma_r(T)$ then $\bar{\lambda} \in \sigma_p(T^*) = \sigma_p(T)$ by Theorem 11.4. But λ must be real by the first part of this proof, so $\lambda \in \sigma_p(T) \cap \sigma_r(T)$, which is impossible.

Finally, if λ_1, λ_2 are distinct eigenvalues, pick u_1, u_2 such that $Tu_1 = \lambda_1 u_1$ and $Tu_2 = \lambda_2 u_2$. There follows

$$\lambda_1\langle u_1, u_2\rangle = \langle \lambda_1 u_1, u_2\rangle = \langle Tu_1, u_2\rangle = \langle u_1, Tu_2\rangle = \langle u_1, \lambda_2 u_2\rangle = \bar{\lambda}_2\langle u_1, u_2\rangle \quad (11.3.27)$$

Since λ_1, λ_2 must be real we see that $(\lambda_1 - \lambda_2)\langle u_1, u_2\rangle = 0$ so $u_1 \perp u_2$ as needed. □

Theorem 11.6. *If T is a unitary operator then $\sigma_r(T) = \emptyset$ and $\sigma(T) \subset \{\lambda \colon |\lambda| = 1\}$.*

Proof. Recall that $\|Tu\| = \|u\|$ for all u when T is unitary. Thus if $Tu = \lambda u$ we then have

$$\|u\| = \|Tu\| = \|\lambda u\| = |\lambda|\|u\| \quad (11.3.28)$$

so $|\lambda| = 1$ must hold for any $\lambda \in \sigma_p(T)$. If $\lambda \in \sigma_r(T)$ then $\bar{\lambda} \in \sigma_p(T^*)$ by Theorem 11.4. Since T^* is also unitary we get $|\lambda| = |\bar{\lambda}| = 1$. Also $T^*u = \bar{\lambda}u$ implies that $u = TT^*u = \bar{\lambda}Tu$ so that $\lambda = 1/\bar{\lambda} \in \sigma_p(T)$, which is a contradiction to the assumption that $\lambda \in \sigma_r(T)$. Thus the residual spectrum of T is empty.

To complete the proof, first note that since $\|T\| = 1$ we must have $|\lambda| \leq 1$ if $\lambda \in \sigma(T)$ by Theorem 9.4. If $|\lambda| < 1$ then $(I - \lambda T^*)^{-1} \in \mathcal{B}(\mathbf{H})$ by the same theorem, and for any $f \in \mathbf{H}$ we can obtain a solution of $\lambda u - Tu = f$ by setting $u = -T^*(I - \lambda T^*)^{-1}f$. Since we already know $\lambda \notin \sigma_p(T)$ it follows that $\lambda I - T$ is one-to-one and onto, and $\|(\lambda I - T)^{-1}\| = \|T^*(I - \lambda T^*)^{-1}\|$ which is finite, and so $\lambda \in \rho(T)$, as needed. □

Example 11.7. Let $T = \mathcal{F}$, the Fourier transform on $\mathbf{H} = L^2(\mathbb{R}^N)$, as defined in Eq. (7.4.52), which we have already established is unitary, see Eq. (9.5.64). From the inversion formula for the Fourier transform it is immediate that $\mathcal{F}^4 = I$. If $\mathcal{F}u = \lambda u$ we would also have $u = \mathcal{F}^4 u = \lambda^4 u$ so that any eigenvalue λ of $\mathcal{F}$ satisfies $\lambda^4 = 1$, that is, $\sigma_p(\mathcal{F}) \subset \{\pm 1, \pm i\}$. We already know that $\lambda = 1$ must be an eigenvalue with a Gaussian $e^{-\frac{|x|^2}{2}}$ as a corresponding eigenfunction. In fact all four values $\pm 1, \pm i$ are eigenvalues with infinite dimensional eigenspaces spanned by products of Gaussians and so-called Hermite polynomials. See Section 2.5 of [10] for more details. In Exercise 11.7 you are asked to show that all other values of λ are in the resolvent set of $\mathcal{F}$. □

Example 11.8. The Hilbert transform $\mathcal{H}$ introduced in Example 9.7 is also unitary on $\mathbf{H} = L^2(\mathbb{R})$. Since also $\mathcal{H}^2 = -I$ it follows that the only possible eigenvalues of $\mathcal{H}$ are $\pm i$. It is readily checked that these are both eigenvalues with the eigenspaces for $\lambda = \pm i$ being the subspaces $M_\pm$ defined in Exercise 9.5. Let us check that any $\lambda \neq \pm i$ is in the resolvent set. If $\lambda u - \mathcal{H}u = f$ then applying $\mathcal{H}$ to both sides we get $\lambda \mathcal{H}u + u = \mathcal{H}f$. Eliminating $\mathcal{H}u$ between these two equations we can solve for

$$u = \frac{\lambda f + \mathcal{H}f}{\lambda^2 + 1} \tag{11.3.29}$$

Conversely by direct substitution we can verify that this formula defines a solution of $\lambda u - \mathcal{H}u = f$, so that $(\lambda I - \mathcal{H})^{-1} = \frac{\lambda I + \mathcal{H}}{\lambda^2 + 1}$ which is obviously bounded for $\lambda \neq \pm i$. □

Finally we discuss an important example of an unbounded operator.

Example 11.9. Let $\mathbf{H} = L^2(\mathbb{R}^N)$ and $Tu = -\Delta u$ on $D(T) = H^2(\mathbb{R}^N)$. If we apply the Fourier transform, then for $f, u \in \mathbf{H}$ the resolvent equation $\lambda u - Tu = f$ is seen to be equivalent to

$$(\lambda - |y|^2)\widehat{u}(y) = \widehat{f}(y) \tag{11.3.30}$$

If $\lambda \in \mathbb{C}\backslash[0, \infty)$ it is straightforward to check that $\widehat{u}(y) = \widehat{f}(y)/(\lambda - |y|^2)$ defines a unique $u \in H^2(\mathbb{R}^N)$ which is a solution of the resolvent equation (it may be convenient here to use the characterization (Eq. 8.6.68) of $H^2(\mathbb{R}^N)$). It is also immediate from Eq. (11.3.30) that $\sigma_p(T) = \emptyset$. On the other hand a solution $\widehat{u}$, and hence u, exists in $\mathbf{H}$ as long as $\widehat{f}$ vanishes in a neighborhood of $|y| = \sqrt{\lambda}$. Such f form a dense subset of $\mathbf{H}$ so $\sigma_r(T) = \emptyset$ also. This could also be shown by verifying that T is self-adjoint. Finally, it is clear that for $\lambda > 0$ there exists a function u such that $\widehat{u} \notin L^2(\mathbb{R}^N)$ but $g := (\lambda - |y|^2)\widehat{u} \in L^2(\mathbb{R}^N)$. If $f \in L^2(\mathbb{R}^N)$ is defined by $\widehat{f} = g$ then it follows that f is not in the range of $\lambda I - T$, so $\lambda \in \sigma_c(T)$ must hold. In summary, $\sigma(T) = \sigma_c(T) = [0, \infty)$. □

11.4 EXERCISES

11.1. Let T be the integral operator

$$Tu(x) = \int_0^1 (x+y)u(y)\,dy$$

on $L^2(0,1)$. Find $\sigma_p(T), \sigma_c(T)$, and $\sigma_r(T)$ and the multiplicity of each eigenvalue.

11.2. Let M be a closed subspace of a Hilbert space $\mathbf{H}$, $M \neq \{0\}, \mathbf{H}$ and let P_M be the usual orthogonal projection onto M. Show that if $\lambda \neq 0, 1$ then $\lambda \in \rho(P_M)$ and

$$\|(\lambda I - P_M)^{-1}\| \le \frac{1}{|\lambda|} + \frac{1}{|1-\lambda|}$$

11.3. Recall that the resolvent operator of T is defined to be $R_\lambda = (\lambda I - T)^{-1}$ for $\lambda \in \rho(T)$.

(a) Prove the resolvent identity (11.1.3).
(b) Deduce from this that R_λ, R_μ commute.
(c) Show also that T, R_λ commute for $\lambda \in \rho(T)$.

11.4. Show that $\lambda \to R_\lambda$ is a continuously differentiable, regarded as a mapping from $\rho(T) \subset \mathbb{C}$ into $\mathcal{B}(\mathbf{H})$, with

$$\frac{dR_\lambda}{d\lambda} = -R_\lambda^2$$

11.5. If in Definition 11.1 of resolvent and spectrum we do not require that T be closed, show that $\rho(T) = \emptyset$ for any nonclosed linear operator T.

11.6. Let T denote the right shift operator on ℓ^2. Show that

(a) $\sigma_p(T) = \emptyset$
(b) $\sigma_c(T) = \{\lambda \colon |\lambda| = 1\}$
(c) $\sigma_r(T) = \{\lambda \colon |\lambda| < 1\}$

11.7. If $\lambda \neq \pm 1, \pm i$ show that λ is in the resolvent set of the Fourier transform $\mathcal{F}$. (Suggestion: Assuming that a solution of $\mathcal{F}u - \lambda u = f$ exists, derive an explicit formula for it by justifying and using the identity

$$\mathcal{F}^4 u = \lambda^4 u + \lambda^3 f + \lambda^2 \mathcal{F} f + \lambda \mathcal{F}^2 f + \mathcal{F}^3 f$$

together with the fact that $\mathcal{F}^4 = I$.)

11.8. Let $\mathbf{H} = L^2(0,1)$, $T_1 u = T_2 u = T_3 u = u'$ on the domains

$$D(T_1) = H^1(0,1)$$
$$D(T_2) = \{u \in H^1(0,1) \colon u(0) = 0\}$$
$$D(T_3) = \{u \in H^1(0,1) \colon u(0) = u(1) = 0\}$$

Show that

(a) $\sigma(T_1) = \sigma_p(T_1) = \mathbb{C}$
(b) $\sigma(T_2) = \emptyset$
(c) $\sigma(T_3) = \sigma_r(T_3) = \mathbb{C}$

11.9. Define the translation operator $Tu(x) = u(x-1)$ on $L^2(\mathbb{R})$.
(a) Find T^*.
(b) Show that T is unitary.
(c) Show that $\sigma(T) = \sigma_c(T) = \{\lambda \in \mathbb{C}: |\lambda| = 1\}$.

11.10. Let $Tu(x) = \int_0^x K(x,y)u(y)\,dy$ be a Volterra integral operator on $L^2(0,1)$ with a bounded kernel, $|K(x,y)| \le M$. Show that $\sigma(T) = \{0\}$. (There are several ways to show that T has no nonzero eigenvalues. Here is one approach: Define the equivalent norm on $L^2(0,1)$

$$\|u\|_\theta^2 = \int_0^1 |u(x)|^2 e^{-2\theta x}\,dx$$

and show that the supremum of $\frac{\|Tu\|_\theta}{\|u\|_\theta}$ can be made arbitrarily small by choosing θ sufficiently large.)

11.11. If T is a symmetric operator, show that

$$\sigma_p(T) \cup \sigma_c(T) \subset \mathbb{R}$$

(It is almost the same as showing that $\sigma(T) \subset \mathbb{R}$ for a self-adjoint operator.)

11.12. The *approximate spectrum* $\sigma_a(T)$ of a linear operator T is the set of all $\lambda \in \mathbb{C}$ such that there exists a sequence $\{u_n\}$ in $\mathbf{H}$ such that $\|u_n\| = 1$ for all n and $\|Tu_n - \lambda u_n\| \to 0$ as $n \to \infty$. Show that

$$\sigma_p(T) \cup \sigma_c(T) \subset \sigma_a(T) \subset \sigma(T)$$

(so that $\sigma_a(T) = \sigma(T)$ in the case of a self-adjoint operator) Show by example that $\sigma_r(T)$ need not be contained in $\sigma_a(T)$.

11.13. The *essential spectrum* $\sigma_e(T)$ of a linear operator T is the set of all $\lambda \in \mathbb{C}$ such that $\lambda I - T$ is not a Fredholm operator[3] (recall Definition 9.5). Show that $\sigma_e(T) \subset \sigma(T)$. Characterize the essential spectrum for the following operators: (i) a linear operator on $\mathbb{C}^n$, (ii) an orthogonal projection on a Hilbert space, (iii) the Fourier transform on $L^2(\mathbb{R}^N)$, and (iv) a multiplication operator on $L^2(\Omega)$.

11.14. If T is a bounded, self-adjoint operator on a Hilbert space $\mathbf{H}$, show that $\langle Tu, u\rangle \ge 0$ for all $u \in \mathbf{H}$ if and only if $\sigma(T) \subset [0,\infty)$.

3. Actually there are several nonequivalent definitions of essential spectrum which can be found in the literature. We are using one of the common ones.

Chapter 12

Compact Operators

12.1 COMPACT OPERATORS

One type of operator, which has not yet been mentioned much in connection with spectral theory, is integral operators. This is because they typically belong to a particular class of operators, known as compact operators, for which there is a well-developed special theory, and whose main points will be presented in this chapter.

If $\mathbf{X}$ is a Banach space, then as usual $K \subset \mathbf{X}$ is compact if any open cover of K has a finite subcover. Equivalently any infinite bounded sequence in K has a subsequence convergent to an element of K. If $\dim(\mathbf{X}) < \infty$ then K is compact if and only if it is closed and bounded, but this is false if $\dim(\mathbf{X}) = \infty$.

Example 12.1. Let $\mathbf{H}$ be an infinite dimensional Hilbert space and $K = \{u \in \mathbf{H}\colon \|u\| \le 1\}$, which is obviously closed and bounded. If we let $\{e_n\}_{n=1}^{\infty}$ be an infinite orthonormal sequence (which we know must exist), there cannot be any convergent subsequence since $\|e_n - e_m\| = \sqrt{2}$ for any $n \ne m$. Thus K is not compact. □

Recall also that $E \subset \mathbf{X}$ is precompact, or relatively compact, if $\overline{E}$ is compact.

Definition 12.1. If $\mathbf{X}, \mathbf{Y}$ are Banach spaces then a linear operator $T\colon \mathbf{X} \to \mathbf{Y}$ is compact if for any bounded set $E \subset \mathbf{X}$ the image $T(E)$ is precompact in $\mathbf{Y}$.

This definition makes sense even if T is nonlinear, but in this book the terminology will only be used in the linear case. We will use the notation $\mathcal{K}(\mathbf{X}, \mathbf{Y})$ to denote the set of compact linear operators from $\mathbf{X}$ to $\mathbf{Y}$ and $\mathcal{K}(\mathbf{X})$ if $\mathbf{Y} = \mathbf{X}$.

Proposition 12.1. *If $\mathbf{X}, \mathbf{Y}$ are Banach spaces then*

1. *$\mathcal{K}(\mathbf{X}, \mathbf{Y})$ is a subspace of $\mathcal{B}(\mathbf{X}, \mathbf{Y})$.*
2. *If $T \in \mathcal{B}(\mathbf{X}, \mathbf{Y})$ and $\dim(R(T)) < \infty$ then $T \in \mathcal{K}(\mathbf{X}, \mathbf{Y})$.*
3. *The identity map I belongs to $\mathcal{K}(\mathbf{X})$ if and only if $\dim(\mathbf{X}) < \infty$.*

Techniques of Functional Analysis for Differential and Integral Equations
http://dx.doi.org/10.1016/B978-0-12-811426-1.00012-X

Proof. If T is compact then $\overline{T(B(0,1))}$ is compact in $\mathbf{Y}$ and in particular is bounded in $\mathbf{Y}$. Thus there exists $M < \infty$ such that $\|Tu\| \le M$ if $\|u\| \le 1$, which means $\|T\| \le M$. It is straightforward to check that a linear combination of compact operators is also compact, hence $\mathcal{K}(\mathbf{X},\mathbf{Y})$ is a vector subspace of $\mathcal{B}(\mathbf{X},\mathbf{Y})$.

If $E \subset \mathbf{X}$ is bounded and $T \in \mathcal{B}(\mathbf{X},\mathbf{Y})$ then $T(E)$ is bounded in $\mathbf{Y}$. Therefore under the assumption of 2., $T(E)$ is a bounded subset of the finite dimensional set $R(T)$, so is relatively compact by the Heine-Borel theorem. This proves 2. and the "if" part of 3. The other half of 3. is equivalent to the statement that the unit ball $B(0,1)$ is not compact if $\dim(\mathbf{X}) = \infty$. This was shown in Example 12.1 in the Hilbert space case, and we refer to Theorem 6.5 of [5] for the general case of a Banach space. □

Recall that when $\dim(R(T)) < \infty$ we say that T is of finite rank. Any degenerate integral operator $Tu(x) = \int_\Omega K(x,y)u(y)\,dy$ with $K(x,y) = \sum_{j=1}^n \phi_j(x)\psi_j(y)$, $\phi_j, \psi_j \in L^2(\Omega)$ for $j = 1,\dots,n$, is therefore of finite rank, and so in particular is compact.

A convenient alternate characterization of compact operators involves the notion of *weak convergence*. Although the following discussion can mostly be carried out in a Banach space setting, we will consider only the Hilbert space case.

Definition 12.2. If $\mathbf{H}$ is a Hilbert space and $\{u_n\}_{n=1}^\infty$ is an infinite sequence in $\mathbf{H}$, we say u_n converges weakly to u in $\mathbf{H}$ ($u_n \overset{w}{\rightarrow} u$), provided that $\langle u_n, v\rangle \rightarrow \langle u, v\rangle$ for every $v \in \mathbf{H}$.

Note by the Riesz Representation Theorem that this is the same as requiring $\ell(u_n) \rightarrow \ell(u)$ for every $\ell \in H^*$—this is the definition to use when generalizing to the Banach space situation. The weak limit, if it exists, is unique, see Exercise 12.3.

In case it is necessary to emphasize the difference between weak convergence and the ordinary notion of convergence in $\mathbf{H}$ we may refer to the latter as *strong convergence*. It is elementary to show that strong convergence always implies weak convergence, but the converse is false, as the following example shows.

Example 12.2. Assume that $\mathbf{H}$ is infinite dimensional and let $\{e_n\}_{n=1}^\infty$ be any orthonormal set in $\mathbf{H}$. From Bessel's inequality we have

$$\sum_{n=1}^\infty |\langle e_n, v\rangle|^2 \le \|v\|^2 < \infty \quad \text{for all } v \in \mathbf{H} \tag{12.1.1}$$

which implies in particular that $\langle e_n, v\rangle \rightarrow 0$ for every $v \in \mathbf{H}$. This means $e_n \overset{w}{\rightarrow} 0$, even though we know it is not strongly convergent, by Example 12.1. □

To make the connection to properties of compact operators, let $\{e_n\}_{n=1}^\infty$ again denote an infinite orthonormal set in an infinite dimensional Hilbert space $\mathbf{H}$ and suppose T is compact on $\mathbf{H}$. If $u_n = Te_n$ then $\{u_n\}_{n=1}^\infty$ is evidently relatively compact in $\mathbf{H}$ so we can find a convergent subsequence $u_{n_k} \to u$. For any $v \in \mathbf{H}$ we then have

$$\langle u_{n_k}, v\rangle = \langle Te_{n_k}, v\rangle = \langle e_{n_k}, T^* v\rangle \to 0 \tag{12.1.2}$$

so that $u_{n_k} = Te_{n_k} \overset{w}{\to} 0$. But since also $u_{n_k} \to u$ we must have $u_{n_k} \to 0$. Since the original sequence could be replaced by any of its subsequences we conclude that for any subsequence e_{n_k} there must exist a further subsequence $e_{n_{k_j}}$ such that $Te_{n_{k_j}} \to 0$. We now claim that $u_n \to 0$, that is, the entire sequence converges, not just the subsequence. If not, then there must exist $\delta > 0$ and a subsequence e_{n_k} such that $\|Te_{n_k}\| \ge \delta$, which contradicts the fact just established that Te_{n_k} must have a subsequence convergent to zero. We have therefore established that any compact operator maps the weakly convergent sequence e_n to a strongly convergent sequence. We will see below that compact operators always map weakly convergent sequences to strongly convergent sequences and that this property characterizes compact operators.

Let us first present some more elementary but important facts about weak convergence in a Hilbert space.[1]

Proposition 12.2. *Let $u_n \overset{w}{\to} u$ in a Hilbert space $\mathbf{H}$. Then*

1. $\|u\| \le \liminf_{n\to\infty} \|u_n\|$.
2. *If $\|u_n\| \to \|u\|$ then $u_n \to u$.*

Proof. We have

$$0 \le \|u_n - u\|^2 = \|u_n\|^2 - 2\mathrm{Re}\langle u_n, u\rangle + \|u\|^2 \tag{12.1.3}$$

or

$$2\mathrm{Re}\langle u_n, u\rangle - \|u\|^2 \le \|u_n\|^2 \tag{12.1.4}$$

Now take the lim inf of both sides to get the conclusion of 1. If $\|u_n\| \to \|u\|$ then the right-hand identity of Eq. (12.1.3) shows that $\|u_n - u\| \to 0$. □

The property in part (a) of the proposition is often referred to as the *weak lower semicontinuity of the norm.* Note that strict inequality can occur, for example, in the case that u_n is an infinite orthonormal set.

Various familiar topological notions may be based on weak convergence.

Definition 12.3. A set $E \subset \mathbf{H}$ is weakly closed if

$$u_n \in E \quad u_n \overset{w}{\to} u \qquad \text{implies } u \in E \tag{12.1.5}$$

and E is weakly open if its complement is weakly closed. We say E is weakly compact if any infinite sequence in E has a subsequence which is weakly convergent to an element $u \in E$.

Clearly a weakly closed set is closed, but the converse is false in general.

1. See also Exercise 5.18 in Chapter 5.

Example 12.3. If $E = \{u \in \mathbf{H}\colon \|u\| = 1\}$ then E is closed but is not weakly closed, since an example where Eq. (12.1.5) fails is provided by any infinite orthonormal sequence. On the other hand, $E = \{u \in \mathbf{H}\colon \|u\| \le 1\}$ is weakly closed by Proposition 12.2. □

Several key facts relating to the weak convergence concept, which we will not prove here but will make extensive use of, are given in the next theorem.

Theorem 12.1. *Let* $\mathbf{H}$ *be a Hilbert space. Then*

1. *Any weakly convergent sequence is bounded.*
2. *Any bounded sequence has a weakly convergent subsequence.*
3. *If* $E \subset \mathbf{H}$ *is convex and closed then it is also weakly closed. In particular any closed subspace is weakly closed.*

The three parts of this theorem are all special cases of some very general results in functional analysis. The first statement is a special case of the Banach-Steinhaus theorem (or Uniform Boundedness Principle), which is more generally a theorem about sequences of bounded linear functionals on a Banach space. See Corollary 1 in Section 23 of [2] or Theorem 5.8 of [32] for the more general Banach space result. The second statement is a special case of the Banach-Alaoglu theorem, which asserts a weak compactness property of bounded sets in the dual space of any Banach space, see Theorem 1 in Section 24 of [2] or Theorem 3.15 of [33] for generalizations. The third part is a special case of Mazur's theorem, also valid in a more general Banach space setting, see Theorem 3.7 of [5].

Now let us return to the main development and prove the following very important characterization of compact linear operators.

Theorem 12.2. *Let* $T \in \mathcal{B}(\mathbf{H})$. *Then* T *is compact if and only if* T *has the property that* $u_n \xrightarrow{w} u$ *implies* $Tu_n \to Tu$.

Proof. Suppose that T is compact and $u_n \xrightarrow{w} u$. Then $\{u_n\}$ is bounded by part (a) of Theorem 12.1. The compactness of T then implies that the image sequence $\{Tu_n\}$ has convergent subsequence. Note also that $Tu_n \xrightarrow{w} Tu$ since for any $v \in \mathbf{H}$ we have

$$\langle Tu_n, v\rangle = \langle u_n, T^* v\rangle \to \langle u, T^* v\rangle = \langle Tu, v\rangle \tag{12.1.6}$$

Thus there must exist a subsequence u_{n_k} such that $Tu_{n_k} \to Tu$ strongly in $\mathbf{H}$. By the same argument, any subsequence of u_n has a further subsequence for which the image sequence converges to Tu and so $Tu_n \to Tu$.

To prove the converse, let $E \subset \mathbf{H}$ be bounded and $\{v_n\}_{n=1}^\infty \subset \overline{T(E)}$. We must then have $v_n = z_n + \epsilon_n$, where $z_n = Tu_n$ for some $u_n \in E$ and $\epsilon_n \to 0$ in $\mathbf{H}$. By the boundedness of E and part 2 of Theorem 12.1 there must exist a weakly convergent subsequence $u_{n_k} \xrightarrow{w} u$. Therefore $v_{n_k} = Tu_{n_k} + \epsilon_{n_k} \to Tu$, and it follows that $T(E)$ is relatively compact, as needed. □

The following theorem will turn out to be a key tool in developing the theory of integral equations with L^2 kernels.

Theorem 12.3. *$\mathcal{K}(\mathbf{H})$ is a closed subspace of $\mathcal{B}(\mathbf{H})$.*

Proof. We have already observed that $\mathcal{K}(\mathbf{H})$ is a subspace of $\mathcal{B}(\mathbf{H})$. To verify that it is closed, pick $T_n \in \mathcal{K}(\mathbf{H})$ such that $\|T_n - T\| \to 0$ for some $T \in \mathcal{B}(\mathbf{H})$. We are done if we show $T \in \mathcal{K}(\mathbf{H})$, and this in turn will follow if we show that for any bounded sequence $\{u_n\}$ there exists a convergent subsequence of the image sequence $\{Tu_n\}$.

Since $T_1 \in \mathcal{K}(\mathbf{H})$ there must exist a subsequence $\{u_n^1\} \subset \{u_n\}$ such that $\{T_1 u_n^1\}$ is convergent. Likewise, since $T_2 \in \mathcal{K}(\mathbf{H})$ there must exist a further subsequence $\{u_n^2\} \subset \{u_n^1\}$ such that $\{T_2 u_n^2\}$ is convergent. Continuing in this way we get $\{u_n^j\}$ such that $\{u_n^{j+1}\} \subset \{u_n^j\}$ and $\{T_j u_n^j\}$ is convergent, for any fixed j.

Now let $z_n = u_n^n$, so that $\{z_n\}_{n\ge j} \subset \{u_n^j\}$ for any j, and is obviously a subsequence of the original sequence $\{u_n\}$. We claim that $\{Tz_n\}$ is convergent, which will complete the proof.

Fix some $\epsilon > 0$. We may first choose M such that $\|u_n\| \le M$ for every n, and then some fixed j such that $\|T_j - T\| < \frac{\epsilon}{4M}$. For this choice of j we can pick N so that $\|T_j z_n - T_j z_m\| < \frac{\epsilon}{2}$ when $m, n \ge N$. We then have, for $n, m \ge N$, that

$$\begin{aligned} \|Tz_n - Tz_m\| &\le \|Tz_n - T_j z_n\| + \|T_j z_n - T_j z_m\| + \|T_j z_m - Tz_m\| \\ &\le \|T - T_j\|(\|z_n\| + \|z_m\|) + \|T_j z_n - T_j z_m\| \le \epsilon \end{aligned} \tag{12.1.7}$$

It follows that $\{Tz_n\}$ is Cauchy, hence convergent, in $\mathbf{H}$. □

Recall that an integral operator

$$Tu(x) = \int_\Omega K(x,y)u(y)\,dy \tag{12.1.8}$$

is of Hilbert-Schmidt type if $K \in L^2(\Omega \times \Omega)$, and we have earlier established that such operators are bounded on $L^2(\Omega)$. We will now show that any Hilbert-Schmidt integral operator is actually compact. The basic idea is to show that T can be approximated by finite rank operators, which we know to be compact, and then apply the previous theorem. First we need a lemma.

Lemma 12.1. *If $\{\phi_n\}_{n=1}^\infty$ is an orthonormal basis of $L^2(\Omega)$ then $\{\phi_n(x)\phi_m(y)\}_{n,m=1}^\infty$ is an orthonormal basis of $L^2(\Omega \times \Omega)$.*

Proof. By direct calculation we see that

$$\int_\Omega \int_\Omega \phi_n(x)\phi_m(y)\overline{\phi_{n'}(x)\phi_{m'}(y)}\,dxdy = \begin{cases} 1 & n = n', m = m' \\ 0 & \text{otherwise} \end{cases} \tag{12.1.9}$$

so that they are orthonormal in $L^2(\Omega \times \Omega)$. To show completeness, then by Theorem 5.4 it is enough to verify the Bessel equality. That is, we show

$$\|f\|^2_{L^2(\Omega\times\Omega)} = \sum_{n,m=1}^{\infty} |c_{n,m}|^2 \tag{12.1.10}$$

where

$$c_{n,m} = \int_\Omega\int_\Omega f(x,y)\overline{\phi_n(x)\phi_m(y)}\,dxdy \tag{12.1.11}$$

and it is enough to do this for $f \in C(\overline{\Omega\times\Omega})$.

By applying the Bessel equality in x for fixed y, and then integrating with respect to y we get

$$\int_\Omega\int_\Omega |f(x,y)|^2\,dxdy = \int_\Omega \sum_{n=1}^{\infty} |c_n(y)|^2\,dy \tag{12.1.12}$$

where $c_n(y) = \int_\Omega f(x,y)\overline{\phi_n(x)}\,dx$. Since we can exchange the sum and integral, it follows by applying the Bessel equality to $c_n(\cdot)$ that we get

$$\int_\Omega\int_\Omega |f(x,y)|^2\,dxdy = \sum_{n=1}^{\infty}\int_\Omega |c_n(y)|^2\,dy = \sum_{n=1}^{\infty}\sum_{m=1}^{\infty} |c_{n,m}|^2 \tag{12.1.13}$$

where

$$c_{n,m} = \int_\Omega c_n(y)\overline{\phi_m(y)}\,dy = \int_\Omega\int_\Omega f(x,y)\overline{\phi_n(x)\phi_m(y)}\,dxdy \tag{12.1.14}$$

as needed. □

Theorem 12.4. *If $K \in L^2(\Omega\times\Omega)$ then the integral operator* (12.1.8) *is compact on $L^2(\Omega)$.*

Proof. Let $\{\phi_n\}$ be an orthonormal basis of $L^2(\Omega)$ and set

$$K_N(x,y) = \sum_{n,m=1}^{N} c_{n,m}\phi_n(x)\phi_m(y) \tag{12.1.15}$$

with $c_{n,m}$ as above, so we know that $\|K_N - K\|_{L^2(\Omega\times\Omega)} \to 0$ as $N \to \infty$. Let T_N be the corresponding integral operator with kernel K_N, which is compact since it has finite rank. Finally since $\|T - T_N\| \le \|K_N - K\|_{L^2(\Omega\times\Omega)} \to 0$ (recall Eq. 9.2.20) it follows from Theorem 12.3 that T is compact. □

12.2 RIESZ-SCHAUDER THEORY

In this section we first establish a fundamental abstract result about the solvability of operator equations of the form $\lambda u - Tu = f$ when T is compact and $\lambda \ne 0$. We will then specialize to the case of a compact integral operator.

Theorem 12.5. *Let $T \in \mathcal{K}(\mathbf{H})$ and $\lambda \ne 0$. Then*

1. *$\lambda I - T$ is a Fredholm operator of index zero.*
2. *If $\lambda \in \sigma(T)$ then $\lambda \in \sigma_p(T)$.*

Recall that the first statement means that $N(\lambda I - T)$ and $N(\overline{\lambda} I - T^*)$ are of the same finite dimension and that $R(\lambda I - T)$ is closed. When the conclusions of the theorem are valid, it follows that

$$R(\lambda I - T) = N(\overline{\lambda} I - T^*)^\perp \tag{12.2.16}$$

and the Fredholm alternative holds:

Either

- $\lambda I - T$ and $\overline{\lambda} I - T^*$ are both one to one, and $\lambda u - Tu = f$ has a unique solution for every $f \in \mathbf{H}$, or
- $\dim N(\lambda I - T) = \dim N(\overline{\lambda} I - T^*) < \infty$ and $\lambda u - Tu = f$ has a solution if and only if $f \perp v$ for any v satisfying $T^* v = \overline{\lambda} v$.

If T is compact then so is T^* (Exercise 12.2), thus all of the same conclusions hold for T^*.

The proof of Theorem 12.5 proceeds by means of a number of intermediate steps, some of which are of independent interest. Without loss of generality we may assume $\lambda = 1$, since we could always write $\lambda I - T = \lambda(I - \lambda^{-1}T)$. For the rest of the section we denote $S = I - T$ with the assumption that $T \in \mathcal{K}(\mathbf{H})$.

Lemma 12.2. *There exists $C > 0$ such that $\|Su\| \geq C\|u\|$ for all $u \in N(S)^\perp$.*

Proof. If no such constant exists then we can find a sequence $\{u_n\}_{n=1}^\infty$ such that $u_n \in N(S)^\perp$, $\|u_n\| = 1$, and $\|Su_n\| \to 0$. By weak compactness there exists a subsequence u_{n_k} such that $u_{n_k} \overset{w}{\to} u$ for some u with $\|u\| \leq 1$. Since T is compact it follows that $Tu_{n_k} \to Tu$, so $u_{n_k} = Su_{n_k} + Tu_{n_k} \to Tu$. By uniqueness of the weak limit $Tu = u$, in other words $u \in N(S)$. On the other hand $u_n \in N(S)^\perp$ implies that $u \in N(S)^\perp$ so that $u = 0$ must hold. Finally we also have $\|u\| = 1$, since $u_{n_k} \to u$ strongly, which is a contradiction. ☐

Lemma 12.3. *$R(S)$ is closed.*

Proof. Let $v_n \in R(S)$, $v_n \to v$ and choose u_n such that $Su_n = v_n$. Let P denote the orthogonal projection onto the closed subspace $N(S)$. If $w_n = u_n - Pu_n$ then $w_n \in N(S)^\perp$ and $Sw_n = Su_n = v_n$. By the previous lemma $\|v_n - v_m\| \geq C\|w_n - w_m\|$ for some $C > 0$, so that $\{w_n\}$ must be a Cauchy sequence. Letting $w = \lim_{n\to\infty} w_n$ we then have $Sw = \lim_{n\to\infty} Sw_n = v$, so that $v \in R(S)$ as needed. ☐

Lemma 12.4. *$R(S) = \mathbf{H}$ if and only if $N(S) = \{0\}$.*

Proof. First suppose that $R(S) = \mathbf{H}$ and that there exists $u_1 \in N(S)$, $u_1 \neq 0$. There must exist $u_2 \in \mathbf{H}$ such that $Su_2 = u_1$, since we have assumed that S is onto. Similarly we can find u_p for $p = 3, 4, \ldots$ such that $Su_p = u_{p-1}$ and evidently $S^{p-1}u_p = u_1, S^p u_p = 0$. Let $N_p = N(S^p)$ so that $N_{p-1} \subset N_p$ and the inclusion is strict, since $u_p \in N_p$ but $u_p \notin N_{p-1}$. Now apply the Gram-Schmidt procedure to the sequence $\{u_p\}$ to get a sequence $\{w_p\}$ such that $w_p \in N_p$, $\|w_p\| = 1$ and $w_p \perp N_{p-1}$. We will be done if we show that $\{Tw_p\}$ has no convergent subsequence, since this will contradict the compactness of T.

Fix $p > q$, let $g = Sw_q - Sw_p - w_q$ and observe that

$$\|Tw_p - Tw_q\| = \|w_p - w_q - Sw_p + Sw_q\| = \|w_p + g\| \tag{12.2.17}$$

We must have $w_p \perp g$ since $Sw_q, Sw_p, w_q \in N_{p-1}$, therefore

$$\|Tw_p - Tw_q\|^2 = \|w_p\|^2 + \|g\|^2 \ge \|w_p\|^2 = 1 \tag{12.2.18}$$

and it follows that there can be no convergent subsequence of $\{Tw_p\}$, as needed.

To prove the converse implication, assume that $N(S) = \{0\}$ so that $\overline{R(S^*)} = N(S)^\perp = \mathbf{H}$ by Corollary 9.1. But as remarked above T^* is also compact, so by Lemma 12.3 $R(S^*)$ is closed, hence $R(S^*) = \mathbf{H}$. By the first half of this lemma $N(S^*) = \{0\}$ so that $\overline{R(S)} = \mathbf{H}$ and therefore finally $R(S) = \mathbf{H}$ by one more application of Lemma 12.3. □

Lemma 12.5. *$N(S)$ is of finite dimension.*

Proof. If not, then there exists an infinite orthonormal basis $\{e_n\}_{n=1}^\infty$ of $N(S)$, and in particular $\|Te_n\| = \|e_n\| = 1$. But since T is compact we also know that $Te_n \to 0$, a contradiction. □

Lemma 12.6. *The null spaces $N(S)$ and $N(S^*)$ are of the same finite dimension.*

Proof. Denote $m = \dim N(S)$, $m^* = \dim N(S^*)$ and suppose that $m^* > m$. Let $w_1, \dots, w_m$, $v_1, \dots, v_{m^*}$ be orthonormal bases of $N(S), N(S^*)$, respectively, and define the operator

$$Au = Su - \sum_{j=1}^{m} \langle u, w_j \rangle v_j \tag{12.2.19}$$

Since $\langle Su, v_j \rangle = 0$ for $j = 1, \dots, m^*$ it follows that

$$\langle Au, v_k \rangle = \begin{cases} -\langle u, w_k \rangle & k = 1, \dots, m \\ 0 & k = m+1, \dots, m^* \end{cases} \tag{12.2.20}$$

Next we claim that $N(A) = \{0\}$. To see this, if $Au = 0$ we would have $\langle u, w_k \rangle = 0$ for $k = 1, \dots, m$, so that $u \in N(S)^\perp$. But it would also follow that $u \in N(S)$ by Eq. (12.2.19), and so $u = 0$.

We may obviously write $A = I - \tilde{T}$ for some $\tilde{T} \in \mathcal{K}(\mathbf{H})$, so by Lemma 12.4 we may conclude that $R(A) = \mathbf{H}$. But $v_{m+1} \notin R(A)$ since if $Au = v_{m+1}$ it would follow that $1 = \|v_{m+1}\|^2 = \langle Au, v_{m+1} \rangle = 0$, a contradiction. This shows that $m^* \le m$, and since $S = S^{**}$ the same argument implies $m \le m^*$. □

Corollary 12.1. *If $0 \in \sigma(S)$ then $0 \in \sigma_p(S)$.*

Proof. If $0 \notin \sigma_p(S)$ then $N(S) = \{0\}$ so that $R(S) = \mathbf{H}$ by Lemma 12.4. But then $0 \in \rho(S)$. □

By combining the conclusions of Lemmas 12.3 and 12.6 and Corollary 12.1, we have completed the proof of Theorem 12.5. Further important information about the spectrum of a compact operator is contained in the next theorem.

Theorem 12.6. *If $T \in \mathcal{K}(\mathbf{H})$ then $\sigma(T)$ is at most countably infinite, with 0 as the only possible accumulation point.*

Proof. Since $\sigma(T)\backslash\{0\} = \sigma_p(T)$, it is enough to show that for any $\epsilon > 0$ there exists at most a finite number of linearly independent eigenvectors of T corresponding to eigenvalues λ with $|\lambda| > \epsilon$. Assuming to the contrary, there must exist $\{u_n\}_{n=1}^\infty$, linearly independent, such that $Tu_n = \lambda_n u_n$ and $|\lambda_n| > \epsilon$. Applying the Gram-Schmidt procedure to the sequence $\{u_n\}_{n=1}^\infty$ we obtain an orthonormal sequence $\{v_n\}_{n=1}^\infty$ such that

$$v_k = \sum_{j=1}^{k} \beta_{kj} u_j \quad \beta_{kk} \neq 0 \tag{12.2.21}$$

Therefore

$$Tv_k - \lambda_k v_k = \sum_{j=1}^{k} \beta_{kj}(\lambda_j - \lambda_k) u_j \tag{12.2.22}$$

implying that

$$Tv_k = \lambda_k v_k + \sum_{j=1}^{k-1} \alpha_{kj} v_j \tag{12.2.23}$$

for some α_{kj}. But then

$$|\lambda_k|^2 \leq |\lambda_k|^2 + \sum_{j=1}^{k-1} |\alpha_{kj}|^2 = \|Tv_k\|^2 \to 0 \tag{12.2.24}$$

since $\{v_n\}_{n=1}^\infty$ is orthonormal and T is compact, contradicting $|\lambda_n| > \epsilon$. □

We emphasize that nothing stated so far implies that a compact operator has any eigenvalues at all. For example, we have already observed that the simple Volterra operator $Tu(x) = \int_0^x u(s)\,ds$, which is certainly compact, has spectrum $\sigma(T) = \sigma_c(T) = \{0\}$ (Example 11.4). In the next section we will see that if the operator T is also self-adjoint, then this sort of behavior cannot happen (i.e., eigenvalues must exist).

We could also use Theorem 12.5 or 12.6 to prove that certain operators are not compact. For example, a nonzero multiplication operator cannot be compact since it has either an uncountable spectrum or an infinite dimensional eigenspace, or both.

We conclude this section by summarizing in the form of a theorem, the implications of the abstract results in this section for the solvability of integral equations

$$\lambda u(x) - \int_\Omega K(x,y)u(y)\,dy = f(x) \quad x \in \Omega \tag{12.2.25}$$

Theorem 12.7. *If $K \in L^2(\Omega \times \Omega)$ then there exists a finite or countably infinite set $\{\lambda_n \in \mathbb{C}\}$ with zero as its only possible accumulation point, such that*

1. *If $\lambda \neq \lambda_n, \lambda \neq 0$ then for every $f \in L^2(\Omega)$ there exists a unique solution $u \in L^2(\Omega)$ of Eq. (12.2.25).*
2. *If $\lambda = \lambda_n \neq 0$ then there exist an integer $m \geq 1$, linearly independent solutions $\{v_1, \dots, v_m\}$ of the homogeneous equation*

$$\lambda v(x) - \int_\Omega K(x,y)v(y)\,dy = 0 \tag{12.2.26}$$

and linearly independent solutions $\{w_1, \dots, w_m\}$ of the adjoint homogeneous equation

$$\bar{\lambda} w(x) - \int_\Omega \overline{K(y,x)} w(y)\,dy = 0 \tag{12.2.27}$$

such that for $f \in L^2(\Omega)$ a solution of Eq. (12.2.25) exists if and only if f satisfies the m solvability conditions $\langle f, w_j \rangle = 0$ for $j = 1, \dots, m$. In such case, Eq. (12.2.25) has the m parameter family of solutions

$$u = u_p + \sum_{j=1}^m c_j v_j \tag{12.2.28}$$

where u_p denotes any particular solution of Eq. (12.2.25).

3. *If $\lambda = 0$ then either existence or uniqueness may fail. The condition that $\langle f, w \rangle = 0$ for any solution w of*

$$\int_\Omega \overline{K(y,x)} w(y)\,dy = 0 \tag{12.2.29}$$

is necessary, but in general not sufficient, for the existence of a solution of Eq. (12.2.25).

12.3 THE CASE OF SELF-ADJOINT COMPACT OPERATORS

In this section we continue with the study of the spectral properties of compact operators, but now make the additional assumption that the operator is self-adjoint. As motivation, let us recall that in the finite dimensional case a Hermitian matrix is always diagonalizable, and in particular there exists an orthonormal basis of eigenvectors of the matrix. If $Tx = Ax$, where A is an $N \times N$ Hermitian matrix with eigenvalues $\{\lambda_1, \dots, \lambda_N\}$ (repeated according to multiplicity) and corresponding orthonormal eigenvectors $\{u_1, \dots, u_N\}$, and we let U denote the $N \times N$ matrix whose columns are $u_1, \dots, u_N$, then $U^*U = I$ and $U^*AU = D$, where D is a diagonal matrix with diagonal entries $\lambda_1, \dots, \lambda_N$. It follows that

$$Ax = UDU^*x = \sum_{j=1}^N \lambda_j \langle u_j, x \rangle u_j \tag{12.3.30}$$

or equivalently

$$T = \sum_{j=1}^{N} \lambda_j P_j \tag{12.3.31}$$

where P_j is the orthogonal projection onto the span of u_j. The property that an operator may have of being expressible as a linear combination of projections is a useful one when true, and as we will see in this section is generally correct for compact self-adjoint operators.

Definition 12.4. If T is a linear operator on a Hilbert space $\mathbf{H}$, the *Rayleigh quotient* for T is

$$J(u) = \frac{\langle Tu, u\rangle}{\|u\|^2} \tag{12.3.32}$$

Clearly J: $D(T)\backslash\{0\} \to \mathbb{C}$ and $|J(u)| \le \|T\|$ for $T \in \mathcal{B}(\mathbf{H})$. If T is self-adjoint then J is real valued since

$$\langle Tu, u\rangle = \langle u, Tu\rangle = \overline{\langle Tu, u\rangle} \tag{12.3.33}$$

The range of the function J is sometimes referred to as the numerical range of T, and we may occasionally use the notation $Q(u) = \langle Tu, u\rangle$, the so-called quadratic form associated with T. Note also that $\sigma_p(T)$ is contained in the numerical range of T, since $J(u) = \lambda$ if $Tu = \lambda u$.

Theorem 12.8. *If $T \in \mathcal{B}(\mathbf{H})$ and $T = T^*$ then*

$$\|T\| = \sup_{u \ne 0} |J(u)| \tag{12.3.34}$$

Proof. If $M = \sup_{u\ne 0} |J(u)|$ then we have already observed that $M \le \|T\|$. To derive the reverse inequality, first observe that since J is real valued,

$$\langle T(u+v), u+v\rangle \le M\|u+v\|^2 \tag{12.3.35}$$

$$-\langle T(u-v), u-v\rangle \le M\|u-v\|^2 \tag{12.3.36}$$

for any $u, v \in \mathbf{H}$. Adding these inequalities and using the self-adjointness give

$$2\text{Re}\,\langle Tu, v\rangle = \langle Tu, v\rangle + \langle Tv, u\rangle \le M(\|u\|^2 + \|v\|^2) \tag{12.3.37}$$

If $u \notin N(T)$ choose $v = (\|u\|/\|Tu\|)Tu$ so that $\|v\| = \|u\|$ and $\langle Tu, v\rangle = \|Tu\|\,\|v\|$. It follows that

$$2\|Tu\|\,\|u\| \le 2M\|u\|^2 \tag{12.3.38}$$

and therefore $\|Tu\| \le M\|u\|$ holds for $u \notin N(T)$. Since the same conclusion is obvious for $u \in N(T)$, we must have $\|T\| \le M$, and the proof is completed. □

We note that the conclusion of theorem is false without the self-adjointness assumption; for example, $J(u) = 0$ for all u if T is the operator of rotation by $\pi/2$ in $\mathbb{R}^2$.

Now consider the function $\alpha \to J(u+\alpha v)$ for fixed $u, v \in \mathbf{H}\backslash\{0\}$ and $\alpha \in \mathbb{R}$. As a function of α it is simply a quotient of quadratic functions, hence differentiable at any α for which $\|u+\alpha v\| \neq 0$. In particular

$$\left.\frac{d}{d\alpha} J(u+\alpha v)\right|_{\alpha=0} \tag{12.3.39}$$

is well defined for any $u \neq 0$. This expression is the directional derivative of J at u in the v direction, and we say that u is a critical point of J if Eq. (12.3.39) is zero for every direction v.

We may evaluate Eq. (12.3.39) by elementary calculus rules and we find that

$$\left.\frac{d}{d\alpha} J(u+\alpha v)\right|_{\alpha=0} = \frac{\langle u, u\rangle(\langle Tu, v\rangle + \langle Tv, u\rangle) - \langle Tu, u\rangle(\langle u, v\rangle + \langle v, u\rangle)}{\langle u, u\rangle^2} \tag{12.3.40}$$

so at a critical point it must hold that

$$\text{Re}\,\langle Tu, v\rangle = J(u)\text{Re}\,\langle u, v\rangle \quad \forall v \in \mathbf{H} \tag{12.3.41}$$

Replacing v by iv we obtain

$$\text{Im}\,\langle Tu, v\rangle = J(u)\,\text{Im}\,\langle u, v\rangle \quad \forall v \in \mathbf{H} \tag{12.3.42}$$

and since J is real valued,

$$\langle Tu, v\rangle = J(u)\langle u, v\rangle \quad \forall v \in \mathbf{H} \tag{12.3.43}$$

If $\lambda = J(u)$ then $\langle Tu - \lambda u, v\rangle = 0$ for all $v \in \mathbf{H}$, so that $Tu = \lambda u$ must hold. We therefore see that eigenvalues of a self-adjoint operator T may be obtained from critical points of the corresponding Rayleigh quotient, and it is also clear that the right-hand side of Eq. (12.3.40) evaluates to be zero for any v if $Tu = \lambda u$. We have therefore established the following.

Proposition 12.3. *Let T be a bounded self-adjoint operator on* $\mathbf{H}$. *Then $u \in \mathbf{H}\backslash\{0\}$ is a critical point of J if and only if u is an eigenvector of T corresponding to eigenvalue $\lambda = J(u)$.*

At this point we have not yet proved that any such critical points exist, and indeed we know that a bounded self-adjoint operator can have an empty point spectrum; for example, a multiplication operator if the multiplier is real valued and all of its level sets have measure zero. Nevertheless we have identified a strategy that will succeed in proving the existence of eigenvalues, once some additional assumptions are made. The main such additional assumption we will now make is that T is compact.

Theorem 12.9. *If $T \in \mathcal{K}(\mathbf{H})$ and $T = T^*$ then either J or $-J$ achieves its maximum on $\mathbf{H}\backslash\{0\}$. In particular, either $\|T\|$ or $-\|T\|$ (or both) belong to $\sigma_p(T)$.*

Proof. If $T = 0$ then $J(u) \equiv 0$ and the conclusion is obvious. Otherwise, if $M := \|T\| > 0$ then by Theorem 12.8 either

$$\sup_{u \neq 0} J(u) = M \quad \text{or} \quad \inf_{u \neq 0} J(u) = -M \tag{12.3.44}$$

or both. For definiteness we assume that the first of these is true, in which case there must exist a sequence $\{u_n\}_{n=1}^\infty$ in $\mathbf{H}$ such that $J(u_n) \to M$. Without loss of generality we may assume $\|u_n\| = 1$ for all n, so that $\langle Tu_n, u_n \rangle \to M$. By weak compactness there is a subsequence $u_{n_k} \overset{w}{\to} u$, for some $u \in \mathbf{H}$, and since T is compact we also have $Tu_{n_k} \to Tu$. Thus

$$0 \le \|Tu_{n_k} - Mu_{n_k}\|^2 = \|Tu_{n_k}\|^2 + M^2 \|u_{n_k}\|^2 - 2M\langle Tu_{n_k}, u_{n_k} \rangle \tag{12.3.45}$$

Letting $k \to \infty$ the right-hand side tends to $\|Tu\|^2 - M^2 \le 0$, and thus $\|Tu\| = M$.

Furthermore, $Tu_{n_k} - Mu_{n_k} \to 0$, and since $M \neq 0$, $\{u_{n_k}\}$ must be strongly convergent to u, in particular $\|u\| = 1$. Thus we have $Tu = Mu$ for some $u \neq 0$, so that $J(u) = M$. This means that J achieves its maximum at u and u is an eigenvector corresponding to eigenvalue $\|T\| = M$, as needed. □

According to this theorem, any nonzero, compact, self-adjoint operator has at least one eigenvector u_1 corresponding to an eigenvalue $\lambda_1 \neq 0$. If another such eigenvector exists which is not a scalar multiple of u_1, then it must be possible to find one which is orthogonal to u_1, since eigenvectors corresponding to distinct eigenvalues are automatically orthogonal (Theorem 11.5) while the eigenvectors corresponding to λ_1 form a subspace which we can find an orthogonal basis of. This suggests that we seek another eigenvector by maximizing or minimizing the Rayleigh quotient over the subspace $\mathbf{H}_1 = \{u_1\}^\perp$.

Let us first make a definition and a simple observation.

Definition 12.5. If T is a linear operator on $\mathbf{H}$ then a subspace $E \subset D(T)$ is *invariant for* T if $T(E) \subset E$.

It is obvious that any eigenspace of T is invariant for T, and in the case of a self-adjoint operator we have also the following.

Lemma 12.7. *If $T \in \mathcal{B}(\mathbf{H})$ is a self-adjoint and E is an invariant subspace for T, then $E^\perp$ is also invariant for T.*

Proof. If $v \in E$ and $u \in E^\perp$ then

$$\langle Tu, v \rangle = \langle u, Tv \rangle = 0 \tag{12.3.46}$$

since $Tv \in E$. Thus $Tu \in E^\perp$. □

Now defining $\mathbf{H}_1 = \{u_1\}^\perp$ as previously, we have immediately that $T \in \mathcal{B}(\mathbf{H}_1)$ and clearly inherits the properties of compactness and self-adjointness from $\mathbf{H}$. Theorem 12.9 is therefore immediately applicable, so that the restriction of T to $\mathbf{H}_1$ has an eigenvector u_2, which is also an eigenvector of T and which is automatically orthogonal to u_1. The corresponding eigenvalue is $\lambda_2 = \pm\|T_1\|$, where T_1 is the restriction of T to $\mathbf{H}_1$, and so obviously $|\lambda_2| \le |\lambda_1|$.

Continuing this way we obtain orthogonal eigenvectors $u_1, u_2, \ldots$ corresponding to real eigenvalues $|\lambda_1| \geq |\lambda_2| \geq \cdots$, where

$$|\lambda_{n+1}| = \max_{\substack{u \in \mathbf{H}_n \\ u \neq 0}} |J(u)| = \|T_n\| \tag{12.3.47}$$

with $\mathbf{H}_n = \{u_1, \ldots, u_n\}^\perp$ and T_n being the restriction of T to $\mathbf{H}_n$. Without loss of generality $\|u_n\| = 1$ for all n obtained this way.

There are now two possibilities, either (i) the process continues indefinitely with $\lambda_n \neq 0$ for all n, or (ii) $\lambda_{n+1} = 0$ for some n. In the first case we must have $\lim_{n\to\infty} \lambda_n = 0$ by Theorem 12.6 and the fact that every eigenspace is of finite dimension. In case (ii), T has only finitely many linearly independent eigenvectors corresponding to nonzero eigenvalues $\lambda_1, \ldots, \lambda_n$ and $T = 0$ on $\mathbf{H}_n$. Assuming for definiteness that $\mathbf{H}$ is separable and of infinite dimension, then $\mathbf{H}_n = N(T)$ is the eigenspace for $\lambda = 0$ which must itself be infinite dimensional.

Theorem 12.10. *Let* $\mathbf{H}$ *be a separable Hilbert space. If* $T \in \mathcal{K}(\mathbf{H})$ *is self-adjoint then*

1. $R(T)$ *has an orthonormal basis consisting of eigenvectors* $\{u_n\}$ *of* T *corresponding to eigenvalues* $\lambda_n \neq 0$.
2. $\mathbf{H}$ *has an orthonormal basis consisting of eigenvectors of* T.

Proof. Let $\{u_n\}$ be the finite or countably infinite set of orthonormal eigenvectors corresponding to the nonzero eigenvalues of T as constructed above. For $u \in \mathbf{H}$ let $v = u - \sum_{j=1}^n \langle u, u_j \rangle u_j$ for some n. Then v is the orthogonal projection of u onto $\mathbf{H}_n$, so $\|v\| \leq \|u\|$ and $\|Tv\| \leq |\lambda_{n+1}|\, \|v\|$. In particular

$$\left\| Tu - \sum_{j=1}^n \langle Tu, u_j \rangle u_j \right\|^2 = \left\| Tu - \sum_{j=1}^n \langle u, u_j \rangle Tu_j \right\|^2 \leq |\lambda_{n+1}|^2 \|u\|^2 \tag{12.3.48}$$

where we have used that

$$\langle u, u_j \rangle Tu_j = \langle u, u_j \rangle \lambda_j u_j = \langle u, \lambda_j u_j \rangle u_j = \langle u, Tu_j \rangle u_j = \langle Tu, u_j \rangle u_j \tag{12.3.49}$$

Letting $n \to \infty$, or taking n sufficiently large in the case of a finite number of nonzero eigenvalues, we therefore see that Tu is in the span of $\{u_n\}$. This completes the proof of 1.

If we now let $\{z_n\}$ be any orthonormal basis of the closed subspace $N(T)$, then each z_n is an eigenvector of T corresponding to eigenvector $\lambda = 0$ and $z_n \perp u_m$ for any m, n since $N(T) = R(T)^\perp$. For any $u \in \mathbf{H}$ let $v = \sum_n \langle u, u_n \rangle u_n$- the series must be convergent by Proposition 5.3 and the fact that $\sum_n |\langle u, u_n \rangle|^2 \leq \|u\|^2$. It is immediate that $u - v \in N(T)$ since

$$Tu = Tv = \sum_n \lambda_n \langle u, u_n \rangle u_n \tag{12.3.50}$$

and so u has a unique representation

$$u = v + (u - v) = \sum_n \langle u, u_n \rangle u_n + \sum_n c_n z_n \tag{12.3.51}$$

for some constants c_n. Thus $\{u_n\} \cup \{z_n\}$ is an orthonormal basis of $\mathbf{H}$. □

We note that either sum in Eq. (12.3.51) can be finite or infinite, but of course they cannot both be finite unless $\mathbf{H}$ is finite dimensional. In the case of a nonseparable Hilbert space it is only necessary to allow for an uncountable basis of $N(T)$. From Eq. (12.3.50) we also get the diagonalization formula

$$T = \sum_n \lambda_n P_n \tag{12.3.52}$$

where $P_n u = \langle u, u_n \rangle u_n$ is the orthogonal projection onto the span of u_n.

The existence of an eigenfunction basis provides a convenient tool for the study of corresponding operator equations. Let us consider the problem

$$\lambda u - Tu = f \tag{12.3.53}$$

where T is a compact, self-adjoint operator on a separable, infinite dimensional Hilbert space $\mathbf{H}$. Let $\{u_n\}_{n=1}^{\infty}$ be an orthonormal basis of eigenvectors of T. We may therefore expand f, and solution u if it exists, in this basis,

$$u = \sum_{n=1}^{\infty} a_n u_n \quad f = \sum_{n=1}^{\infty} b_n u_n \quad a_n = \langle u, u_n \rangle \quad b_n = \langle f, u_n \rangle \tag{12.3.54}$$

Inserting these into the equation and using $Tu_n = \lambda_n u_n$ there results

$$\sum_{n=1}^{\infty} ((\lambda - \lambda_n) a_n - b_n) u_n = 0 \tag{12.3.55}$$

Thus it is a necessary condition that $(\lambda - \lambda_n) a_n = b_n$ for all n, in order that a solution u exists.

Now let us consider several cases.

Case 1. If $\lambda \neq \lambda_n$ for every n and $\lambda \neq 0$, then $\lambda \in \rho(T)$ so a unique solution u of Eq. (12.3.53) exists, which must be given by

$$u = \sum_{n=1}^{\infty} \frac{\langle f, u_n \rangle}{\lambda - \lambda_n} u_n \tag{12.3.56}$$

Note that there exists a constant C such that $1/|\lambda - \lambda_n| \leq C$ for all n, from which it follows directly that the series is convergent in $\mathbf{H}$ and $\|u\| \leq C\|f\|$.

Case 2. Suppose $\lambda = \lambda_m$ for some m and $\lambda \neq 0$. It is then necessary that $b_n = 0$ for all n for which $\lambda_n = \lambda_m$, which amounts precisely to the solvability condition on f already derived, that $f \perp z$ for all $z \in N(\lambda I - T)$. When this holds the constants a_n may be chosen arbitrarily for these n values, while $a_n = b_n/(\lambda - \lambda_n)$ must hold otherwise. Thus the general solution may be written

$$u = \sum_{\{n:\, \lambda_n \neq \lambda_m\}} \frac{\langle f, u_n \rangle}{\lambda - \lambda_n} u_n + \sum_{\{n:\, \lambda_n = \lambda_m\}} c_n u_n \tag{12.3.57}$$

for any $f \in R(\lambda I - T)$.

Case 3. If $\lambda = 0$ and $\lambda_n \neq 0$ for all n then the unique solution is given by

$$u = -\sum_{n=1}^{\infty} \frac{\langle f, u_n \rangle}{\lambda_n} u_n \tag{12.3.58}$$

provided the series is convergent in $\mathbf{H}$. Since $\lambda_n \to 0$ must hold in this case, there will always exist $f \in \mathbf{H}$ for which the series is not convergent, as must be the case since $R(T)$ is dense but not equal to all of $\mathbf{H}$. In fact we obtain the precise characterization that $f \in R(T)$ if and only if

$$\sum_{n=1}^{\infty} \frac{|\langle f, u_n \rangle|^2}{\lambda_n^2} < \infty \tag{12.3.59}$$

Case 4. If $\lambda = 0 \in \sigma_p(T)$ let $\{u_n\} \cup \{z_n\}$ be an orthonormal basis of eigenvectors as above, with the z_n's being a basis of $N(T)$. If a solution u exists, then by matching coefficients in the basis expansions of Tu and f we get that a solution exists if f has the properties

$$\langle f, z_n \rangle = 0 \quad \forall n \quad \text{and} \quad \sum_n \frac{|\langle f, u_n \rangle|^2}{\lambda_n^2} < \infty \tag{12.3.60}$$

in which case the general solution is

$$u = -\sum_n \frac{\langle f, u_n \rangle}{\lambda_n} u_n + \sum_n c_n z_n \quad \sum_n c_n^2 < \infty \tag{12.3.61}$$

12.4 SOME PROPERTIES OF EIGENVALUES

When T is a self-adjoint compact operator, we have seen in the previous section that solution formulas for the equation $\lambda u - Tu = f$ can be given purely in terms of the eigenvalues and eigenvectors of T, along with f itself. This means that all of the properties of T are encoded by these eigenvalues and eigenvectors. We will briefly pursue some consequences of this in the case that T is an integral operator, in which case we may anticipate that properties of the kernel of the operator are directly connected to those of the eigenvalues and eigenvectors. Thus let

$$Tu(x) = \int_\Omega K(x, y) u(y)\, dy \tag{12.4.62}$$

where $K \in L^2(\Omega \times \Omega)$ and $K(x, y) = \overline{K(y, x)}$. Considered as an operator on $L^2(\Omega)$, Theorem 12.10 is then applicable, so we know there must exist an orthonormal basis of eigenfunctions $\{u_n\}_{n=1}^{\infty}$ and real eigenvalues λ_n such that $Tu_n = \lambda_n u_n$, that is

$$\int_{\Omega} K(x,y)u_n(y)\,dy = \lambda_n u_n(x) \tag{12.4.63}$$

or equivalently

$$\int_{\Omega} K(y,x)\overline{u_n(y)}\,dy = \lambda_n \overline{u_n(x)} \tag{12.4.64}$$

This may be regarded as the statement that for almost every $x \in \Omega$, $\lambda_n \overline{u_n(x)}$ is the nth generalized Fourier coefficient of $K(\cdot, x)$ with respect to the u_n basis. In particular, by the Bessel equality

$$\int_{\Omega} |K(x,y)|^2\,dy = \sum_{n=1}^{\infty} \lambda_n^2 |u_n(x)|^2 \quad \text{for a.e. } x \in \Omega \tag{12.4.65}$$

and integrating with respect to x gives

$$\iint_{\Omega\times\Omega} |K(x,y)|^2\,dydx = \sum_{n=1}^{\infty} \lambda_n^2 \int_{\Omega} |u_n(x)|^2\,dx = \sum_{n=1}^{\infty} \lambda_n^2 \tag{12.4.66}$$

It also follows from the above considerations that

$$K(y,x) = \sum_{n=1}^{\infty} \lambda_n \overline{u_n(x)} u_n(y) \tag{12.4.67}$$

or

$$K(x,y) = \sum_{n=1}^{\infty} \lambda_n u_n(x) \overline{u_n(y)} \tag{12.4.68}$$

in the sense that the convergence takes place in $L^2(\Omega)$ with respect to y for a.e. x and vice versa. Formally at least, it follows by setting $y = x$ that

$$K(x,x) = \sum_{n=1}^{\infty} \lambda_n |u_n(x)|^2 \tag{12.4.69}$$

and integrating in x that

$$\int_{\Omega} K(x,x)\,dx = \sum_{n=1}^{\infty} \lambda_n \tag{12.4.70}$$

This last identity, however, cannot be proved to be correct without further assumptions, if for no other reason than that $K(x,x)$, being a restriction of K to a set of measure zero in $\Omega \times \Omega$, could be changed in an arbitrary way without changing the spectrum of T. Likewise, the sum on the right need not be convergent without further restrictions. Here we state without proof *Mercer's theorem*, which gives sufficient conditions for Eq. (12.4.70) to hold—see for example [9, p. 138].

Theorem 12.11. *Let T be the compact self-adjoint integral operator* (12.4.62). *Assume that Ω is bounded, K is continuous on $\overline{\Omega \times \Omega}$ and that all but finitely many of the nonzero eigenvalues of T are of the same sign. Then Eq. (12.4.68) is valid, where the convergence is absolute and uniform, and in particular Eq. (12.4.70) holds.*

12.5 SINGULAR VALUE DECOMPOSITION AND NORMAL OPERATORS

If T is a compact operator we know from explicit examples that the point spectrum of T may be empty. However, if we let $S = T^*T$, the so-called *normal operator* of T, then S is compact and self-adjoint (see Exercise 12.1), so that Theorem 12.10 applies to S. If we assume for simplicity that $\mathbf{H}$ is separable, there must therefore exist an orthonormal basis $\{u_n\}_{n=1}^{\infty}$ of $\mathbf{H}$ consisting of eigenvectors of S, that is

$$T^*Tu_n = \lambda_n u_n \tag{12.5.71}$$

Note that if J is the Rayleigh quotient for S then

$$\lambda_n = J(u_n) = \langle Su_n, u_n\rangle = \|Tu_n\|^2 \geq 0 \tag{12.5.72}$$

We define $\sigma_n = \sqrt{\lambda_n}$ to be the nth *singular value of* T. If $T \neq 0$ and we list the nonzero eigenvalues of S in decreasing order, $\lambda_1 \geq \lambda_2 \geq \cdots$ (this is possibly a finite list) then from Theorem 12.9 it is immediate that $\lambda_1 = \|T\|^2$. Thus we have the following simple but important result.

Proposition 12.4. *If $T \in \mathcal{K}(\mathbf{H})$ then $\|T\| = \sigma_1$, the largest singular value of T.*

Now for any n for which $\lambda_n > 0$, let $v_n = Tu_n/\sigma_n$. We then have

$$Tu_n = \sigma_n v_n \quad T^*v_n = \sigma_n u_n \tag{12.5.73}$$

The u_n's are orthonormal by construction, and

$$\langle v_n, v_m\rangle = \frac{1}{\sigma_n\sigma_m}\langle Tu_n, Tu_m\rangle = \frac{\lambda_n}{\sigma_n\sigma_m}\langle u_n, u_m\rangle \tag{12.5.74}$$

so that the v_n's are also orthonormal. We say that u_n is the nth *right singular vector* of T and v_n is the nth *left singular vector*. The collection $\{\sigma_n, u_n, v_n\}$ is a *singular system* for T.

From Eq. (12.3.51) we then have

$$Tu = \sum_n \langle u, u_n\rangle Tu_n = \sum_n \sigma_n\langle u, u_n\rangle v_n \tag{12.5.75}$$

or

$$T = \sum_n \sigma_n Q_n, \quad \text{where} \quad Q_n u = \langle u, u_n\rangle v_n \tag{12.5.76}$$

Here Q_n is not a projection unless $u_n = v_n$, but is a so-called *rank one operator*. The representation Eq. (12.5.76) of T as a sum of rank one operators is the *singular value decomposition* of T.

Next, let us consider a normal operator $T \in \mathcal{K}(\mathbf{H})$, which we recall means that $T^*T = TT^*$. For simplicity let us also assume that all eigenvalues of the compact self-adjoint operator $S = T^*T$ are nonzero and simple. In that case, if $Su_n = \lambda_n u_n$ it follows that

$$STu_n = T^*T^2u_n = TT^*Tu_n = TSu_n = \lambda_n Tu_n \tag{12.5.77}$$

which means either $Tu_n = 0$ or Tu_n is an eigenvector of S corresponding to λ_n. The first case cannot occur since then $Su_n = 0$ would hold, so it must be that u_n and Tu_n are nonzero and linearly dependent, $Tu_n = \theta_n u_n$ for some $\theta_n \in \mathbb{C}\backslash\{0\}$. Thus $\mathbf{H}$ has an orthonormal basis consisting of eigenvectors of T since these are the same as the eigenvectors of S. With a somewhat more complicated proof, the same can be shown for any normal operator T, see Section 56 of [2].

12.6 EXERCISES

12.1. Show that if $S \in \mathcal{B}(\mathbf{H})$ and T is compact, then TS and ST are also compact. (In algebraic terms this means that the set of compact operators is an *ideal* in $\mathcal{B}(\mathbf{H})$.)

12.2. If $T \in \mathcal{B}(\mathbf{H})$ and T^*T is compact, show that T must be compact. Use this to show that if T is compact then T^* must also be compact.

12.3. Prove that a sequence $\{x_n\}_{n=1}^{\infty}$ in a Hilbert space can have at most one weak limit.

12.4. If $T \in \mathcal{B}(\mathbf{H})$ is compact and $\mathbf{H}$ is of infinite dimension, show that $0 \in \sigma(T)$.

12.5. Let $\{\phi_j\}_{j=1}^n, \{\psi_j\}_{j=1}^n$ be linearly independent sets in $L^2(\Omega)$,

$$K(x,y) = \sum_{j=1}^{n} \phi_j(x)\psi_j(y)$$

be the corresponding degenerate kernel and T be the corresponding integral operator. Show that the problem of finding the nonzero eigenvalues of T always amounts to a matrix eigenvalue problem. In particular, show that T has at most n nonzero eigenvalues. Find $\sigma_p(T)$ in the case that $K(x,y) = 6 + 12xy + 60x^2y^3$ and $\Omega = (0,1)$. (Feel free to use Matlab or some such thing to solve the resulting matrix eigenvalue problem.)

12.6. Let

$$Tu(x) = \frac{1}{x}\int_0^x u(y)\,dy \quad u \in L^2(0,1)$$

Show that $(0,2) \subset \sigma_p(T)$ and that T is not compact. (Suggestion: Look for eigenfunctions in the form $u(x) = x^{\alpha}$.)

12.7. Let $\{\lambda_j\}_{j=1}^{\infty}$ be a sequence of nonzero real numbers satisfying

$$\sum_{j=1}^{\infty} \lambda_j^2 < \infty$$

Construct a symmetric Hilbert-Schmidt kernel K such that the corresponding integral operator has eigenvalues $\lambda_j, j = 1, 2, \dots$ and for which 0 is an eigenvalue of infinite multiplicity. (Suggestion: Look for such a K in the form $K(x,y) = \sum_{j=1}^{\infty} \lambda_j u_j(x) \overline{u_j(y)}$, where $\{u_j\}$ are orthonormal, but not complete, in $L^2(\Omega)$.)

12.8. On the Hilbert space $\mathbf{H} = \ell^2$ define the operator T by

$$T\{x_1, x_2, \dots\} = \{a_1 x_1, a_2 x_2, \dots\}$$

for some sequence $\{a_n\}_{n=1}^{\infty}$. Show that T is compact if and only if $\lim_{n\to\infty} a_n = 0$.

12.9. Let T be the integral operator with kernel $K(x,y) = e^{-|x-y|}$ on $L^2(-1,1)$. Find all of the eigenvalues and eigenfunctions of T. (Suggestion: $Tu = \lambda u$ is equivalent to an ODE problem. Do not forget about boundary conditions. The eigenvalues may need to be characterized in terms of the roots of a certain nonlinear function.)

12.10. We say that $T \in \mathcal{B}(\mathbf{H})$ is a positive operator if $\langle Tx, x\rangle \ge 0$ for all $x \in \mathbf{H}$. If T is a positive self-adjoint compact operator show that T has a square root, more precisely there exists a compact self-adjoint operator S such that $S^2 = T$. (Suggestion: If $T = \sum_{n=1}^{\infty} \lambda_n P_n$ try $S = \sum_{n=1}^{\infty} \sqrt{\lambda_n} P_n$. In a similar manner, one can define other fractional powers of T.)

12.11. Suppose that $S \in \mathcal{B}(\mathbf{H})$, $0 \in \rho(S)$, T is a compact operator on $\mathbf{H}$, and $N(S+T) = \{0\}$. Show that the operator equation

$$Sx + Tx = y$$

has a unique solution for every $y \in \mathbf{H}$.

12.12. Compute the singular value decomposition of the Volterra operator

$$Tu(x) = \int_0^x u(s)\, ds$$

in $L^2(0,1)$ and use it to find $\|T\|$. Is T normal? (Suggestion: The equation $T^*Tu = \lambda u$ is equivalent to an ODE eigenvalue problem, which you can solve explicitly.)

12.13. The concept of a Hilbert-Schmidt operator can be defined abstractly as follows. If $\mathbf{H}$ is a separable Hilbert space, we say that $T \in \mathcal{B}(\mathbf{H})$ is Hilbert-Schmidt if

$$\sum_{n=1}^{\infty} \|Tu_n\|^2 < \infty \tag{12.6.78}$$

for some orthonormal basis $\{u_n\}_{n=1}^{\infty}$ of $\mathbf{H}$.

(a) Show that if T is Hilbert-Schmidt then the sum (12.6.78) must be finite for *any* orthonormal basis of $\mathbf{H}$. (Suggestion: If $\{v_n\}_{n=1}^{\infty}$ is another orthonormal basis, then

$$\sum_{n=1}^{\infty} \|Tv_n\|^2 = \sum_{n,m=1}^{\infty} |\langle Tv_n, u_m\rangle|^2 = \sum_{n,m=1}^{\infty} |\langle v_n, T^* u_m\rangle|^2 = \sum_{n,m=1}^{\infty} |\langle u_n, T^* u_m\rangle|^2$$

etc.)

(b) Show that a Hilbert-Schmidt operator is compact.

12.14. If $Q \in \mathcal{B}(\mathbf{H})$ is a Fredholm operator of index zero, show that there exists a one-to-one operator $S \in \mathcal{B}(\mathbf{H})$ and $T \in \mathcal{K}(\mathbf{H})$ such that $Q = S + T$. (Hint: Define $T = AP$, where P is the orthogonal projection onto $N(Q)$ and A: $N(Q) \to N(Q^*)$ is one-to-one and onto.)

Chapter 13

Spectra and Green's Functions for Differential Operators

In this chapter we will focus more on spectral properties of unbounded operators, about which we have had little to say up to this point. Two simple but key observations are that (i) many interesting unbounded linear operators have an inverse which is compact, or more generally a resolvent which is compact for most values of the spectral parameter, and (ii) if $\lambda \neq 0$ is an eigenvalue of some operator then λ^{-1} is an eigenvalue of the inverse operator, with the same eigenvector. Thus we may be able to obtain a great deal of information about the spectrum of an unbounded operator by looking at its inverse, if the inverse exists. We will carry this plan out in detail for two important special cases. The first is the case of a second-order differential operator in one space dimension (Sturm-Liouville theory), and the second is the case of the Laplacian operator in a bounded domain of $\mathbb{R}^N$.

13.1 GREEN'S FUNCTIONS FOR SECOND-ORDER ODEs

Let us reconsider the operator on $L^2(0,1)$ from Example 11.6, namely

$$Tu = -u'' \quad D(T) = \{u \in H^2(0,1) \colon u(0) = u(1) = 0\} \tag{13.1.1}$$

Any $u \in N(T)$ is a linear function vanishing at the endpoints, so the associated problem

$$-u'' = f \quad 0 < x < 1 \quad u(0) = u(1) = 0 \tag{13.1.2}$$

has at most one solution for any $f \in L^2(0,1)$. In fact an explicit solution formula was given in Exercise 1.7 of Chapter 1, at least for $f \in C([0,1])$, and it is not hard to check that it remains valid for $f \in L^2(0,1)$ in the sense that if

$$G(x,y) = \begin{cases} y(1-x) & 0 < y < x < 1 \\ x(1-y) & 0 < x < y < 1 \end{cases} \tag{13.1.3}$$

Techniques of Functional Analysis for Differential and Integral Equations
http://dx.doi.org/10.1016/B978-0-12-811426-1.00013-1

then

$$u(x) = \int_0^1 G(x,y)f(y)\,dy \tag{13.1.4}$$

satisfies $-u'' = f$ in the sense of distributions on $(0,1)$, as well as the given boundary conditions.

Let us next consider how Eqs. (13.1.3), (13.1.4) might be derived in the first place. Formally, if Eq. (13.1.4) holds, then

$$u''(x) = \int_0^1 G_{xx}(x,y)f(y)\,dy = -f(x) \tag{13.1.5}$$

which suggests $G_{xx}(x,y) = -\delta(x-y)$ for all $y \in (0,1)$. This in turn means, in particular, that

$$G(x,y) = \begin{cases} Ax+B & 0<x<y \\ Cx+D & y<x<1 \end{cases} \tag{13.1.6}$$

for some constants $A, B, C, D \dots$ depending on y as a parameter. In order that u satisfies the required boundary conditions, we should have $B = C + D = 0$. Recalling the discussion leading up to Eq. (6.3.53) we expect that $x \to G(x,y)$ should be continuous at $x = y$ and $x \to G_x(x,y)$ should have a jump of magnitude -1 at $x = y$. These four conditions uniquely determine the four coefficients determining G in Eq. (13.1.3). We call G the *Green's function* for the problem (13.1.2). The integral operator with kernel $G(x,y)$ is the inverse of T and is clearly compact.

Now let us consider a more general situation of this type. Define a differential expression

$$Lu = a_2(x)u'' + a_1(x)u' + a_0(x)u \tag{13.1.7}$$

where we require the coefficients to satisfy $a_j \in C([a,b])$ for $j = 1,2,3$ and $a_2(x) \neq 0$ on $[a,b]$, together with boundary operators

$$\begin{aligned} &B_1u = c_1u(a) + c_2u'(a) \quad && B_2u = c_3u(b) + c_4u'(b) \\ &|c_1| + |c_2| \neq 0 && |c_3| + |c_4| \neq 0 \end{aligned} \tag{13.1.8}$$

We seek a solution for the problem

$$Lu(x) = f(x) \quad a<x<b \quad B_1u = B_2u = 0 \tag{13.1.9}$$

in the form

$$u(x) = \int_a^b G(x,y)f(y)\,dy \tag{13.1.10}$$

for some suitable kernel function $G(x,y)$.

Proceeding again as above we compute formally that

$$Lu(x) = \int_a^b L_xG(x,y)f(y)\,dy \tag{13.1.11}$$

where the subscript on L reminds us that L operates in the x variable for fixed y. Thus

$$L_x G = \delta(x - y) \tag{13.1.12}$$

should hold, and

$$B_{1x} G = B_{2x} G = 0 \tag{13.1.13}$$

in order that the boundary conditions for u be satisfied. In particular G should satisfy $L_x G = 0$ for $a < x < y < b$ and $a < y < x < b$, plus certain matching conditions at $x = y$, which may be stated as follows:

- G should be continuous at $x = y$ since otherwise $L_x G$ would contain a term of the form $C\delta'(x - y)$.
- G_x should experience a jump at $x = y$ of the correct magnitude such that $a_2(x)G_{xx}(x, y) = \delta(x - y)$, in other words the jump in G_x should be $1/a_2(y)$.

The same conclusion could be (formally) derived by integrating both sides of Eq. (13.1.12) from $y - \epsilon$ to $y + \epsilon$ and letting $\epsilon \to 0+$. Thus our conditions may be summarized as

$$G(y+, y) - G(y-, y) = 0, \quad G_x(y+, y) - G_x(y-, y) = \frac{1}{a_2(y)} \tag{13.1.14}$$

$$B_{1x} G = B_{2x} G = 0 \tag{13.1.15}$$

We now claim that such a function $G(x, y)$ can be found, under the additional assumption that the homogeneous problem (13.1.9) with $f \equiv 0$ has only the zero solution.

First observe that we can find nontrivial solutions $\phi_1, \phi_2 \in H^2(a, b)$ of

$$L\phi_1 = 0 \quad a < x < b, \quad B_1\phi_1 = 0 \tag{13.1.16}$$

$$L\phi_2 = 0 \quad a < x < b, \quad B_2\phi_2 = 0 \tag{13.1.17}$$

since each amounts to a second-order ODE with only one initial condition. Now look for G in the form

$$G(x, y) = \begin{cases} C_1(y)\phi_1(x) & a < x < y < b \\ C_2(y)\phi_2(x) & a < y < x < b \end{cases} \tag{13.1.18}$$

It is then automatic that $L_x G = 0$ for $x \neq y$, and that the boundary conditions (13.1.15) hold. In order that the remaining conditions (13.1.14) be satisfied we need to have that

$$C_1(y)\phi_1(y) - C_2(y)\phi_2(y) = 0 \tag{13.1.19}$$

$$C_1(y)\phi_1'(y) - C_2(y)\phi_2'(y) = -\frac{1}{a_2(y)} \tag{13.1.20}$$

Thus unique constants $C_1(y), C_2(y)$ exist provided the coefficient matrix is nonsingular, or equivalently the Wronskian of ϕ_1, ϕ_2 is nonzero for every y. But it is known from ODE theory that if the Wronskian is zero at any point then ϕ_1, ϕ_2 must be linearly dependent, in which case either one is a nontrivial solution of the homogeneous problem. This contradicts the assumption we made, and so the first part of the following theorem has been established.

Theorem 13.1. *Assume that Eq. (13.1.9) with $f \equiv 0$ has only the zero solution. Then*

1. *There exists a unique function $G(x, y)$ defined for $a \le x, y \le b$ such that $L_x G(x, y) = \delta(x - y)$ in the sense of distributions on (a, b) for fixed y, and Eqs. (13.1.14),* (13.1.15) *hold.*
2. *G is bounded on $[a, b] \times [a, b]$.*
3. *If $f \in L^2(a, b)$ and*

$$u(x) = Sf := \int_a^b G(x, y) f(y)\, dy \tag{13.1.21}$$

then u is the unique solution of Eq. (13.1.9).

Proof. We have proved the first part, and the third part is left for the exercises. We can explicitly solve for $C_1(y), C_2(y)$, obtaining

$$C_1(y) = \frac{\phi_2(y)}{a_2(y)W(y)} \quad C_2(y) = \frac{\phi_1(y)}{a_2(y)W(y)} \tag{13.1.22}$$

where W is the Wronskian determinant $W(y) = \phi_1(y)\phi_2'(y) - \phi_2(y)\phi_1'(y)$. Since $\phi_1, \phi_2 \in C^1([a, b])$ and a_2, W cannot vanish, it follows that there exists an upper bound for ϕ_1, ϕ_2, C_1, C_2, and so for G. □

In particular, if we define the unbounded linear operator

$$Tu = Lu \quad D(T) = \{u \in H^2(a, b): B_1 u = B_2 u = 0\} \tag{13.1.23}$$

then T^{-1}, given by Eq. (13.1.21) clearly satisfies the Hilbert-Schmidt condition and so is compact operator on $L^2(a, b)$.[1]

Corollary 13.1. *Assume that Eq. (13.1.9) with $f \equiv 0$ has only the zero solution and define T by Eq. (13.1.23). Then $\sigma(T)$ consists of at most countably many nonzero simple eigenvalues with no finite accumulation point.*

Proof. By Theorem 13.1 $0 \in \rho(T)$. If $\lambda \in \sigma(T)$ then $\mu = \lambda^{-1} \in \sigma(T^{-1})$ since if $\mu \in \rho(T^{-1})$ it would follow that the equation $Tu - \lambda u = f$ has the unique solution

$$u = \mu(\mu I - T^{-1})^{-1} T^{-1} f \quad \mu = \lambda^{-1} \tag{13.1.24}$$

1. Note that we observe a careful distinction between the operator T and the differential expression defined by L—the operator T corresponds to the triple (L, B_1, B_2).

which implies that $\lambda \in \rho(T)$. Thus $\sigma(T)$ is contained in the set $\{\lambda\colon \lambda^{-1} \in \sigma(T^{-1})\}$, which is at most countable by Theorem 12.6. In addition every such point must be in $\sigma_p(T^{-1})$ and so $\sigma(T) = \sigma_p(T)$. Since $\sigma(T^{-1})$ is bounded with zero as its only accumulation point it follows that $\sigma(T)$ can have no finite accumulation point. Finally, all eigenvalues of T must be simple, since if there existed two linearly independent functions in $N(T - \lambda I)$ these would form a fundamental set for the ODE $Lu = \lambda u$. But then every solution of $Lu = \lambda u$ would have to be in $D(T)$, in particular also satisfying the boundary conditions $B_1 u = B_2 u = 0$, which is clearly false. □

Example 13.1. For the case

$$Lu = u'' - u \quad B_1 u = u'(0) \quad B_2 u = u(1) \tag{13.1.25}$$

we can choose

$$\phi_1(x) = \cosh x \quad \phi_2(x) = \sinh(x-1) \tag{13.1.26}$$

The matching and jump conditions at $x = y$ then amount to

$$C_1(y)\cosh(y) - C_2(y)\sinh(y-1) = 0 \tag{13.1.27}$$

$$C_1(y)\sinh(y) - C_2(y)\cosh(y-1) = -1 \tag{13.1.28}$$

The solution pair is $C_1(y) = \sinh(y-1)/\cosh(1)$, $C_2(y) = \cosh(y)/\cosh(1)$ giving the Green's function

$$G(x,y) = \begin{cases} \frac{\sinh(y-1)\cosh(x)}{\cosh(1)} & 0 < x < y < 1 \\ \frac{\sinh(x-1)\cosh(y)}{\cosh(1)} & 0 < y < x < 1 \end{cases} \tag{13.1.29}$$

If T is the operator corresponding to L, B_1, B_2 then it may be checked by explicit calculation that

$$\sigma(T) = \left\{-1 - \left(\left(n + \frac{1}{2}\right)\pi\right)^2\right\}_{n=0}^{\infty} \tag{13.1.30}$$

Note that the Green's function is real and symmetric, so that the corresponding operator integral operator S in Eq. (13.1.21), and hence also $T = S^{-1}$ is self-adjoint. □

13.2 ADJOINT PROBLEMS

In this section we consider in more detail the adjoint of the operator T defined in Eq. (13.1.23). First we observe that formally, for $\phi, \psi \in C_0^\infty(a,b)$ we have

$$\langle L\phi, \psi\rangle = \int_a^b (a_2\phi'' + a_1\phi' + a_0\phi)\overline{\psi} \tag{13.2.31}$$

$$= \int_a^b \phi((a_2\overline{\psi})'' - (a_1\overline{\psi})' + a_0\overline{\psi})dx \tag{13.2.32}$$

That is to say,

$$\langle L\phi, \psi \rangle = \langle \phi, L^* \psi \rangle \tag{13.2.33}$$

where

$$L^* \psi = (\overline{a_2} \psi)'' - (\overline{a_1} \psi)' + \overline{a_0} \psi \tag{13.2.34}$$

For simplicity we will make the additional assumptions on the coefficients that

$$a_j \in C^j([a,b]) \quad \text{and is real valued for} \quad j = 0, 1, 2 \tag{13.2.35}$$

so in particular the integration by parts is valid. Furthermore since

$$L^* \psi = a_2 \psi'' + (2a_2' - a_1)\psi' + (a_2'' - a_1' + a_0)\psi \tag{13.2.36}$$

we see that $L^* \psi = L\psi$ precisely if $a_1 = a_2'$. We say that the expression L is *formally self-adjoint* in this case, but note that this is not the same as having the corresponding operator T be self-adjoint, since so far there has been no taking account of the boundary conditions, which are part of the definition of T.

To pursue this point, we see from an integration by parts that for any $\phi, \psi \in C^2([a,b])$ we have

$$\langle L\phi, \psi \rangle - \langle \phi, L^* \psi \rangle = J(\phi, \psi)|_a^b \tag{13.2.37}$$

where

$$J(\phi, \psi) = a_2(\phi' \overline{\psi} - \phi \overline{\psi}') + (a_1 - a_2')\phi \overline{\psi} \tag{13.2.38}$$

is the *boundary functional*. Since we can choose ϕ, ψ to have compact support, in which case the boundary term is zero, the expression for T^* must be given by L^*. It then follows that $D(T^*)$ must be such that $J(\phi, \psi)|_a^b = 0$ whenever $\phi \in D(T)$ and $\psi \in D(T^*)$. As we will see, this amounts to the specification of two more homogeneous boundary conditions to be satisfied by ψ.

Example 13.2. As in Example 13.1 consider $L\phi = \phi'' - \phi$ on $(0,1)$, which is formally self-adjoint, together with the boundary operators $B_1\phi = \phi'(0)$, $B_2\phi = \phi(1)$. By direct calculation we see that

$$J(\phi, \psi)|_0^1 = \phi(0)\psi'(0) + \phi'(1)\psi(1) \tag{13.2.39}$$

if $B_1\phi = B_2\phi = 0$. But otherwise $\phi(0), \phi'(1)$ can take on arbitrary values, and so the only way it can be guaranteed that $J(\phi, \psi)|_0^1 = 0$ is always true is if $\psi'(0) = \psi(1) = 0$, that is, ψ satisfies the same boundary conditions as ϕ. Thus we expect that $T^* = T$, consistent with the earlier observation that T^{-1} is self-adjoint. □

Example 13.3. Let

$$L\phi = x^2\phi'' + x\phi' - \phi \quad 1 < x < 2 \quad B_1\phi = \phi'(1) \quad B_2\phi = \phi(2) + \phi'(2) \tag{13.2.40}$$

In this case we find that the expression for the adjoint operator is

$$L^*\psi = (x^2\psi)'' - (x\psi)' - \psi = x^2\psi'' + 3x\psi' \tag{13.2.41}$$

Next, the boundary functional is

$$J(\phi, \psi) = x^2(\phi'\psi - \phi\psi') - x\phi\psi \tag{13.2.42}$$

so that if $B_1\phi = B_2\phi = 0$ it follows that

$$J(\phi, \psi)|_1^2 = \phi(2)(-6\psi(2) - 4\psi'(2)) + \phi(1)(\psi'(1) + \psi(1)) \tag{13.2.43}$$

Since $\phi(1), \phi(2)$ can be chosen arbitrarily, it must be that

$$2\psi'(2) + 3\psi'(2) = 0 \quad \psi'(1) + \psi(1) = 0 \tag{13.2.44}$$

for $\psi \in D(T^*)$. □

Definition 13.1. We say that a set of boundary operators $\{B_1^*, B_2^*\}$ are adjoint to $\{B_1, B_2\}$, with respect to L, if

$$J(\phi, \psi)|_a^b = 0 \tag{13.2.45}$$

whenever $B_1\phi = B_2\phi = B_1^*\psi = B_2^*\psi = 0$. The conditions $B_1^*\psi = B_2^*\psi = 0$ are referred to as the adjoint boundary conditions (with respect to L).

Thus, for example, in Examples 13.2 and 13.3 we found adjoint boundary operators $\{\psi'(0), \psi(1)\}$ and $\{\psi'(1) + \psi(1), 2\psi'(2) + 3\psi(2)\}$, respectively. The operators B_1^*, B_2^* are not themselves unique, since for example they could always be interchanged or multiplied by constants. However, the subspace $\{\psi\colon B_1^*\psi = B_2^*\psi = 0\}$ is uniquely determined. If we now define $T^*\psi = L^*\psi$ on the domain

$$D(T^*) = \{\psi \in H^2(a,b)\colon B_1^*\psi = B_2^*\psi = 0\} \tag{13.2.46}$$

then $\langle T\phi, \psi\rangle = \langle\phi, T^*\psi\rangle$ if $\phi \in D(T)$ and $\psi \in D(T^*)$ and so T^* is the adjoint operator of T.

It can be shown (Exercise 13.5) that if $a_1 = a_2'$ (i.e., L is formally self-adjoint), and the boundary conditions are of the form (13.1.8), then the adjoint boundary conditions coincide with the original boundary conditions, so that T is self-adjoint. It is possible to also consider nonseparated boundary conditions of the form

$$B_1u = c_1u(a) + c_2u'(a) + c_3u(b) + c_4u'(b) = 0 \tag{13.2.47}$$

$$B_2u = d_1u(a) + d_2u'(a) + d_3u(b) + d_4u'(b) = 0 \tag{13.2.48}$$

to allow, for example, for periodic boundary conditions, see Exercise 13.7.

If T satisfies the assumptions of Theorem 13.1 then $N(T^*) = R(T)^\perp = \{0\}$. Thus T^* also satisfies these assumptions, and so has a corresponding Green's function which we denote by $G^*(x, y)$. Let us observe, at least formally, the important property

$$G(x, y) = G^*(y, x) \quad x, y \in (a, b) \tag{13.2.49}$$

To see this, use $L_zG(z,y) = \delta(z-y)$, $L_z^*G^*(z,x) = \delta(z-x)$ to get

$$G^*(y,x) - G(x,y) = \int_a^b G^*(z,x)L_zG(z,y)\,dz - \int_a^b G(z,y)L_z^*G^*(z,x)\,dz \tag{13.2.50}$$

$$= J(G^*(z,x), G(z,y))\Big|_{z=a}^{z=b} = 0 \tag{13.2.51}$$

where the last equality follows from the fact that G, G^* satisfy, respectively, the $\{B_1, B_2\}$ and $\{B_1^*, B_2^*\}$ boundary conditions as a function of their first variable. This confirms the expected result that $G(x,y) = G(y,x)$ if $T = T^*$. Furthermore it shows that as a function of the second variable, $G(x,y)$ satisfies the homogeneous adjoint equation for $x \neq y$ and the adjoint boundary conditions.

13.3 STURM-LIOUVILLE THEORY

If the operator T in Eq. (13.1.23) is self-adjoint, then the existence of real eigenvalues and eigenfunctions can be directly proved as a consequence of the fact that T^{-1} is compact and self-adjoint. But even if T is not self-adjoint, it is still possible to obtain such results by using a special device known as the Liouville transformation. Essentially we will produce a compact self-adjoint operator in a slightly different space, whose spectrum must agree with that of T. The resulting conclusions about the spectral properties of second-order ordinary differential operators, together with a number of other closely related facts, is generally referred to as *Sturm-Liouville theory*.

As in Eq. (13.1.7), let

$$L_0\phi = a_2(x)\phi'' + a_1(x)\phi' + a_0(x)\phi \tag{13.3.52}$$

with the assumptions that $a_j \in C([a,b])$, and now for definiteness $a_2(x) < 0$. Define

$$p(x) = \exp\left(\int_a^x \frac{a_1(s)}{a_2(s)}\,ds\right) \quad \rho(x) = -\frac{p(x)}{a_2(x)} \quad q(x) = a_0(x)\rho(x) \tag{13.3.53}$$

so that p, ρ are both positive and continuous on $[a,b]$. We then observe by simple calculus identities that $L_0\phi = \lambda\phi$ is equivalent to

$$-(p\phi')' + q\phi = \lambda\rho\phi \tag{13.3.54}$$

If we now define L_1, L by

$$L_1\phi = -(p\phi')' + q\phi \quad L\phi = \frac{L_1\phi}{\rho} \tag{13.3.55}$$

then we see that

$$L_0\phi = \lambda\phi \quad \text{if and only if} \quad L\phi = \lambda\phi \tag{13.3.56}$$

Note that L_1 is formally self-adjoint. In order to realize L itself as a self-adjoint operator we introduce the weighted space

$$L^2_\rho(a,b) = \left\{\phi\colon \int_a^b |\phi(x)|^2 \rho(x)\, dx < \infty\right\} \tag{13.3.57}$$

Since ρ is continuous and positive on $[a,b]$, this space may be regarded as the Hilbert space equipped with inner product

$$\langle \phi, \psi \rangle_\rho := \int_a^b \phi(x)\overline{\psi(x)}\rho(x)\, dx \tag{13.3.58}$$

for which the corresponding norm $\|\phi\|^2_\rho = \int_a^b |\phi(x)|^2 \rho(x)\, dx$ is equivalent to the usual $L^2(a,b)$ norm. We have obviously

$$\langle L\phi, \psi \rangle_\rho - \langle \phi, L\psi \rangle_\rho = \langle L_1\phi, \psi \rangle - \langle \phi, L_1\psi \rangle = 0 \tag{13.3.59}$$

for $\phi, \psi \in C_0^\infty(a,b)$. For $\phi, \psi \in C^2([a,b])$ we have instead, just as before, that

$$\langle L\phi, \psi \rangle_\rho - \langle \phi, L\psi \rangle_\rho = J(\phi, \psi)|_a^b \tag{13.3.60}$$

where $J(\phi,\psi) = p(\psi'\overline{\phi} - \psi\overline{\phi}')$. In the case of separated boundary conditions (13.1.8) we still have the property remarked earlier that $\{B_1^*, B_2^*\} = \{B_1, B_2\}$ so that the operator T_1 corresponding to $\{L_1, B_1, B_2\}$ is self-adjoint.

It follows in particular that the solution of

$$L_1\phi = f \quad B_1\phi = B_2\phi = 0 \tag{13.3.61}$$

may be given as

$$\phi(x) = \int_a^b G_1(x,y) f(y)\, dy \tag{13.3.62}$$

as long as there is no nontrivial solution of the homogeneous problem. The Green's function G_1 will have the properties stated in Theorem 13.1 and $G_1(x,y) = G_1(y,x)$ by the self-adjointness. The eigenvalue condition $L_1\phi = \lambda\rho\phi$ then amounts to

$$\phi(x) = \lambda \int_a^b G_1(x,y)\rho(y)\phi(y)\, dy \tag{13.3.63}$$

If we let $\psi(x) = \sqrt{\rho(x)}\phi(x)$, $\mu = 1/\lambda$ and

$$G(x,y) = \sqrt{\rho(x)}\sqrt{\rho(y)}G_1(x,y) \tag{13.3.64}$$

then we see that

$$\int_a^b G(x,y)\psi(y)\, dy = \mu\psi(x) \tag{13.3.65}$$

must hold. Conversely, any nontrivial solution of Eq. (13.3.65) gives rise, via all of the same transformations, to an eigenfunction of L_0 with the $\{B_1, B_2\}$

boundary conditions. The integral operator S with kernel G is clearly compact and self-adjoint, and 0 is not an eigenvalue, since $S\psi = 0$ would imply that zero is a solution of $L_1 u = \sqrt{\rho(x)}\psi(x)$. In particular, if $S\psi = \mu\psi$, then $\phi = \psi/\sqrt{\rho}$ satisfies

$$L_0\phi = \lambda\phi \quad B_1\phi = B_2\phi = 0 \tag{13.3.66}$$

with $\lambda = 1/\mu$.

The choice we made above that $a_2(x) < 0$ implies that the set of eigenvalues is bounded below. Consider for example the case that the boundary conditions are $\phi(a) = \phi(b) = 0$. From the fact that λ_n and any corresponding eigenfunction ϕ_n satisfy $L_1\phi_n = \lambda_n\rho\phi_n$, it follows, upon multiplying by ϕ_n and integrating by parts, that

$$\int_a^b (p|\phi_n'|^2 + q|\phi_n|^2)\,dx = \lambda_n \int_a^b \rho|\phi_n|^2\,dx \tag{13.3.67}$$

Since $p > 0$ we get in particular that

$$\lambda_n = \frac{\int_a^b (p|\phi_n'|^2 + q|\phi_n|^2)\,dx}{\int_a^b \rho|\phi_n|^2\,dx} \geq \frac{\int_a^b q|\phi_n|^2\,dx}{\int_a^b \rho|\phi_n|^2\,dx} \geq C \tag{13.3.68}$$

where $C = \min q/\max\rho$. The same conclusion holds for the case of more general boundary conditions, see Exercise 13.11.

Next we can say a little more about the eigenfunctions $\{\phi_n\}$. We know by Theorem 12.10 that the eigenfunctions $\{\psi_n\}$ of the operator S may be chosen as an orthonormal basis of $L^2(a,b)$. Since ϕ_n may be taken to be $\psi_n/\sqrt{\rho}$, by the preceding discussion, it follows that

$$\int_a^b \phi_n\phi_m\rho\,dx = \int_a^b \psi_n\psi_m\,dx = \begin{cases} 0 & n \neq m \\ 1 & n = m \end{cases} \tag{13.3.69}$$

Thus the eigenfunctions are orthonormal in the weighted space $L^2_\rho(a,b)$. We can also easily verify the completeness of these eigenfunctions as follows. For any $f \in L^2_\rho(a,b)$ we have that $\sqrt{\rho}f \in L^2(a,b)$, so

$$f\sqrt{\rho} = \sum_{n=1}^\infty c_n\psi_n \quad c_n = \langle f\sqrt{\rho}, \psi_n\rangle \tag{13.3.70}$$

in the sense of L^2 convergence. Equivalently, this means

$$f = \sum_{n=1}^\infty c_n\phi_n \quad c_n = \langle f\rho, \phi_n\rangle = \langle f, \phi_n\rangle_\rho \tag{13.3.71}$$

also in the sense of L^2 or L^2_ρ convergence, and so the completeness follows from Theorem 5.4.

From these observations, together with Theorem 12.10 and Corollary 13.1 we obtain the following.

Theorem 13.2. *Assume that* $a_0, a_1, a_2 \in C([a,b])$, $a_2(x) < 0$ *on* $[a,b]$, *and that* $|c_1| + |c_2| \neq 0, |c_3| + |c_4| \neq 0$. *Then the problem*

$$a_2\phi'' + a_1\phi' + a_0\phi = \lambda\phi \quad a < x < b \quad c_1\phi(a) + c_2\phi'(a) = c_3\phi(b) + c_4\phi'(b) = 0 \tag{13.3.72}$$

has a countable sequence of simple real eigenvalues $\{\lambda_n\}_{n=1}^{\infty}$, *with* $\lambda_n \to \infty$. *The corresponding eigenfunctions may be chosen to form an orthonormal basis of* $L^2_\rho(a,b)$.

There is one other notable property of the eigenfunctions, which we mention without proof: *The eigenfunction* ϕ_n *has exactly* $n-1$ *roots in* (a,b). See for example Theorem 2.1, Chapter 8 of [7].

13.4 LAPLACIAN WITH HOMOGENEOUS DIRICHLET BOUNDARY CONDITIONS

In this section we develop some theory for the very important eigenvalue problem

$$-\Delta u = \lambda u \quad x \in \Omega \tag{13.4.73}$$

$$u = 0 \quad x \in \partial\Omega \tag{13.4.74}$$

Here Ω is a bounded open set in $\mathbb{R}^N$, $N \geq 2$, with sufficiently smooth boundary. The general approach will again be to obtain the existence of eigenvalues and eigenfunctions by first looking at an appropriately defined inverse operator. To begin making precise the definitions of the operators involved, set

$$Tu = -\Delta u \quad \text{on } D(T) = \{u \in H_0^1(\Omega) \colon \Delta u \in L^2(\Omega)\} \tag{13.4.75}$$

to be regarded as an unbounded operator on $L^2(\Omega)$.

Recall that in Section 8.1 we have defined the Sobolev spaces $H^1(\Omega)$ and $H_0^1(\Omega)$, and it was mentioned there that it is appropriate to regard $u \in H_0^1(\Omega)$ as meaning that $u \in H^1(\Omega)$ and $u = 0$ on $\partial\Omega$. The precise meaning of this needs to be clarified, since in general a function $u \in H^1(\Omega)$ need not be continuous on $\overline{\Omega}$, so that its restriction to the lower dimensional set $\partial\Omega$ is not defined in an obvious way. The following theorem is proved in [5, Lemma 9.9 and following discussion] or [11, Theorem 1, Section 5.5].

Theorem 13.3. *If* Ω *is a bounded domain in* $\mathbb{R}^N$ *with a* C^1 *boundary, then there exists a bounded linear operator* $\tau \colon H^1(\Omega) \to L^2(\partial\Omega)$ *such that*

$$\tau u = u|_{\partial\Omega} \quad \text{if } u \in H^1(\Omega) \cap C(\overline{\Omega}) \tag{13.4.76}$$

$$\tau u = 0 \quad \text{if } u \in H_0^1(\Omega) \tag{13.4.77}$$

The mapping τ in this theorem is the *trace operator*, that is, the operator of restriction to $\partial\Omega$, and τu is called the trace of u on $\partial\Omega$. According to the theorem, the trace is well defined for any $u \in H^1(\Omega)$, it coincides with the usual notion of restriction if u happens to be continuous on $\overline{\Omega}$, and any function $u \in H_0^1(\Omega)$ has trace equal to 0. It can be furthermore be shown that the expected integration by parts formula (see Eq. A.3.31)

$$\int_\Omega u \frac{\partial v}{\partial x_j}\, dx = -\int_\Omega \frac{\partial u}{\partial x_j} v\, dx + \int_{\partial\Omega} uvn_j\, dS \tag{13.4.78}$$

remains valid as long as $u, v \in H^1(\Omega)$, where in the boundary integral u, v must be understood as meaning τu and τv. The boundary integral is well defined since these traces belong to $L^2(\partial\Omega)$, according to the theorem.

For any $f \in L^2(\Omega)$, the condition that $u \in D(T)$ and $Tu = f$ means

$$u \in H_0^1(\Omega) \quad -\int_\Omega u\Delta v\, dx = \int_\Omega fv\, dx \quad \forall v \in C_0^\infty(\Omega) \tag{13.4.79}$$

The first integral may be equivalently written as $\int_\Omega \nabla u \cdot \nabla v\, dx$, using the integration by parts formula, and then by the density of $C_0^\infty(\Omega)$ in $H_0^1(\Omega)$, we see that

$$u \in H_0^1(\Omega) \quad \int_\Omega \nabla u \cdot \nabla v\, dx = \int_\Omega fv\, dx \quad \forall v \in H_0^1(\Omega) \tag{13.4.80}$$

must hold. Conversely, any function u satisfying Eq. (13.4.80) must also satisfy $Tu = f$.

In particular, if λ is an eigenvalue of T then

$$u \in H_0^1(\Omega) \quad \int_\Omega \nabla u \cdot \nabla v\, dx = \lambda \int_\Omega uv\, dx \quad \forall v \in H_0^1(\Omega) \tag{13.4.81}$$

We note that $\lambda \geq 0$ must hold, since we can choose $v = u$. As we will see below $\lambda = 0$ is impossible also.

Another tool we will make good use of is the so-called *Poincaré inequality*.

Proposition 13.1. *If Ω is a bounded open set in $\mathbb{R}^N$ then there exists a constant C, depending only on Ω, such that*

$$\|u\|_{L^2(\Omega)} \leq C\|\nabla u\|_{L^2(\Omega)} \quad \forall u \in H_0^1(\Omega) \tag{13.4.82}$$

Proof. It is enough to prove the stated inequality for $u \in C_0^\infty(\Omega)$. If we let R be large enough so that $\Omega \subset Q_R = \{x \in \mathbb{R}^N\colon |x_j| < R, j = 1, \dots, N\}$ then defining $u = 0$ outside of Ω we may also regard u as an element of $C_0^\infty(Q_R)$, with identical norms whether considered on Ω or Q_R. Therefore

$$\|u\|^2_{L^2(\Omega)} = \int_{Q_R} u^2\, dx = -\int_{Q_R} x_1 \frac{\partial}{\partial x_1} u^2 dx \tag{13.4.83}$$

$$= -2\int_{Q_R} x_1 u \frac{\partial u}{\partial x_1} dx \tag{13.4.84}$$

$$\leq 2R\|u\|_{L^2(\Omega)}\|\nabla u\|_{L^2(\Omega)} \tag{13.4.85}$$

Thus the conclusion holds with $C = 2R$. □

Note that we do not really need Ω to be bounded here, only that it be contained between two parallel hyperplanes. It is an immediate consequence of Poincaré's inequality that

$$\|u\|_{H_0^1(\Omega)} := \|\nabla u\|_{L^2(\Omega)} \tag{13.4.86}$$

defines a norm on $H_0^1(\Omega)$ which is equivalent to the original norm it inherits as a subspace of $H^1(\Omega)$, since

$$1 \le \frac{\|u\|^2_{H^1(\Omega)}}{\|u\|^2_{H_0^1(\Omega)}} = \frac{\int_\Omega (u^2 + |\nabla u|^2)\,dx}{\int_\Omega |\nabla u|^2\,dx} \le C^2 + 1 \tag{13.4.87}$$

Unless otherwise stated we always assume that the norm on $H_0^1(\Omega)$ is that given by Eq. (13.4.86), which of course corresponds to the inner product

$$\langle u, v\rangle_{H_0^1(\Omega)} = \int_\Omega \nabla u \cdot \overline{\nabla v}\,dx \tag{13.4.88}$$

A simple but important connection between the eigenvalues of T and the Poincaré inequality, obtained by choosing $v = u$ in the right-hand equality of Eq. (13.4.81), is that any such eigenvalue λ satisfies

$$\lambda \ge \frac{1}{C^2} > 0 \tag{13.4.89}$$

where C is any constant for which Poincaré's inequality is valid. We will see later that there is a "best constant," namely a value $C = C_P$ for which Eq. (13.4.82) is true, but is false for any smaller value, and the smallest positive eigenvalue of T is precisely $1/C_P^2$.

Any constant which works in the Poincaré inequality also provides a lower bound for the operator T, as follows: If $Tu = f$ then choosing $v = u$ in Eq. (13.4.81) we get

$$\int_\Omega |\nabla u|^2\,dx = \int_\Omega fu\,dx \le \|f\|_{L^2(\Omega)}\|u\|_{L^2(\Omega)} \le C\|f\|_{L^2(\Omega)}\|\nabla u\|_{L^2(\Omega)} \tag{13.4.90}$$

Therefore

$$\|u\|_{H_0^1(\Omega)} \le C\|f\|_{L^2(\Omega)} \quad \text{and} \quad \|u\|_{L^2(\Omega)} \le C^2\|f\|_{L^2(\Omega)} \tag{13.4.91}$$

or equivalently $\|Tu\|_{L^2(\Omega)} \ge C^{-2}\|u\|_{L^2(\Omega)}$.

Proposition 13.2. *Considered as a linear operator on $L^2(\Omega)$, T is one-to-one, onto, has a bounded inverse and is self-adjoint.*

Proof. If $f \in L^2(\Omega)$ define the linear functional ϕ by

$$\phi(v) = \int_\Omega fv\,dx \tag{13.4.92}$$

Then ϕ is continuous on $H_0^1(\Omega)$ since

$$|\phi(v)| \le \|f\|_{L^2(\Omega)}\|v\|_{L^2(\Omega)} \le C\|f\|_{L^2(\Omega)}\|v\|_{H_0^1(\Omega)} \tag{13.4.93}$$

By the Riesz Representation Theorem (Theorem 5.6) there exists a unique $u \in H_0^1(\Omega)$ such that

$$\langle u, v\rangle_{H_0^1(\Omega)} = \int_\Omega \nabla u \cdot \nabla v \, dx = \phi(v) \tag{13.4.94}$$

which is equivalent to $Tu = f$, as explained above. Thus T is onto. The property that T is one-to-one with a bounded inverse is now immediate from Eq. (13.4.91).

Finally, from Eq. (13.4.81) it follows that $\langle Tu, v\rangle = \int_\Omega \nabla u \cdot \nabla v \, dx = \langle u, Tv\rangle$ (i.e., T is symmetric) and a linear operator which is symmetric and onto must be self-adjoint, see Exercise 10.7 of Chapter 10. □

Next we consider the construction of an inverse operator to T, in the form of an integral operator

$$Sf(x) = \int_\Omega G(x, y)f(y) \, dy \tag{13.4.95}$$

where G will again be called the Green's function for $Tu = f$, assuming it exists. Thus $u(x) = Sf(x)$ should be the solution of

$$-\Delta u = f(x) \quad x \in \Omega \quad u(x) = 0 \quad x \in \partial\Omega \tag{13.4.96}$$

Analogously to the ODE case discussed in the previous section, we expect that G should formally satisfy

$$-\Delta_x G(x, y) = \delta(x - y) \quad x \in \Omega \quad G(x, y) = 0 \quad x \in \partial\Omega \tag{13.4.97}$$

for every fixed $y \in \Omega$. Recall that we already know that there exists $\Gamma(x)$ such that $-\Delta\Gamma = \delta$ in the sense of distributions, so if we set $h(x, y) = G(x, y) - \Gamma(x - y)$ then it is necessary for h to satisfy

$$-\Delta_x h(x, y) = 0 \quad x \in \Omega \quad h(x, y) = -\Gamma(x - y) \quad x \in \partial\Omega \tag{13.4.98}$$

for every fixed $y \in \Omega$. Note that since $x - y \ne 0$ for $x \in \partial\Omega$ and $y \in \Omega$, the boundary function for h is infinitely differentiable. Thus we have a parameterized set of boundary value problems, each having the form of finding a function harmonic in Ω satisfying a prescribed smooth Dirichlet type boundary condition. Such a problem is known to have a unique solution, assuming only very minimal hypotheses on the smoothness of $\partial\Omega$, see for example Theorem 2, Section 4.3 of [24]. In a few special cases it is possible to compute $h(x, y)$, and hence $G(x, y)$, explicitly, see Exercise 13.18 for the case when Ω is a ball.

Note, however, that whatever h may be, $G(x, y)$ is singular when $x = y$, and possesses the same local integrability properties as $\Gamma(x - y)$. It is not hard to check that $\int_{\Omega\times\Omega} |\Gamma(x - y)|^2 \, dxdy$ is finite for $N = 2, 3$ but not for $N \ge 4$. Thus

G is not of Hilbert-Schmidt type in general, so we cannot directly conclude in this way that $S = T^{-1}$ is compact on $L^2(\Omega)$. Nevertheless the operator is indeed compact. One approach to showing this comes from the general theory of singular integral operators, see Chapter 14. A simple alternative, which we will use here, is based on the following result, which is of independent importance.

Theorem 13.4. *(Rellich-Kondrachov) A bounded set in $H_0^1(\Omega)$ is precompact in $L^2(\Omega)$.*

For a proof we refer to [11, Section 5.7, Theorem 1] or [5, Theorem 9.16], where somewhat more general statements are given. With some minimal smoothness assumption on $\partial\Omega$ we can replace $H_0^1(\Omega)$ by $H^1(\Omega)$. It is an equivalent statement to say that the identity map i: $H_0^1(\Omega) \to L^2(\Omega)$ is a compact linear operator. Other terminology such as that $H_0^1(\Omega)$ is compactly embedded, compactly included, or compactly injected in $L^2(\Omega)$ (or $H_0^1(\Omega) \hookrightarrow L^2(\Omega)$) are also commonly used.

Corollary 13.2. *If $S = T^{-1}$ then S is a compact self-adjoint operator on $L^2(\Omega)$.*

Proof. If $E \subset L^2(\Omega)$ is bounded, then by Eq. (13.4.91) the image $S(E) = \{u = Sf : f \in E\}$ is bounded in $H_0^1(\Omega)$. The Rellich-Kondrachov theorem then implies $S(E)$ is precompact as a subset of $L^2(\Omega)$, so S: $L^2(\Omega) \to L^2(\Omega)$ is compact. The self-adjointness of S follows immediately from that of T. □

Thus S possesses an infinite sequence of real eigenvalues $\{\mu_n\}_{n=1}^\infty$, $\lim_{n\to\infty} \mu_n = 0$, and corresponding eigenfunctions $\{\psi_n\}_{n=1}^\infty$, which may be chosen as an orthonormal basis of $L^2(\Omega)$. As usual, the reciprocals $\lambda_n = 1/\mu_n$ are eigenvalues of $T = S^{-1}$, and recall that all eigenvalues of T are strictly positive. We have established the following.

Theorem 13.5. *The operator*

$$Tu = -\Delta u \quad D(T) = \{u \in H_0^1(\Omega) : \Delta u \in L^2(\Omega)\} \tag{13.4.99}$$

has an infinite sequence of real eigenvalues of finite multiplicity,

$$0 < \lambda_1 \leq \lambda_2 \leq \lambda_3 \leq \cdots \leq \lambda_n \to +\infty \tag{13.4.100}$$

and corresponding eigenfunctions $\{\psi_n\}_{n=1}^\infty$, which may be chosen as an orthonormal basis of $L^2(\Omega)$.

The convention here is that an eigenvalue in this sequence is repeated according to its multiplicity. In comparison with the Sturm-Liouville case, an eigenvalue need not be simple, although the multiplicity must still be finite, thus repetitions in the sequence (13.4.100) may occur. It does turn out to be the case, however, that λ_1 is always simple, that is to say $\lambda_1 < \lambda_2$. We refer to λ_n, ψ_n as Dirichlet eigenvalues and eigenfunctions for the domain Ω. Among many other things, knowledge of the existence of these eigenvalues and eigenfunctions allows us to greatly expand the scope of the separation of variables method.

Example 13.4. Consider the initial and boundary value problem for the heat equation in a bounded domain $\Omega \subset \mathbb{R}^N$,

$$u_t - \Delta u = 0 \qquad x \in \Omega \qquad t > 0 \tag{13.4.101}$$

$$u(x,t) = 0 \qquad x \in \partial\Omega \qquad t > 0 \tag{13.4.102}$$

$$u(x,0) = f(x) \quad x \in \Omega \tag{13.4.103}$$

Employing the separation of variables method, we begin by looking for solutions in the product form $\psi(x)\phi(t)$, which satisfy the PDE and the homogeneous boundary condition. Substituting we see that $\phi'(t)\psi(x) = \phi(t)\Delta\psi(x)$ should hold, and therefore

$$\phi' + \lambda\phi = 0 \quad t > 0 \qquad \Delta\psi + \lambda\psi = 0 \quad x \in \Omega \tag{13.4.104}$$

In addition the boundary condition implies that $\psi(x) = 0$ for $x \in \partial\Omega$. In order to have a nonzero solution, we conclude that λ, ψ must be a Dirichlet eigenvalue/eigenfunction pair for the domain Ω, and then correspondingly $\phi(t) = Ce^{-\lambda t}$. By linearity we therefore see that if λ_n, ψ_n denote the Dirichlet eigenvalues and $L^2(\Omega)$ orthonormalized eigenfunctions, then

$$u(x,t) = \sum_{n=1}^{\infty} c_n e^{-\lambda_n t}\psi_n(x) \tag{13.4.105}$$

is a solution of Eqs. (13.4.101), (13.4.102), as long as the coefficients c_k are sufficiently rapidly decaying.

In order that Eq. (13.4.103) also holds, we must have

$$f(x) = u(x,0) = \sum_{n=1}^{\infty} c_n\psi_n(x) \tag{13.4.106}$$

and so from the orthonormality, $c_n = \langle f, \psi_n\rangle$. We have thus obtained the (formal) solution

$$u(x,t) = \sum_{n=1}^{\infty} \langle f, \psi_n\rangle e^{-\lambda_n t}\psi_n(x) \tag{13.4.107}$$

of Eqs. (13.4.101)–(13.4.103). □

Making use of estimates which may be found in more advanced PDE textbooks, it can be shown that for any $f \in L^2(\Omega)$ the series (13.4.107) is uniformly convergent to an infinitely differentiable limit $u(x,t)$ for $t > 0$, where u is a classical solution of Eqs. (13.4.101), (13.4.102), and the initial condition (13.4.103) is satisfied at least in the sense that $\lim_{t\to 0}\int_\Omega (u(x,t) - f(x))^2\,dx = 0$. Under stronger conditions on f, the nature of the convergence at $t = 0$ can be shown to be correspondingly stronger. We refer, for example, to [11] for more details. At the very least, since $\sum_{n=1}^{\infty} |c_n|^2 < \infty$ must hold, the obvious estimate $\sum_{n=1}^{\infty} |c_n e^{-\lambda_n t}|^2 < \infty$ for $t \ge 0$ implies that the series is convergent in $L^2(\Omega)$ for every fixed $t \ge 0$.

Note, again at a formal level at least, that the expression for the solution u can be rewritten as

$$u(x,t) = \sum_{n=1}^{\infty} \left(\int_\Omega f(y)\psi_n(y)\,dy \right) e^{-\lambda_n t}\psi_n(x) \tag{13.4.108}$$

$$= \int_\Omega f(y) \left(\sum_{n=1}^{\infty} e^{-\lambda_n t}\psi_n(x)\psi_n(y) \right) dy \tag{13.4.109}$$

$$= \int_\Omega f(y)G(x,y,t)dy \tag{13.4.110}$$

suggesting that

$$G(x,y,t) := \sum_{n=1}^{\infty} e^{-\lambda_n t}\psi_n(x)\psi_n(y) \tag{13.4.111}$$

should be regarded as the Green's function for Eqs. (13.4.101)–(13.4.103).

13.5 EXERCISES

13.1. Let $Lu = (x-2)u'' + (1-x)u' + u$ on $(0,1)$.

(a) Find the Green's function for

$$Lu = f \quad u'(0) = 0 \quad u(1) = 0$$

(Hint: First show that $x-1, e^x$ are linearly independent solutions of $Lu = 0$.)

(b) Find the adjoint operator and boundary conditions.

13.2. Let

$$Tu = -\frac{d}{dx}\left(x\frac{du}{dx} \right)$$

on the domain

$$D(T) = \{u \in H^2(1,2) \colon u(1) = u(2) = 0\}$$

(a) Show that $N(T) = \{0\}$.

(b) Find the Green's function for the boundary value problem $Tu = f$.

(c) State and prove a result about the continuous dependence of the solution u on f in part (b).

13.3. Let ϕ, ψ be solutions of $Lu = a_2(x)u'' + a_1(x)u' + a_0(x)u = 0$ on (a,b) and $W(\phi,\psi)(x) = \phi(x)\psi'(x) - \phi'(x)\psi(x)$ be the corresponding Wronskian determinant.

(a) Show that W is either zero everywhere or zero nowhere. (Suggestion: Find a first-order ODE satisfied by W.)
(b) If $a_1(x) = 0$ show that the W is constant.

13.4. Prove the validity of Eq. (13.1.21). (Suggestions: Start by writing $u(x)$ in the form

$$u(x) = \phi_2(x) \int_a^x C_2(y)f(y)\,dy + \phi_1(x) \int_x^b C_1(y)f(y)\,dy$$

and note that some of the terms that arise in the expression for $u'(x)$ will cancel.)

13.5. Let $Lu = a_2(x)u'' + a_1(x)u' + a_0(x)u$ with $a_2' = a_1$, so that L is formally self-adjoint. If $B_1u = C_1u(a) + C_2u'(a), B_2u = C_3u(b) + C_4u'(b)$, show that $\{B_1^*, B_2^*\} = \{B_1, B_2\}$.

13.6. Find the Green's function for

$$u'' + 2u' - 3u = f(x) \quad 0 < x < \infty \quad u(0) = 0 \quad \lim_{x\to\infty} u(x) = 0$$

(think of the last condition as a "boundary condition at infinity"). Using the Green's function, find $u(2)$ if $f(x) = e^{-6x}$.

13.7. Consider the second-order operator

$$Lu = a_2(x)u'' + a_1(x)u' + a_0(x)u \quad a < x < b$$

with *nonseparated* boundary conditions

$$B_1u = \alpha_{11}u(a) + \alpha_{12}u'(a) + \beta_{11}u(b) + \beta_{12}u'(b) = 0$$
$$B_2u = \alpha_{21}u(a) + \alpha_{22}u'(a) + \beta_{21}u(b) + \beta_{22}u'(b) = 0$$

where the vectors $(\alpha_{11}, \alpha_{12}, \beta_{11}, \beta_{12})$, $(\alpha_{21}, \alpha_{22}, \beta_{21}, \beta_{22})$ are linearly independent. We again say that two other nonseparated boundary conditions B_1^*, B_2^* are adjoint to B_1, B_2 with respect to L if $J(u, v)|_a^b = 0$ whenever $B_1u = B_2u = B_1^*v = B_2^*v = 0$.

Find the adjoint operator and boundary conditions in the case that

$$Lu = u'' + xu'$$
$$B_1u = u'(0) - 2u(1) \quad B_2u = u(0) + u(1)$$

13.8. When we rewrite $a_2(x)u'' + a_1(x)u' + a_0(x)u = \lambda u$ as

$$-(p(x)u')' + q(x)u = \lambda\rho(x)u$$

the latter is often referred to as the *Liouville normal form*. Consider the eigenvalue problem

$$x^2u'' + xu' + u = \lambda u \quad 1 < x < 2$$
$$u(1) = u(2) = 0$$

(a) Find the Liouville normal form.

(b) What is the orthogonality relationship satisfied by the eigenfunctions?

(c) Find the eigenvalues and eigenfunctions. (You may find the original form of the equation easier to work with than the Liouville normal form when computing the eigenvalues and eigenfunctions.)

13.9. Consider the Sturm-Liouville equation in the Liouville normal form,

$$-(p(x)u')' + q(x)u = \lambda\rho(x)u \quad a < x < b$$

where $p, \rho \in C^2([a,b])$, $q \in C([a,b])$, $p, \rho > 0$ on $[a,b]$. Let

$$\sigma(x) = \sqrt{\frac{\rho(x)}{p(x)}} \quad \eta(x) = (p(x)\rho(x))^{1/4} \quad L = \int_a^b \sigma(s)\,ds$$

$$\phi(x) = \frac{1}{L}\int_a^x \sigma(s)\,ds$$

If $\psi = \phi^{-1}$ (the inverse function of ϕ) and $v(z) = \eta(\psi(z))u(\psi(z))$ show that v satisfies

$$-v'' + Q(z)v = \mu v \quad 0 < z < 1 \tag{13.5.112}$$

for some Q depending on p, ρ, q, and $\mu = L^2\lambda$. (This is mainly a fairly tedious exercise with the chain rule. Focus on making the derivation as clean as possible and be sure to say exactly what $Q(z)$ is. The point of this is that *every* eigenvalue problem for a second-order ODE is equivalent to one with an equation of the form (13.5.112), provided that the coefficients have sufficient smoothness. The map $u(x) \to v(z)$ is sometimes called the *Liouville transformation*, and the ODE (13.5.112) is the *canonical form* for a second-order ODE eigenvalue problem.)

13.10. Consider the Sturm-Liouville problem

$$u'' + \lambda u = 0 \qquad 0 < x < 1$$
$$u(0) - u'(0) = u(1) = 0$$

(a) Multiply the equation by u and integrate by parts to show that any eigenvalue is positive.

(b) Show that the eigenvalues are the positive solutions of

$$\tan\sqrt{\lambda} = -\sqrt{\lambda}.$$

(c) Show graphically that such roots exist, and form an infinite sequence λ_k such that $\left(k - \frac{1}{2}\right)\pi < \sqrt{\lambda_k} < k\pi$ and

$$\lim_{k\to\infty}\left(\sqrt{\lambda_k} - \left(k - \frac{1}{2}\right)\pi\right) = 0$$

13.11. Complete the proof that $\lambda_n \to +\infty$ under the assumptions of Theorem 13.2. (Suggestion: You can obtain an inequality like Eq. (13.3.68), except it may also contain boundary terms. It may be helpful to establish and use the following inequality: for any $\epsilon > 0$ there exists $M < \infty$ such that

$$\|u\|^2_{L^\infty(a,b)} \le \epsilon \|u'\|^2_{L^2(a,b)} + M\|u\|^2_{L^2(a,b)}$$

for any $u \in C^1([a,b])$.)

13.12. Using separation of variables, compute explicitly the Dirichlet eigenvalues and eigenfunctions of $-\Delta$ when the domain is a rectangle $(0,A) \times (0,B)$ in $\mathbb{R}^2$. Verify directly that the first eigenvalue is simple, and that the first eigenfunction is of constant sign. Can there be other eigenvalues of multiplicity greater than 1? (Hint: Your answer should depend on whether the ratio A/B is rational or irrational.)

13.13. Find Dirichlet eigenvalues and eigenfunctions of $-\Delta$ in the unit ball $B(0,1) \subset \mathbb{R}^2$. (Suggestion: Express the PDE and do separation of variables in polar coordinates. Your answer should involve Bessel functions.)

13.14. If $\{\psi_n\}_{n=1}^\infty$ are Dirichlet eigenfunctions of the Laplacian making up an orthonormal basis of $L^2(\Omega)$, let $\zeta_n = \psi_n/\sqrt{\lambda_n}$ (λ_n the corresponding eigenvalue).

(a) Show that $\{\zeta_n\}_{n=1}^\infty$ is an orthonormal basis of $H_0^1(\Omega)$.

(b) Show that $f \in H_0^1(\Omega)$ if and only if $\sum_{n=1}^\infty \lambda_n |\langle f, \psi_n\rangle|^2 < \infty$.

13.15. If $\Omega \subset \mathbb{R}^n$ is a bounded open set with smooth enough boundary, find a solution of the wave equation problem

$$u_{tt} - \Delta u = 0 \quad x \in \Omega \quad t > 0$$
$$u(x,t) = 0 \quad x \in \partial\Omega \ t > 0$$
$$u(x,0) = f(x) \quad u_t(x,0) = g(x) \ x \in \Omega$$

in the form

$$u(x,t) = \sum_{n=1}^\infty c_n(t)\psi_n(x)$$

where $\{\psi_n\}_{n=1}^\infty$ are the Dirichlet eigenfunctions of $-\Delta$ in Ω.

13.16. Derive formally that

$$G(x,y) = \sum_{n=1}^\infty \frac{\psi_n(x)\psi_n(y)}{\lambda_n} \tag{13.5.113}$$

where λ_n, ψ_n are the Dirichlet eigenvalues and normalized eigenfunctions for the domain Ω, and $G(x,y)$ is the corresponding Green's function in Eq. (13.4.95). (Suggestion: If $-\Delta u = f$, expand both u and f in the ψ_n basis.)

13.17. Formulate and prove a result which says that under appropriate conditions

$$u(x,t) \approx Ce^{-\lambda_1 t}\psi_1(x) \tag{13.5.114}$$

as $t \to \infty$, where u is the solution of Eqs. (13.4.101)–(13.4.103).

13.18. If $\Omega = B(0,1) \subset \mathbb{R}^N$ show that the function $h(x,y)$ appearing in Eq. (13.4.98) is given by

$$h(x,y) = -\Gamma(|x|y - x/|x|) \tag{13.5.115}$$

13.19. Prove the Rellich-Kondrachov theorem, Theorem 13.4, directly in the case of one space dimension, by using the Arzela-Ascoli theorem.

Chapter 14

Further Study of Integral Equations

In this chapter we present a brief discussion of several loosely related topics in the theory of integral equations.

14.1 SINGULAR INTEGRAL OPERATORS

In the very broadest sense, an integral operator

$$Tu(x) = \int_\Omega K(x,y)u(y)dy \tag{14.1.1}$$

is said to be singular if the kernel $K(x,y)$ fails to be C^∞ at one or more points. Of course this need not significantly affect the general properties of T, but there are certain more specific kinds of singularity which occur in natural and important ways, which do affect the general behavior of the operator, and so call for some specific study.

First of all let us observe that singularity is not necessarily a bad thing. For example, the problem of solving $Tu = f$ with a C^∞ kernel $K(x,y)$ is a first kind integral equation, for which a solution only exists, in general, for very restricted f. By comparison, the corresponding second kind integral equation $\lambda u - Tu = f$ may be regarded, at least formally, as a first kind equation with the "very singular" kernel $\lambda\delta(x-y) - K(x,y)$, and will have a unique solution for a much larger class of f's, typically all $f \in L^2(\Omega)$ in fact.

As a second kind of example, recall that if $\Omega = (a,b) \subset \mathbb{R}$, a Volterra type integral equation is generally easier to analyze and solve than a corresponding non-Volterra type equation. The more amenable nature of the Volterra equation may be understood as the fact that the Volterra operator $Tu(x) = \int_a^x K(x,y)u(y)dy$ could be rewritten as $\int_a^b \tilde{K}(x,y)u(y)dy$, where

$$\tilde{K}(x,y) = \begin{cases} K(x,y) & a < y < x < b \\ 0 & a < x < y < b \end{cases} \tag{14.1.2}$$

That is to say, $\tilde{K}$ is singular when $y = x$ no matter how smooth K itself is so singularity is built in to the very structure of a Volterra type integral equation.

Techniques of Functional Analysis for Differential and Integral Equations
http://dx.doi.org/10.1016/B978-0-12-811426-1.00014-3

Let us also mention that it often appropriate to regard T as being singular if the underlying domain Ω is unbounded. One might expect this from the fact that if we were to make a change of variable to map the unbounded domain Ω onto a convenient bounded domain, the price to be paid normally is that the transformed kernel will become singular at those points which are the image of ∞. The Fourier transform could be regarded in this light, and its very nice behavior viewed as due to, rather than despite, the presence of singularity.

In this section we will focus on a specific class of singular integral operators, in which the kernel K is assumed to satisfy

$$|K(x,y)| \le \frac{M}{|x-y|^\alpha} \quad x,y \in \Omega \tag{14.1.3}$$

for some constant M and exponent $\alpha > 0$, with Ω a bounded domain in $\mathbb{R}^N$. If $\alpha < N$ then K is said to be *weakly singular*. The main result to be proved below is that an integral operator with weakly singular kernel is compact on $L^2(\Omega)$. Note that such an operator may or may not be of Hilbert-Schmidt type. For example, if $K(x,y) = 1/|x-y|^\alpha$ then $K \in L^2(\Omega \times \Omega)$ if and only if $\alpha < N/2$. The Green's function $G(x,y)$ for the Laplacian (see Eq. 13.4.95) is always weakly singular, and the compactness result below provides an alternative to the Rellich-Kondrachov theorem (Theorem 13.4) for proving compactness of the corresponding integral operator.

We begin with the following lemma.

Lemma 14.1. *Suppose $K \in L^1(\Omega \times \Omega)$ and there exists a constant C such that*

$$\int_\Omega |K(x,y)|dx \le C \ \ \forall y \in \Omega \qquad \int_\Omega |K(x,y)|dy \le C \ \ \forall x \in \Omega \tag{14.1.4}$$

Then the corresponding integral operator T is a bounded linear operator on $L^2(\Omega)$ with $\|T\| \le C$.

Proof. Using the Schwarz inequality we get

$$\int_\Omega |K(x,y)|\,|u(y)|dy \le \sqrt{\int_\Omega |K(x,y)|dy}\sqrt{\int_\Omega |K(x,y)|\,|u(y)|^2dy} \tag{14.1.5}$$

$$\le \sqrt{C}\sqrt{\int_\Omega |K(x,y)|\,|u(y)|^2dy} \tag{14.1.6}$$

and therefore

$$\int_\Omega |Tu(x)|^2dx \le C\int_\Omega \left[\int_\Omega |K(x,y)|\,|u(y)|^2dy\right]dx \tag{14.1.7}$$

$$= C\int_\Omega |u(y)|^2\left[\int_\Omega |K(x,y)|dx\right]dy \tag{14.1.8}$$

$$\le C^2\int_\Omega |u(y)|^2dy \tag{14.1.9}$$

as needed. □

We can now prove the compactness result mentioned above.

Theorem 14.1. *If Ω is a bounded domain in $\mathbb{R}^N$ and K is a weakly singular kernel, then the integral operator* (14.1.1) *is compact on $L^2(\Omega)$.*

Proof. First observe that

$$\int_\Omega |K(x,y)|dy \le M \int_\Omega \frac{dy}{|x-y|^\alpha} \le M \int_{B(x,R)} \frac{dy}{|x-y|^\alpha} \tag{14.1.10}$$

$$= M\Omega_{N-1} \int_0^R r^{N-1-\alpha} dr = \frac{M\Omega_{N-1}R^{N-\alpha}}{N-\alpha} \tag{14.1.11}$$

for some large enough R depending on Ω. Here Ω_{N-1} denotes the surface area of the unit sphere in $\mathbb{R}^N$, see Eq. (A.4.37). The same is true if we integrate with respect to x instead of y, and so by Lemma 14.1, T is bounded. Now let

$$K_m(x,y) = \begin{cases} K(x,y) & |x-y| > \frac{1}{m} \\ 0 & |x-y| \le \frac{1}{m} \end{cases} \tag{14.1.12}$$

and note that $K - K_m$ satisfies the same estimate as K discussed above, except that R may be replaced by $1/m$. That is,

$$\int_\Omega |K(x,y) - K_m(x,y)|dy \le \frac{M\Omega_{N-1}}{(N-\alpha)m^{N-\alpha}} \tag{14.1.13}$$

and likewise for the integral with respect to x. Thus, if T_m is the integral operator with kernel K_m, then using Lemma 14.1 once more we get

$$\|T - T_m\| \le \frac{M\Omega_{N-1}}{(N-\alpha)m^{N-\alpha}} \to 0 \tag{14.1.14}$$

as $m \to \infty$. Since $K_m \in L^\infty(\Omega \times \Omega)$, the operator T_m is compact for each m, by Theorem 12.4, and so the compactness of T follows from Theorem 12.3. □

Theorem 14.2. *Let Ω be a bounded domain in $\mathbb{R}^N$ and assume K is a weakly singular kernel, which is continuous on $\overline{\Omega \times \Omega}$ for $x \ne y$. If $u \in L^\infty(\Omega)$ then Tu is uniformly continuous on Ω.*

Proof. Fix $\epsilon > 0$, pick $\alpha \in (0, N)$ such that Eq. (14.1.3) holds, and set

$$H(x,y) = K(x,y)|x-y|^\alpha \tag{14.1.15}$$

so H is bounded and continuous for $x \ne y$. Assuming $u \in L^\infty(\Omega)$ and $x \in \Omega$, we have for $z \in B(x,\delta) \cap \Omega$

$$|Tu(z) - Tu(x)| = \left| \int_\Omega (K(z,y) - K(x,y))u(y)dy \right| \tag{14.1.16}$$

$$\le \int_{\Omega \cap B(x,2\delta)} (|K(z,y)| + |K(x,y)|)|u(y)|dy \tag{14.1.17}$$

$$+ \int_{\Omega \backslash B(x,2\delta)} |(K(z,y) - K(x,y))u(y)|dy \tag{14.1.18}$$

The integral in Eq. (14.1.17) may be estimated by

$$\|H\|_\infty \|u\|_\infty \int_{B(x,2\delta)} \frac{1}{|z-y|^\alpha} + \frac{1}{|x-y|^\alpha} dy \tag{14.1.19}$$

and so tends to zero as $\delta \to 0$ at a rate which is independent of x, z. We fix $\delta > 0$ such that this term is less than ϵ.

In the remaining integral, assuming that $|x-z| < \delta$ we have $|y-x| > 2\delta$ and so also $|y-z| > \delta$. If $E_\delta = \{(x,y) \in \overline{\Omega} \times \overline{\Omega}\colon |x-y| \geq \delta\}$ then K is uniformly continuous on E_δ, so there must exist $\delta' < \delta$ such that for $z \in B(x,\delta') \cap \Omega$ the integral in Eq. (14.1.18) is less than ϵ. This completes the proof. □

In general compactness fails if $\alpha \geq N$. A good example to keep in mind is the Hilbert transform (9.2.32) which is in the borderline case $\alpha = N = 1$, and which we have already noted is not a compact operator. Actually this example does not quite fit in to our discussion since the underlying domain is $\Omega = \mathbb{R}$ which is not bounded. If, however, we consider the so-called *finite* Hilbert transform defined by[1]

$$\mathcal{H}_0 u(x) = \frac{1}{\pi} \int_0^1 \frac{u(y)}{x-y} dy \tag{14.1.20}$$

as an operator on $L^2(0,1)$, it is known (see [20]) that the spectrum $\sigma_p(\mathcal{H}_0)$ consists of the segment of the imaginary axis connecting the points $\pm i$. In particular, since this set is uncountable, $\mathcal{H}_0$ is not compact. See Chapter 5, Section 2 of [16] for discussion of the operator equation $\mathcal{H}_0 u = f$. Note that boundedness of $\mathcal{H}_0$ is automatic from the corresponding property for the Hilbert transform. A thorough investigation of operators which generalize the Hilbert transform may be found in [36].

14.2 LAYER POTENTIALS

A certain type of singular integral operator which has played an important role in the historical development of the theory of elliptic PDEs is the so-called *layer potential*, see for example [21] for a very classical treatment. Layer potentials actually come in two common varieties. If Γ denotes the fundamental solution (8.5.59) of Laplace's equation in $\mathbb{R}^N$ for $N \geq 2$, and $\Sigma \subset \mathbb{R}^N$ is a smooth bounded $N-1$ dimensional surface, set

$$S\phi(x) = \int_\Sigma \Gamma(x-y)\phi(y)ds(y) \tag{14.2.21}$$

$$D\phi(x) = \int_\Sigma \frac{\partial}{\partial n_y} \Gamma(x-y)\phi(y)ds(y) \tag{14.2.22}$$

1. The integral below should be understood in the principal value sense.

which are, respectively, known as single- and double-layer potentials on Σ with density ϕ. To immediately see why such operators might arise naturally in connection with elliptic PDEs, observe that for any ϕ which is well behaved on Σ, $S\phi$ and $D\phi$ are harmonic functions in the complement of Σ. For example, if $u(x) = S\phi(x)$ then

$$\Delta u(x) = \int_\Sigma \Delta_x \Gamma(x-y)\phi(y)ds(y) = 0 \tag{14.2.23}$$

may be easily shown to be legitimate for $x \notin \Sigma$, taking into account that $\Delta\Gamma(x) = 0$ for $x \neq 0$. Likewise if $u(x) = D\phi(x)$ then

$$\Delta u(x) = \int_\Sigma \Delta_x \frac{\partial}{\partial n_y}\Gamma(x-y)\phi(y)ds(y) = \int_\Sigma \frac{\partial}{\partial n_y}\Delta_x\Gamma(x-y)\phi(y)ds(y) = 0 \tag{14.2.24}$$

A wise choice of ϕ may then allow us to find harmonic functions satisfying some desired further properties, such as prescribed boundary behavior.

To clarify the definition of the double-layer potential D, we suppose that a unit vector $n(x)$ normal to Σ is chosen, which is a continuous function of $x \in \Sigma$ (typically this amounts to making a consistent choice of the sign of $n(x)$, since there are two unit normal vectors at each point of Σ). If Σ is a simple closed surface then we will always adopt the usual convention which is to take $n(x)$ to be the outward normal. In any case we have

$$\frac{\partial}{\partial n_y}\Gamma(x-y) = -\sum_{j=1}^N \Gamma_{x_j}(x-y)n_j(y) = -\frac{(x-y)\cdot n(y)}{\Omega_{N-1}|x-y|^N} := K(x,y) \quad y \in \Sigma \tag{14.2.25}$$

Both $S\phi$ and $D\phi$ are obviously well defined for $x \notin \Sigma$, and the kernels are well defined for $x \neq y$ even if $x \in \Sigma$. If we wish to view either S or D as an operator, say, on $L^2(\Sigma)$ then formally at least we should think of Σ as being $N-1$ dimensional, and since the singularity of Γ is like[2] $|x|^{2-N}$, S has the character of a weakly singular integral operator. In the case of D, however, the singularity of Γ_{x_j} is like $|x|^{1-N}$, so K appears to be exactly in the borderline case where compactness is lost. On the other hand, under some reasonable assumptions on Σ we will see that extra decay of K when $x \to y$ is provided by the $n(y)$ factor, so that compactness of D will be restored.

Let us consider now the Dirichlet problem

$$\Delta u = 0 \quad x \in \Omega \quad u = f \quad x \in \Sigma \tag{14.2.26}$$

where Ω is a bounded, connected domain in $\mathbb{R}^N$, $N \geq 2$, and $\Sigma = \partial\Omega$. We will seek a solution in the form of a double-layer potential $u(x) = D\phi(x)$ for some density ϕ defined on Σ. As mentioned above, it is automatic that u is harmonic

2. With the usual modification for $N = 2$.

in Ω, so the condition which ϕ must be chosen to satisfy is that $D\phi = f$ on Σ, or more precisely

$$\lim_{\substack{z\to x\\ z\in\Omega}} \int_\Sigma K(z,y)\phi(y)ds(y) = f(x) \tag{14.2.27}$$

for $x \in \Sigma$.

The distinction between evaluating $D\phi$ on Σ and on the other hand taking the limit of $D\phi$ from inside Ω at a point of Σ is important in the following discussion, and must be observed rigorously—they are in fact not the same in general, and it is mainly the latter which we care about. The simplest possible case, which is contained in the following lemma, illustrates the point.

Lemma 14.2. *If $\phi(x) = 1$ and $\Sigma = \partial\Omega$ is C^2 then*

$$D\phi(x) = \begin{cases} 1 & x \in \Omega \\ \frac{1}{2} & x \in \Sigma \\ 0 & x \in \overline{\Omega}^c \end{cases} \tag{14.2.28}$$

Proof. If $x \in \overline{\Omega}^c$ then $y \to \Gamma(x-y)$ is a harmonic function in all of Ω, so integration by parts gives

$$D\phi(x) = \int_\Sigma \frac{\partial}{\partial n_y}\Gamma(x-y)ds(y) = \int_\Omega \Delta_y \Gamma(x-y)dy = 0 \tag{14.2.29}$$

If $x \in \Omega$, pick $\epsilon > 0$ such that $B(x,\epsilon) \subset \Omega$ and set $\Omega_\epsilon = \Omega \backslash B(x,\epsilon)$. We get

$$0 = \int_{\Omega_\epsilon} \Delta\Gamma(x-y)dy = \int_{\partial\Omega_\epsilon} \frac{\partial}{\partial n_y}\Gamma(x-y)ds(y) \tag{14.2.30}$$

$$= \int_\Sigma \frac{\partial}{\partial n_y}\Gamma(x-y)ds(y) + \int_{|y-x|=\epsilon} \frac{\partial}{\partial n_y}\Gamma(x-y)ds(y) \tag{14.2.31}$$

For $|x-y| = \epsilon$ it is easy to check that $n(y) = (x-y)/|x-y|$ (the outward normal points *toward* x) and so the second integral evaluates to be

$$-\int_{|y-x|=\epsilon} \frac{1}{\Omega_{N-1}\epsilon^{N-1}}ds(y) = -1 \tag{14.2.32}$$

which establishes Eq. (14.2.28) for $x \in \Omega$. Finally for $x \in \Sigma$, we repeat the same calculation and find that the integral in Eq. (14.2.32) is replaced by

$$\int_{\Omega\cap|y-x|=\epsilon} \frac{1}{\Omega_{N-1}\epsilon^{N-1}}ds(y) \tag{14.2.33}$$

Since we assumed that Σ is C^2 it follows that as $\epsilon \to 0$ we get precisely half of surface area (i.e., Σ might as well be a hyperplane), so that the limit of $1/2$ results, as needed. □

Note that if we allowed Σ to have a corner at some point x, then the conclusion that $D\phi(x) = 1/2$ for $x \in \Sigma$ would definitely no longer be valid.

If $u(x) = D\phi(x)$ for some ϕ, let us now define

$$u^+(x) = \lim_{\alpha\to 0+} u(x+\alpha n(x)) \quad u^-(x) = \lim_{\alpha\to 0-} u(x+\alpha n(x)) \tag{14.2.34}$$

Thus u^-, u^+ are, respectively, limiting values of u from inside and outside the domain. In the above example we saw that $u(x) - u^\pm(x) = \pm\frac{1}{2}$ for $x \in \Sigma$, and the normal derivative of $D\phi$ (namely zero) is continuous across Σ. These facts generalize in the following way.

Theorem 14.3. *If* $\Sigma = \partial\Omega$ *is* C^2, $\phi \in C(\Sigma)$, *and* $u = D\phi(x)$ *then*

$$u(x) - u^\pm(x) = \pm\frac{\phi(x)}{2} \quad x \in \Sigma \tag{14.2.35}$$

and

$$\lim_{\alpha\to 0+}\left(\frac{\partial}{\partial n}(u(x+\alpha n(x))) - \frac{\partial}{\partial n}(u(x-\alpha n(x)))\right) = 0 \tag{14.2.36}$$

The proof of this result involves technicalities which are beyond the scope of this book. We refer to Chapter 3 of [12] or Chapter 2 of [8] for details. Note that Eq. (14.2.36) does not imply that the normal derivative of u has a well-defined limit on Σ from the outside or inside, only that the difference has limit zero.

Thus in general $D\phi$ experiences a jump as Σ is crossed, whose magnitude at $x \in \Sigma$ is precisely $\phi(x)$. For the Dirichlet problem (14.2.26) the precise meaning of the boundary condition is that we seek a density ϕ such that $u^-(x) = f(x)$ for $x \in \Sigma$. It then follows from Eq. (14.2.35) that ϕ should satisfy

$$\frac{\phi(x)}{2} + \int_\Sigma K(x,y)\phi(y)ds(y) = f(x) \quad x \in \Sigma \tag{14.2.37}$$

Conversely, if ϕ is a continuous solution of Eq. (14.2.37) and we set $u(x) = D\phi(x)$ then u is harmonic inside Ω and $u^-(x) = u(x) + \frac{\phi(x)}{2} = f(x)$, as required. We therefore have obtained the very interesting and useful result that solvability properties of Eq. (14.2.26) can be analyzed in terms of the second kind integral equation (14.2.37). We can likewise study the corresponding exterior Dirichlet problem, in which we seek u harmonic in $\overline{\Omega}^c$ with prescribed boundary values on Σ, by looking instead at

$$-\frac{\phi(x)}{2} + \int_\Sigma K(x,y)\phi(y)ds(y) = f(x) \quad x \in \Sigma \tag{14.2.38}$$

The strategy now is to show that D is a compact operator on $L^2(\Sigma)$, so that the general theory from Chapter 12 can be applied. Again, the technicalities are lengthy so we will content ourselves with a heuristic discussion, referring to [12] for a detailed treatment.

In the previous section we have established a sufficient condition for a singular integral operator to be compact. Here, the underlying domain Σ is not a domain in $\mathbb{R}^N$ but assuming it is a reasonably smooth surface (e.g., C^2), it is "locally" a domain in $\mathbb{R}^{N-1}$. Thus compactness can be proved, as before, if the singularity of K has an associated exponent $\alpha < N-1$. The explicit expression (14.2.25) for K does not appear to imply this, but it will if we take into account that $x-y$ becomes orthogonal to $n(y)$ if $x, y \in \Sigma$ and $x \to y$. More precisely we have the following lemma.

Lemma 14.3. *If Σ is a C^2 surface then there exists a constant M such that*

$$|(x-y)\cdot n(y)| \le M|x-y|^2 \quad x, y \in \Sigma \tag{14.2.39}$$

Proof. Fix $x \in \Sigma$. Without loss of generality we may assume that $x = 0$ and that $n(0) = (0, 0, \dots, 1)$. Thus in a neighborhood of $x = 0$ the surface Σ is given by $y_n = \Psi(y')$, where $y' = (y_1, \dots, y_{n-1})$, Ψ is C^2 near 0, and $\Psi(0) = \nabla\Psi(0) = 0$. In particular $\Psi(y) = O(|y'|^2)$ as $y' \to 0$. By Taylor's theorem, for $y \in \Sigma$

$$(x-y)\cdot n(y) = -y\cdot(n(0) + n(y) - n(0)) = -y_n + y\cdot(n(0) - n(y)) \tag{14.2.40}$$

$$= -\Psi(y_1, \dots, y_{n-1}) + y\cdot(n(0) - n(y)) \tag{14.2.41}$$

Since Σ is C^2 it follows that $n(y)$ is C^1, and so both terms in Eq. (14.2.41) are $O(|y'|^2)$, which is the needed conclusion at fixed x. The implied constant depends only on bounds for the curvature of Σ and so a constant M exists which is independent of $x \in \Sigma$. □

Corollary 14.1. *The kernel $K(x, y)$ in Eq.* (14.2.25) *satisfies*

$$|K(x, y)| \le M|x-y|^{2-N} \quad x, y \in \Sigma \tag{14.2.42}$$

and in particular D is compact on $L^2(\Sigma)$.

From Theorem 12.5 it now follows that there exists a unique solution of Eq. (14.2.37) for every $f \in C(\Sigma)$ (or even $L^2(\Sigma)$) provided that it can be verified that there is no nontrivial solution of the corresponding homogeneous equation. Assuming for the sake of contradiction that such a solution $\phi \not\equiv 0$ exists then it follows first of all that $u = D\phi$ is a solution of Eq. (14.2.26) with $f \equiv 0$. This must mean $u \equiv 0$ and so in consequence $u^-(x) = \frac{\partial u^-}{\partial n} \equiv 0$ on Σ. By the matching condition for the normal derivative of u stated in Theorem 14.3, u^+ must then satisfy the homogeneous Neumann condition in the exterior domain $\overline{\Omega}^c$. It can then be shown (Exercise 14.8) that $u^+ \equiv 0$ so that using Eq. (14.2.35) one more time we obtain that

$$\phi(x) = u^-(x) - u^+(x) = 0 \tag{14.2.43}$$

Thus $D\phi - \phi/2 = 0$ has only the trivial solution, as needed.

14.3 CONVOLUTION EQUATIONS

Consider the convolution type integral equation

$$\lambda u(x) - \int_{\mathbb{R}^N} K(x-y)u(y)dy = f(x) \quad x \in \mathbb{R}^N \tag{14.3.44}$$

where $K, f \in L^2(\mathbb{R}^N)$. If there exists a solution $u \in L^2(\mathbb{R}^N)$ then by Theorem 7.8 it must hold that

$$\left(\lambda - (2\pi)^{N/2}\widehat{K}(y)\right)\widehat{u}(y) = \widehat{f}(y) \quad \text{a.e. } y \in \mathbb{R}^N \tag{14.3.45}$$

The solution is evidently unique, at least in $L^2(\mathbb{R}^N)$, provided $(2\pi)^{N/2}\widehat{K}(y) \neq \lambda$ a.e. If also there exists $\epsilon > 0$ such that

$$|\lambda - (2\pi)^{N/2}\widehat{K}(y)| \geq \epsilon \quad \text{a.e. } y \in \mathbb{R}^N \tag{14.3.46}$$

then

$$\widehat{u}(y) = \frac{\widehat{f}(y)}{\lambda - (2\pi)^{N/2}\widehat{K}(y)} \tag{14.3.47}$$

defines a solution for every $f \in L^2(\mathbb{R}^N)$.

The requirement $K \in L^2(\mathbb{R}^N)$ can clearly be weakened to some extent. Recall that $K * u$ is well defined under a number of different sets of assumptions which have been made earlier; for example, (i) $K \in \mathcal{D}'(\mathbb{R}^N)$ and $u \in \mathcal{D}(\mathbb{R}^N)$, (ii) $K \in \mathcal{S}'(\mathbb{R}^N)$ and $u \in \mathcal{S}(\mathbb{R}^N)$, or (iii) $K \in L^p(\mathbb{R}^N)$ and $u \in L^q(\mathbb{R}^N)$ with $p^{-1}+q^{-1} \geq 1$, and all of these are subject to further refinement. Thus a separate analysis of existence and uniqueness for Eq. (14.3.44) could be carried out under a wide variety of assumptions. Let us note in particular that Eq. (14.3.47) provides at least a formal solution formula provided that $K \in \mathcal{S}'(\mathbb{R}^N), \widehat{f}, \widehat{K}$ are regular distributions (i.e., functions), and $(2\pi)^{N/2}\widehat{K}(y) \neq \lambda$ a.e.

Example 14.1. In Eq. (14.3.44) let $N = 1$ and $K = \frac{1}{\pi}\text{pv}\frac{1}{x}$ so that $K * u = \mathcal{H}u$, the Hilbert transform of u defined in Eq. (9.2.32). Referring to the formula (7.8.169) for the Fourier transform of K, we obtain

$$(\lambda\widehat{u} - i\text{sgn}(y))\widehat{u}(y) = \widehat{f}(y) \tag{14.3.48}$$

Thus for $f \in L^2(\mathbb{R})$ and $\lambda \neq \pm i$ it is clear that

$$\widehat{u}(y) = \frac{\widehat{f}(y)}{\lambda - i\text{sgn}(y)} \tag{14.3.49}$$

defines the unique solution of Eq. (14.3.44). □

Now let us consider a closely related situation of a so-called *Hankel type* integral equation,

$$\int_{\mathbb{R}^N} K(x+y)u(y)dy = f(x) \quad x \in \mathbb{R}^N \tag{14.3.50}$$

If we let $K_1(x) = K(-x)$ and $f_1(x) = f(-x)$ then Eq. (14.3.50) is equivalent to $K_1 * u = f_1$, and so

$$\widehat{u}(y) = \frac{1}{(2\pi)^{N/2}} \frac{\widehat{f_1}(y)}{\widehat{K_1}(y)} \tag{14.3.51}$$

If we temporarily denote the usual reflection operator by $\mathcal{R}$ (i.e., $\mathcal{R}\phi(x) = \phi(-x)$), note that $\mathcal{R}$ commutes with the Fourier transform. Thus

$$\widehat{u} = \frac{1}{(2\pi)^{N/2}} \mathcal{R}\left(\frac{\widehat{f}}{\widehat{K}}\right) \tag{14.3.52}$$

and so from the inversion theorem the solution u is

$$u = \frac{1}{(2\pi)^{N/2}} \widehat{\left(\frac{\widehat{f}}{\widehat{K}}\right)} \tag{14.3.53}$$

assuming that the expression is meaningful.

Note that using this approach it would not be straightforward to include a λu term on left-hand side of Eq. (14.3.50).

14.4 WIENER-HOPF TECHNIQUE

Consider[3] in one dimension, the integral equation of the special type

$$\lambda u(x) - \int_0^\infty K(x-y)u(y)dy = f(x) \quad x > 0 \tag{14.4.54}$$

Here the kernel depends on the difference of the two arguments, as in a convolution equation, but it is not actually a convolution type equation since the integration only takes place over $(0, \infty)$. Nevertheless we can make some artificial extensions for mathematical convenience. Assuming that there exists a solution u to be found, we let $u(x) = f(x) = 0$ for $x < 0$ and

$$g(x) = \begin{cases} \int_0^\infty K(x-y)u(y)dy & x < 0 \\ 0 & x > 0 \end{cases} \tag{14.4.55}$$

It then follows that

$$\lambda u(x) - \int_{-\infty}^\infty K(x-y)u(y)dy = f(x) - g(x) \quad x \in \mathbb{R} \tag{14.4.56}$$

This resulting equation is of convolution type, but contains the additional unknown term g. On the other hand when considered as a solution on all of $\mathbb{R}$, u should be regarded as constrained by the property that it has support in the positive half line.

---

3. Throughout this section it will be assumed that the reader has some familiarity with basic ideas and techniques of complex analysis.

A pair of operators which are technically useful for dealing with this situation are the so-called Hardy space projection operators $P_\pm$ defined as

$$P_\pm\phi = \frac{1}{2}(\phi \pm i\mathcal{H}\phi) \tag{14.4.57}$$

where $\mathcal{H}$ is the Hilbert transform. To motivate these definitions, recall from the discussion just above Eq. (9.2.32) that $\widehat{(\mathcal{H}\phi)}(y) = -i\text{sgn}(y)\widehat{\phi}(y)$, so

$$\widehat{(P_+\phi)}(y) = \begin{cases} \widehat{\phi}(y) & y > 0 \\ 0 & y < 0 \end{cases} \quad \widehat{(P_-\phi)}(y) = \begin{cases} \widehat{\phi}(y) & y < 0 \\ 0 & y > 0 \end{cases} \tag{14.4.58}$$

It is therefore simple to see that $P_\pm$ are the orthogonal projections of $L^2(\mathbb{R})$ onto the corresponding closed subspaces

$$H^2_+ := \{u \in L^2(\mathbb{R}) \colon \widehat{u}(y) = 0 \ \ \forall y < 0\} \quad H^2_- := \{u \in L^2(\mathbb{R}) \colon \widehat{u}(y) = 0 \ \ \forall y > 0\} \tag{14.4.59}$$

for which $L^2(\mathbb{R}) = H^2_+ \oplus H^2_-$ (see also Exercise 9.5). These are so-called Hardy spaces, which of course may be considered as Hilbert spaces in their own right, see Chapter 3 of [10]. In particular it can be readily seen (Exercise 14.10) that if $\phi \in H^2_+$ then ϕ has an analytic extension to the upper half of the complex plane,

$$\int_{-\infty}^{\infty} |\phi(x+iy)|^2 dx \le \|\phi\|^2_{L^2(\mathbb{R})} \quad \forall y > 0 \tag{14.4.60}$$

and

$$\phi(\cdot + iy) \to \phi \text{ in } L^2(\mathbb{R}) \text{ as } y \to 0+ \tag{14.4.61}$$

Likewise a function $\phi \in H^2_-$ has an analytic extension to the lower half of the complex plane with analogous properties.

A very important converse of the above is given by the following theorem.

Theorem 14.4. *If ϕ is analytic in the upper half of the complex plane and there exists a constant C such that*

$$\sup_{y>0} \int_{-\infty}^{\infty} |\phi(x+iy)|^2 dx = C \tag{14.4.62}$$

then $\phi \in H^2_+$ and

$$\int_{-\infty}^{\infty} |\phi(x)|^2 dx = \int_0^{\infty} |\widehat{\phi}(y)|^2 dy = C \tag{14.4.63}$$

This is one of a number of closely related theorems which together are generally referred to as *Paley-Wiener theory*, see Theorem 19.2 of [32] or Theorem 1, Section 3.4 of [10] for a proof. The spaces $H^2_\pm$ actually belong to the larger family of Hardy spaces $H^p_\pm$, $1 \le p \le \infty$, where for example $\phi \in H^p_+$ if ϕ has an analytic extension to the upper half of the complex plane and

$$\|\phi(\cdot + iy)\|_{L^p(\mathbb{R})} \le \|\phi\|_{L^p(\mathbb{R})} \quad \forall y > 0 \tag{14.4.64}$$

Returning to Eq. (14.4.56) we note that $\widehat{u}, \widehat{f} \in H^2_-$ while $\widehat{g} \in H^2_+$. Suppose now that it is possible to find a pair of functions $q_\pm \in H^\infty_\pm$ such that

$$\lambda - \sqrt{2\pi}\widehat{K}(y) = \frac{q_-(y)}{q_+(y)} \quad y \in \mathbb{R} \tag{14.4.65}$$

Then from Eq. (14.4.56) it follows that

$$q_-\widehat{u} = q_+\widehat{f} - q_+\widehat{g} \tag{14.4.66}$$

From the assumptions made on q_+ and Theorem 14.4 we can conclude that $q_+\widehat{g} \in H^2_+$, and likewise $q_-\widehat{u} \in H^2_-$. In particular $P_-(q_+\widehat{g}) = 0$, so that

$$q_-\widehat{u} = P_-(q_-\widehat{u}) = P_-(q_+\widehat{f}) \tag{14.4.67}$$

We thus obtain at least a formal solution formula for the Fourier transform of the solution, namely

$$\widehat{u} = \frac{P_-(q_+\widehat{f})}{q_-} \tag{14.4.68}$$

In order that this formula be meaningful it is sufficient that $1/q_\pm \in H^\infty_\pm$ along with the other assumptions already made, see Exercise 14.11. The central question which remains to be more thoroughly studied is the existence of the pair of functions $q_\pm$ satisfying all of the above requirements. We refer the reader to Chapter 3 of [16] or Chapter 3 of [10] for further reading about this more advanced topic, and conclude just with an example.

Example 14.2. Consider Eq. (14.4.54) with $K(x) = e^{-|x|}$, that is

$$\lambda u(x) - \int_0^\infty e^{-|x-y|}u(y)dy = f(x) \quad x > 0 \tag{14.4.69}$$

Since

$$\widehat{K}(y) = \sqrt{\frac{2}{\pi}}\frac{1}{y^2+1} \tag{14.4.70}$$

we get

$$\lambda - \sqrt{2\pi}\widehat{K}(y) = \lambda\left(\frac{y^2+b^2}{y^2+1}\right) \tag{14.4.71}$$

where $b^2 = (\lambda - 2)/\lambda$. If we require $\lambda \notin [0,2]$ then b may be chosen as real and positive, and we have

$$\lambda - \sqrt{2\pi}\widehat{K}(y) = \frac{q_-(y)}{q_+(y)} \tag{14.4.72}$$

where

$$q_-(y) = \lambda\left(\frac{y-ib}{y-i}\right) \quad q_+(y) = \left(\frac{y+i}{y+ib}\right) \tag{14.4.73}$$

We see immediately that $q_\pm, q_\pm^{-1} \in H_\pm^\infty$, and so Eq. (14.4.68) provides the unique solution of Eq. (14.4.69) provided $\lambda \notin [0,2]$. Note the significance of this restriction on λ is that it precisely the requirement that λ does not belong to the closure of the numerical range of $\sqrt{2\pi}\widehat{K}(y)$. □

14.5 EXERCISES

14.1. The Abel integral equation is

$$Tu(x) = \int_0^x \frac{u(y)}{\sqrt{x-y}}dy = f(x)$$

a first kind Volterra equation with a weakly singular kernel. Derive the explicit solution formula

$$u(x) = \frac{1}{\pi}\frac{d}{dx}\int_0^x \frac{f(y)}{\sqrt{x-y}}dy$$

(Suggestions: It amounts to showing that $T^2u(x) = \pi\int_0^x u(y)dy$. You will need to evaluate an integral of the form $\int_y^x \frac{dz}{\sqrt{z-y}\sqrt{x-z}}$. Use the change of variable $z = y\cos^2\theta + x\sin^2(\theta)$.)

14.2. Let K_1, K_2 be weakly singular kernels with associated exponents α_1, α_2, and let T_1, T_2 be the associated Volterra integral operators. Show that T_1T_2 is also a Volterra operator with a weakly singular kernel and associated exponent $\alpha_1 + \alpha_2 - 1$.

14.3. If $P(x)$ is any nonzero polynomial, show that the first kind Volterra integral equation

$$\int_a^x P(x-y)u(y)dy = f(x)$$

is equivalent to a second kind Volterra integral equation.

14.4. If T is a weakly singular Volterra integral operator, show that there exists a positive integer n such that T^n is a Volterra integral operator with a bounded kernel.

14.5. Use Eq. (14.3.47) to obtain an explicit solution of Eq. (14.3.44) if

$$N = 1 \quad \lambda = 1 \quad K(x) = e^{-|x|} \quad f(x) = \begin{cases} e^{-x} & x > 0 \\ 0 & x < 0 \end{cases} \tag{14.5.74}$$

14.6. Discuss the solvability of the integral equation

$$\int_0^\infty \frac{u(s)}{s+t}ds = f(t) \quad t > 0 \tag{14.5.75}$$

(Suggestions: Introduce new variables

$$\xi = \frac{1}{2}\log t \quad \eta = \frac{1}{2}\log s \quad \psi(\eta) = e^{\eta}u(e^{2\eta}) \quad g(\xi) = e^{\xi}f(e^{2\xi})$$

You may find it useful to work out, or look up, the Fourier transform of the hyperbolic secant function.)

14.7. Let $\Omega = B(0, R)$.

(a) Show that the kernel $K(x, y)$ in Eq. (14.2.37) is constant for $x, y \in \Sigma = \partial\Omega$.

(b) Solve the integral equation (14.2.37) for ϕ.

(c) Using this expression for ϕ, derive the Poisson integral formula for the solution of Eq. (14.2.26), namely

$$u(x) = \frac{R^2 - |x|^2}{\Omega_{N-1}R}\int_{|y|=R}\frac{f(y)}{|x-y|^N}ds(y) \tag{14.5.76}$$

14.8. If ϕ is a solution of Eq. (14.2.37) with $f = 0$, and $u = D\phi$, show that $u^+(x) = 0$ on Σ. (Suggestions: First show that u^+ is constant in Ω^c and then that the constant must be zero.)

14.9. Suppose that we look for a solution of

$$\Delta u = 0 \quad x \in \Omega$$

$$\frac{\partial u}{\partial n} + u = f \quad x \in \partial\Omega$$

in the form of a single-layer potential

$$u(x) = \int_{\partial\Omega}\Gamma(x-y)\phi(y)dy$$

Look up the jump conditions for the single-layer potential and find an integral equation for the density ϕ.

14.10. If $\phi \in H^2_+$ show that ϕ has an analytic extension to the upper half of the complex plane. To be precise, show that if

$$\tilde{\phi}(z) = \frac{1}{\sqrt{2\pi}}\int_0^{\infty}\widehat{\phi}(t)e^{itz}dt$$

then $\tilde{\phi}$ is defined and analytic on $\{z = x + iy : y > 0\}$ and

$$\lim_{y\to 0+}\tilde{\phi}(\cdot + iy) = \phi \quad \text{in } L^2(\mathbb{R})$$

Along the way show that Eq. (14.4.60) holds.

14.11. Assume that $f \in L^2(0, \infty)$ and that Eq. (14.4.65) is valid for some $q_\pm$ with $q_\pm, q_\pm^{-1} \in H^\infty_\pm$. Verify that Eq. (14.4.68) defines a function $u \in L^2(\mathbb{R})$ such that $u(x) = 0$ for $x < 0$.

14.12. Find $q_\pm$ in Eq. (14.4.65) for the case

$$K(x) = \text{sinc}(x) := \begin{cases} \frac{\sin(\pi x)}{\pi x} & x \neq 0 \\ 1 & x = 0 \end{cases}$$

and $\lambda = -1$. (Suggestion: Look for $q_\pm$ in the form $q_\pm(x) = \lim_{y \to 0\pm} F(x + iy)$, where $F(z) = ((z - \pi)/(z + \pi))^{i\alpha}$.)

Chapter 15

Variational Methods

Throughout this chapter the spaces of functions we work with will be assumed to be real, unless otherwise stated.

15.1 THE DIRICHLET QUOTIENT

We have earlier introduced the concept of the Rayleigh quotient

$$J(u) = \frac{\langle Tu, u\rangle}{\langle u, u\rangle} \tag{15.1.1}$$

for a linear operator T on a Hilbert space $\mathbf{H}$. In the previous discussion we were mainly concerned with the case that T was a bounded or even a compact operator, but now we will allow for T to be unbounded. In such a case, $J(u)$ is defined at least for $u \in D(T)\backslash\{0\}$, and possibly has a natural extension to some larger domain. The principal case of interest to us here is the case of the Dirichlet Laplacian discussed in Section 13.4,

$$Tu = -\Delta u \quad \text{on } D(T) = \{u \in H_0^1(\Omega)\colon \Delta u \in L^2(\Omega)\} \tag{15.1.2}$$

In this case

$$J(u) = \frac{\langle -\Delta u, u\rangle}{\langle u, u\rangle} = \frac{\int_\Omega |\nabla u|^2 dx}{\int_\Omega |u|^2 dx} = \frac{\|u\|^2_{H_0^1(\Omega)}}{\|u\|^2_{L^2(\Omega)}} \tag{15.1.3}$$

which we may evidently regard as being defined on all of $H_0^1(\Omega)$ except the origin. We will refer to any of these equivalent expressions as the *Dirichlet quotient* (or *Dirichlet form*) for $-\Delta$. Throughout this section we take Eq. (15.1.3) as the definition of J, and denote by $\{\lambda_n, \psi_n\}$ the eigenvalues and eigenfunctions of T, where we may choose the ψ_n's to be an orthonormal basis of $L^2(\Omega)$, according to the discussion of Section 13.4. It is immediate that

$$J(\psi_n) = \lambda_n \tag{15.1.4}$$

for all n.

Techniques of Functional Analysis for Differential and Integral Equations
http://dx.doi.org/10.1016/B978-0-12-811426-1.00015-5

If we define a critical point of J to be any $u \in H_0^1(\Omega)\backslash\{0\}$ for which

$$\frac{d}{d\alpha}J(u+\alpha v)\bigg|_{\alpha=0} = 0 \quad \forall v \in H_0^1(\Omega) \tag{15.1.5}$$

then precisely as in Eq. (12.3.40) and the following discussion we find that at any critical point there holds

$$\int_\Omega \nabla u \cdot \nabla v dx = J(u) \int_\Omega uv dx \quad \forall v \in H_0^1(\Omega) \tag{15.1.6}$$

In other words, $Tu = \lambda u$ must hold with $\lambda = J(u)$. Conversely, by straightforward calculation, any eigenfunction of T is a critical point of J. Thus the set of eigenfunctions of T coincides with the set of critical points of the Dirichlet quotient, and by Eq. (15.1.4) the eigenvalues are exactly the critical values of J.

Among these critical points, one might expect to find a point at which J achieves its minimum value, which must then correspond to the critical value λ_1, the least eigenvalue of T. We emphasize, however, that the existence of a minimizer of J must be proved—it is not immediate from anything we have stated so far. We give one such proof here, and indicate another one in Exercise 15.3.

Theorem 15.1. *There exists $\psi \in H_0^1(\Omega)\backslash\{0\}$ such that $J(\psi) \le J(\phi)$ for all $\phi \in H_0^1(\Omega)\backslash\{0\}$.*

Proof. Let

$$\lambda = \inf_{\phi \in H_0^1(\Omega)} J(\phi) \tag{15.1.7}$$

so $\lambda > 0$ by the Poincaré inequality. Therefore there exists $\psi_n \in H_0^1(\Omega)$ such that $J(\psi_n) \to \lambda$. Without loss of generality we may assume $\|\psi_n\|_{L^2(\Omega)} = 1$ for all n, in which case $\|\psi_n\|^2_{H_0^1(\Omega)} \to \lambda$. In particular $\{\psi_n\}$ is a bounded sequence in $H_0^1(\Omega)$, so by Theorem 12.1 there exists $\psi \in H_0^1(\Omega)$ such that $\psi_{n_k} \overset{w}{\to} \psi$ in $H_0^1(\Omega)$, for some subsequence. By Theorem 13.4 it follows that $\psi_{n_k} \to \psi$ strongly in $L^2(\Omega)$, so in particular $\|\psi\|_{L^2(\Omega)} = 1$. Finally, using the lower semicontinuity property of weak convergence (Proposition 12.2)

$$\lambda \le J(\psi) = \|\psi\|^2_{H_0^1(\Omega)} \le \liminf_{n_k\to\infty} \|\psi_{n_k}\|^2_{H_0^1(\Omega)} = \liminf_{n_k\to\infty} J(\psi_{n_k}) = \lambda \tag{15.1.8}$$

so that $J(\psi) = \lambda$, that is, J achieves its minimum at ψ. □

Note that by its very definition, the minimum λ_1 of the Rayleigh quotient J, gives rise to the best constant in the Poincaré inequality, namely Eq. (13.4.82) is valid with $C = \frac{1}{\sqrt{\lambda_1}}$ and no smaller C works.

The above argument provides a proof of the existence of one eigenvalue of T, namely the smallest eigenvalue λ_1, with corresponding eigenfunction ψ_1, which is completely independent from the proof given in Chapter 12. It is natural to ask then whether the existence of the other eigenvalues can be obtained in a similar way. Of course they can no longer be obtained by minimizing the Dirichlet quotient (nor is there any maximum to be found), but we know in fact that J has other critical points, since other eigenfunctions exist. Consider, for example the case of λ_2, for which there must exist an eigenfunction orthogonal in $L^2(\Omega)$ to the eigenfunction already found for λ_1. Thus it is a natural conjecture that λ_2 may be obtained by minimizing J over the orthogonal complement of ψ_1. Specifically, if we set

$$H_1 = \left\{\phi \in H_0^1(\Omega) \colon \int_\Omega \phi\psi_1 dx = 0\right\} \tag{15.1.9}$$

then the existence of a minimizer of J over H_1 can be proved just as in Theorem 15.1. If the minimum occurs at ψ_2, with $\lambda_2 = J(\psi_2)$ then the critical point condition amounts to

$$\int_\Omega \nabla\psi_2 \cdot \nabla v dx = \lambda_2 \int_\Omega \psi_2 v dx \quad \forall v \in H_1 \tag{15.1.10}$$

Furthermore, if $v = \psi_1$ then

$$\int_\Omega \nabla\psi_2 \cdot \nabla\psi_1 dx = -\int_\Omega \psi_2 \Delta\psi_1 = -\lambda_1 \int_\Omega \psi_2\psi_1 = 0 \tag{15.1.11}$$

since $\psi_2 \in H_1$. It follows that Eq. (15.1.10) holds for every $v \in H_0^1(\Omega)$, so ψ_2 is an eigenvalue of T for eigenvalue λ_2. Clearly $\lambda_2 \geq \lambda_1$, since λ_2 is obtained by minimization over a smaller set.

We may continue this way, successively minimizing the Rayleigh quotient over the orthogonal complement in $L^2(\Omega)$ of the previously obtained eigenfunctions, to obtain a variational characterization of all eigenvalues.

Theorem 15.2. *We have*

$$\lambda_n = J(\psi_n) = \min_{u \in H_{n-1}} J(u) \tag{15.1.12}$$

where

$$H_n = \left\{u \in H_0^1(\Omega) \colon \int_\Omega u\psi_k dx = 0, k = 1, 2, \dots, n\right\} \quad H_0 = H_0^1(\Omega) \tag{15.1.13}$$

This proof is essentially a mirror image of the proof of Theorem 12.10, in which a compact operator has been replaced by an unbounded operator, and maximization has been replaced by minimization. One could also look at critical points of the reciprocal of J in order to maintain it as a maximization problem, but it is more common to proceed as we are doing. Similar results can be obtained for a larger class of unbounded self-adjoint operators, see for example [39]. The eigenfunctions may be interpreted as *saddle points* of J, that is, critical points which are not local extrema.

The characterization of eigenvalues and eigenfunctions stated in Theorem 15.2 is unsatisfactory, in the sense that the minimization problem to be solved in order to obtain an eigenvalue λ_n requires knowledge of the eigenfunctions corresponding to smaller eigenvalues. We next discuss two alternative characterizations of eigenvalues, which may be regarded as advantageous from this point of view.

If E is a finite dimensional subspace of $H_0^1(\Omega)$, we define

$$\mu(E) = \max_{u \in E} J(u) \tag{15.1.14}$$

and set

$$S_n = \{E \subset H_0^1(\Omega) \colon \text{E is a subspace, } \dim(E) = n\} \quad n = 0, 1, \dots \tag{15.1.15}$$

Note that $\mu(E)$ exists and is finite for $E \in S_n$, since if we choose any orthonormal basis $\{\zeta_1, \dots, \zeta_n\}$ of S_n then

$$\max_{u \in E} J(u) = \max_{\sum_{k=1}^n |c_k|^2 = 1} \int_\Omega \left| \sum_{k=1}^n c_k \nabla \zeta_k \right|^2 dx \tag{15.1.16}$$

Thus finding $\mu(E)$ amounts to maximizing a continuous function over a compact set.

Theorem 15.3. *(Poincaré Min-Max formula) We have*

$$\lambda_n = \min_{E \in S_n} \mu(E) = \min_{E \in S_n} \max_{u \in E} J(u) \tag{15.1.17}$$

for $n = 1, 2, \dots$.

Proof. J is constant on any one-dimensional subspace, that is, $\mu(E) = J(\phi)$ if $E = \text{span}\{\phi\}$, so the conclusion is equivalent to the statement of Theorem 15.1 for $n = 1$. For $n \geq 2$, if $E \in S_n$ we can find $w \in E$, $w \neq 0$ such that $w \perp \psi_k$ for $k = 1, \dots, n-1$, since this amounts to $n-1$ linear equations for n unknowns (here $\{\psi_n\}$ still denotes the orthonormalized Dirichlet eigenfunctions). Thus $w \in H_{n-1}$ and so by Theorem 15.2.

$$\lambda_n \leq J(w) \leq \max_{u \in E} J(u) = \mu(E) \tag{15.1.18}$$

It follows that

$$\lambda_n \leq \inf_{E \in S_n} \mu(E) \tag{15.1.19}$$

On the other hand, if we choose $E = \text{span}\{\psi_1, \dots, \psi_n\}$ note that

$$J(u) = \frac{\sum_{k=1}^n \lambda_k c_k^2}{\sum_{k=1}^n c_k^2} \tag{15.1.20}$$

for any $u = \sum_{k=1}^{n} c_k \psi_k \in E$. Thus

$$\mu(E) = J(\psi_n) = \lambda_n \tag{15.1.21}$$

and so the infimum in Eq. (15.1.19) is achieved for this E. The conclusion (15.1.17) then follows. □

A companion result with a similar proof (e.g., Theorem 5.2 of [39]) is the following.

Theorem 15.4. *(Courant-Weyl Max-Min formula) We have*

$$\lambda_n = \max_{E \in S_{n-1}} \min_{u \perp E} J(u) \tag{15.1.22}$$

for $n = 1, 2, \ldots$.

An interesting application of the variational characterization of the first eigenvalue is the following monotonicity property. We use temporarily the notation $\lambda_1(\Omega)$ to denote the smallest Dirichlet eigenvalue of $-\Delta$ for the domain Ω.

Theorem 15.5. *If* $\Omega \subset \Omega'$ *then* $\lambda_1(\Omega') \le \lambda_1(\Omega)$.

Proof. By the density of $C_0^\infty(\Omega)$ in $H_0^1(\Omega)$ and Theorem 15.1, for any $\epsilon > 0$ there exists $u \in C_0^\infty(\Omega)$ such that

$$J(u) \le \lambda_1(\Omega) + \epsilon \tag{15.1.23}$$

But extending u to be zero outside of Ω we may regard it as also belonging to $C_0^\infty(\Omega')$, and the value of $J(u)$ is the same whichever domain we have in mind. Therefore

$$\lambda_1(\Omega') \le J(u) \le \lambda_1(\Omega) + \epsilon \tag{15.1.24}$$

and so the conclusion follows by letting $\epsilon \to 0$. □

15.2 EIGENVALUE APPROXIMATION

The variational characterizations of eigenvalues discussed in the previous section lead immediately to certain estimates for the eigenvalues. In the simplest possible situation, if we choose any nonzero function $v \in H_0^1(\Omega)$ (which we call the *trial function* in this context), then from Theorem 15.1 we have that

$$\lambda_1 \le J(v) \tag{15.2.25}$$

an upper bound for the smallest eigenvalue. Furthermore, if we can choose v to "resemble" the corresponding eigenfunction ψ_1, then we will typically find that $J(v)$ is close to λ_1. If, for example in the one-dimensional case $\Omega = (0, 1)$, we choose $v(x) = x(1 - x)$ then by direct calculation we get that $J(v) = 10$, which should be compared to the exact value $\pi^2 \approx 9.87$. The trial function $v(x) = x^2(1-x)$, which is not so much like $\psi_1 = \sin(\pi x)$, provides a correspondingly poorer approximation $J(v) = 14$, which is still of course a valid upper bound.

The so-called *Rayleigh-Ritz method* generalizes this idea, so as to provide inequalities and/or approximations for other eigenvalues besides the first one. Let $v_1, v_2, \dots, v_n$ denote n linearly independent trial functions in $H_0^1(\Omega)$. Then $E = \text{span}\{v_1, v_2, \dots, v_n\}$ is an n-dimensional subspace of $H_0^1(\Omega)$, and so

$$\lambda_1 \le \min_{u\in E} J(u) \quad \lambda_n \le \max_{u\in E} J(u) \tag{15.2.26}$$

by Theorems 15.2 and 15.3.

The problem of computing critical points of J over E is a calculus problem, which may be handled as follows. Any $u \in E$ may be written as $u = \sum_{k=1}^n c_k v_k$, and so

$$J(u) = \frac{\int_\Omega \left|\sum_{k=1}^n c_k \nabla v_k\right|^2}{\int_\Omega \left(\sum_{k=1}^n c_k v_k\right)^2} = J(c_1, \dots, c_n) \tag{15.2.27}$$

The critical point condition $\frac{\partial J}{\partial c_j} = 0, j = 1, \dots, n$ is readily seen to be equivalent to the linear system for $c = (c_1, \dots, c_n)^T$,

$$Ac = \Lambda Bc \tag{15.2.28}$$

where A, B are the $n \times n$ matrices with entries

$$A_{kj} = \int_\Omega \nabla v_k \cdot \nabla v_j dx \quad B_{kj} = \int_\Omega v_k v_j dx \tag{15.2.29}$$

and $\Lambda = J(u)$. In other words, the critical points are obtained as the eigenvalues of the generalized eigenvalue problem (15.2.28) defined by means of the two matrices A, B.

As usual, the set of all eigenvalues of Eq. (15.2.28) are obtained as the roots of the nth degree polynomial $\det(A - \Lambda B) = 0$. We denote these roots (which must be positive and real, by the symmetry of A, B) as $0 < \Lambda_1 \le \Lambda_2 \le \cdots \le \Lambda_n$, with points repeated as needed according to multiplicity. Thus Eq. (15.2.26) amounts to

$$\lambda_1 \le \Lambda_1 \quad \lambda_n \le \Lambda_n \tag{15.2.30}$$

The following theorem states that similar inequalities can be proved for all of the intermediate eigenvalues as well, and we refer to [39] for the proof.

Theorem 15.6. *We have*

$$\lambda_k \le \Lambda_k \quad k = 1, \dots, n \tag{15.2.31}$$

As in the case of a single eigenvalue, a good choice of trial functions $\{v_1, \dots, v_n\}$ will typically result in values of $\Lambda_1, \dots, \Lambda_n$, which are good approximations to $\lambda_1, \dots, \lambda_n$.

15.3 THE EULER-LAGRANGE EQUATION

In Section 15.1 we observed that the problem of minimizing the nonlinear functional J in Eq. (15.1.3), or more generally finding any critical point of J, leads to the eigenvalue problem for T defined in Eq. (15.1.2). This corresponds to the situation found even in elementary calculus, where to solve an optimization problem, we look for points where a derivative is equal to zero. In the Calculus of Variations, we continue to extend this kind of thinking from finite dimensional to infinite dimensional situations.

Suppose $\mathbf{X}$ is a vector space, $\mathcal{X} \subset \mathbf{X}$, $J: \mathcal{X} \to \mathbb{R}$ is a functional, nonlinear in general, and consider the problem

$$\min_{x \in \mathcal{X}} J(x) \tag{15.3.32}$$

There may also be constraints to be satisfied, for example, in the form $H(x) = C$, where $H: \mathcal{X} \to \mathbb{R}$, so that the problem may be given as

$$\min_{\substack{x \in \mathcal{X} \\ H(x)=C}} J(x) \tag{15.3.33}$$

We refer to Eqs. (15.3.32), (15.3.33) as the unconstrained and constrained cases, respectively.[1]

In the unconstrained case, if x is a solution of Eq. (15.3.32) which is also an interior point of $\mathcal{X}$, then $\alpha \to J(x + \alpha y)$ has a minimum at $\alpha = 0$, and so

$$\left. \frac{d}{d\alpha} J(x + \alpha y) \right|_{\alpha=0} = 0 \quad \forall y \in \mathbf{X} \tag{15.3.34}$$

must be satisfied. In the constrained case, a solution must instead have the property that there exists a constant λ such that

$$\left. \frac{d}{d\alpha} (J(x + \alpha y) - \lambda H(x + \alpha y)) \right|_{\alpha=0} = 0 \quad \forall y \in \mathbf{X} \tag{15.3.35}$$

This property, which in the finite dimensional case is the familiar Lagrange multiplier condition, may be justified for the infinite dimensional case in several ways—here is one of them. Suppose we can find a constant λ such that the unconstrained problem of minimizing $J - \lambda H$ has a solution x for which $H(x) = C$, that is $J(z) - \lambda H(z) \geq J(x) - \lambda H(x)$ for all z. But if we require z to satisfy the constraint, then $H(z) = H(x)$, and so $J(z) \geq J(x)$ for all z for which $H(z) = C$. Thus the constrained minimization problem may be regarded as that of solving Eq. (15.3.35) simultaneously with the constraint $H(x) = C$. The special value of λ is called a *Lagrange multiplier* for the problem. In either the constrained or unconstrained case, the equation which results from Eq. (15.3.34) or (15.3.35) is called the *Euler-Lagrange equation*.

1. Even though the definition of $\mathcal{X}$ itself will often amount to the imposition of certain constraints.

The same conditions would be satisfied if we were seeking a maximum rather than a minimum, and may also be satisfied at critical points which are neither. The Euler-Lagrange equation must be viewed as a *necessary* condition for a solution, but it does not follow that any solution of the Euler-Lagrange equation must also be a solution of the original optimization problem. Just as in elementary calculus, we only obtain candidates for the solution in this way, and some further argument will in general be needed to complete the solution.

15.4 VARIATIONAL METHODS FOR ELLIPTIC BOUNDARY VALUE PROBLEMS

We now present the application of variational methods, and obtain the Euler-Lagrange equation in explicit form, for several important PDE problems.

Example 15.1. Let J denote the Dirichlet quotient defined in Eq. (15.1.3) which we regard as defined on $\mathcal{X} = \{u \in H_0^1(\Omega)\colon u \neq 0\} \subset H_0^1(\Omega)$. Precisely as in Eq. (12.3.40) we find that

$$\left.\frac{d}{d\alpha}J(u+\alpha v)\right|_{\alpha=0} = 2\frac{(\int_\Omega u^2dx)(\int_\Omega \nabla u\cdot\nabla v dx) - (\int_\Omega |\nabla u|^2dx)(\int_\Omega uv dx)}{(\int_\Omega u^2dx)^2} \tag{15.4.36}$$

The condition (15.3.34) for an unconstrained minimum of J over $\mathcal{X}$ then amounts to

$$u \in H_0^1(\Omega)\backslash\{0\} \quad \int_\Omega \nabla u\cdot\nabla v dx - \lambda\int_\Omega uv dx = 0 \quad \forall v \in H_0^1(\Omega) \tag{15.4.37}$$

with $\lambda = J(u)$. Thus the Euler-Lagrange equation for this problem is precisely the equation for a Dirichlet eigenfunction in Ω. □

Example 15.2. Let

$$J(u) = \int_\Omega |\nabla u|^2 dx \quad H(u) = \int_\Omega u^2 dx \tag{15.4.38}$$

both regarded as functionals on $\mathcal{X} = \mathbf{X} = H_0^1(\Omega)$. By elementary calculations,

$$\left.\frac{d}{d\alpha}J(u+\alpha v)\right|_{\alpha=0} = 2\int_\Omega \nabla u\cdot\nabla v dx \quad \left.\frac{d}{d\alpha}H(u+\alpha v)\right|_{\alpha=0} = 2\int_\Omega uv dx \tag{15.4.39}$$

The condition (15.3.35) for a minimum of J subject to the constraint $H(u) = 1$ then amounts to Eq. (15.4.37) again, except now the solution is automatically normalized in L^2. Thus we can regard the problem of finding eigenvalues as coming from either a constrained or an unconstrained optimization problem. □

Example 15.3. Define J as in Example 15.1, except replace $H_0^1(\Omega)$ by $H^1(\Omega)$. The condition for a solution of the unconstrained problem is then

$$u \in H^1(\Omega)\backslash\{0\} \quad \int_\Omega \nabla u \cdot \nabla v dx - \lambda \int_\Omega uv dx = 0 \quad \forall v \in H^1(\Omega) \tag{15.4.40}$$

Since we are still free to choose $v \in C_0^\infty(\Omega)$ it again follows that $-\Delta u = \lambda u$ for $\lambda = J(u)$, but there is no longer an evident boundary condition for u to be satisfied. We observe, however, that if we choose v to be, say, in $C^1(\overline{\Omega})$ in Eq. (15.4.40), then an integration by parts yields

$$-\int_\Omega v\Delta u dx + \int_{\partial\Omega} v\frac{\partial u}{\partial n} ds = \lambda \int_\Omega uv dx \tag{15.4.41}$$

and since the Ω integrals must cancel, we get

$$\int_{\partial\Omega} v\frac{\partial u}{\partial n} ds = 0 \quad \forall v \in C^1(\overline{\Omega}) \tag{15.4.42}$$

Since v is otherwise arbitrary, we conclude that $\frac{\partial u}{\partial n} = 0$ on $\partial\Omega$ should hold. Thus, by looking for critical points of the Dirichlet quotient over the larger space $H^1(\Omega)$ we get eigenfunctions of $-\Delta$ subject to the homogeneous Neumann condition, in place of the Dirichlet condition. Since this condition was not imposed explicitly, but rather followed from the choice of function space we used, it is often referred to in this context as the *natural* boundary condition.

Note that the actual minimum in this case is clearly $J = 0$, achieved for any constant function u. Thus it is the fact that infinitely many other critical points can be shown to exist which makes this of interest. □

Example 15.4. Let $f \in L^2(\Omega)$, and set

$$J(u) = \frac{1}{2}\int_\Omega |\nabla u|^2 dx - \int_\Omega fu dx \quad u \in H_0^1(\Omega) \tag{15.4.43}$$

The condition for an unconstrained critical point is readily seen to be

$$u \in H_0^1(\Omega) \quad \int_\Omega \nabla u \cdot \nabla v dx - \int_\Omega fv dx = 0 \ \forall v \in H_0^1(\Omega) \tag{15.4.44}$$

Thus, in the distributional sense at least, a minimizer is a solution of the Poisson problem

$$-\Delta u = f \quad x \in \Omega \quad u = 0 \quad x \in \partial\Omega \tag{15.4.45}$$

The existence of a unique solution is already known from Proposition 13.2, and is explicitly given by the integral operator S appearing in Eq. (13.4.95). The main interest here is that we have obtained a variational characterization of it. Furthermore, we will give a direct proof of the existence of a unique solution of Eq. (15.4.44), which is of interest because it is easily adaptable to some other situations, even if it does not provide a new result in this particular case. The proof illustrates the so-called *direct method of the Calculus of Variations.* □

Theorem 15.7. *The problem of minimizing the functional J defined in Eq. (15.4.43) has a unique solution, which therefore also satisfies Eq. (15.4.44).*

Proof. If C denotes any constant for which the Poincaré inequality (13.4.82) is valid, we obtain

$$\left|\int_\Omega fudx\right| \le \|f\|_{L^2}\|u\|_{L^2} \le C\|f\|_{L^2}\|u\|_{H_0^1} \le \frac{1}{4}\|u\|_{H_0^1}^2 + C^2\|f\|_{L^2}^2 \qquad (15.4.46)$$

so that

$$J(u) \ge \frac{1}{4}\|u\|_{H_0^1}^2 - C^2\|f\|_{L^2}^2 \qquad (15.4.47)$$

In particular, J is bounded below, so

$$d := \inf_{u\in H_0^1(\Omega)} J(u) \qquad (15.4.48)$$

is finite and there exists a sequence $u_n \in H_0^1(\Omega)$ such that $J(u_n) \to d$. Also, since

$$\|u_n\|_{H_0^1}^2 \le 4\left(J(u_n) + C^2\|f\|_{L^2}^2\right) \qquad (15.4.49)$$

the sequence $\{u_n\}$ is bounded in $H_0^1(\Omega)$. By Theorem 12.1 there exists $u \in H_0^1(\Omega)$ and a subsequence weakly convergent in $H_0^1(\Omega)$, $u_{n_k} \overset{w}{\to} u$, which is therefore strongly convergent in $L^2(\Omega)$ by Theorem 13.4. Finally,

$$d \le J(u) = \frac{1}{2}\int_\Omega |\nabla u|^2 dx - \int_\Omega fudx \qquad (15.4.50)$$

$$\le \liminf_{n_k\to\infty}\left(\frac{1}{2}\int_\Omega |\nabla u_{n_k}|^2 dx - \int_\Omega fu_{n_k}dx\right) \qquad (15.4.51)$$

$$= \liminf_{n_k\to\infty} J(u_{n_k}) = d \qquad (15.4.52)$$

Here, the inequality on the second line follows from the first part of Proposition 12.2 and the fact that the $\int_\Omega fu_{n_k}dx$ term is convergent. We conclude that $J(u) = d$ so J achieves its minimum value.

If two such solutions u_1, u_2 exist, then the difference $u = u_1 - u_2$ must satisfy

$$\int_\Omega \nabla u \cdot \nabla v dx = 0 \quad \forall v \in H_0^1(\Omega) \qquad (15.4.53)$$

Choosing $v = u$ we get $\|u\|_{H_0^1} = 0$, so $u_1 = u_2$. □

Here is one immediate generalization about the solvability of Eq. (15.4.45), which is easy to obtain by the above method. Suppose that there exists $p \in [1, 2)$ such that the inequality

$$\left|\int_\Omega fudx\right| \le C\|f\|_{L^p}\|u\|_{H_0^1} \qquad (15.4.54)$$

holds. Then the remainder of the proof remains valid, establishing the existence of a solution for all $f \in L^p(\Omega)$ for this choice of p, corresponding to a class of f's which is larger than $L^2(\Omega)$. It can in fact be shown that Eq. (15.4.54) is correct for $p = \frac{2N}{N+2}$, see Exercise 15.15.

Next consider the functional J in Eq. (15.4.43) except now regarded as defined on all of $H^1(\Omega)$, in which case the critical point condition is

$$u \in H^1(\Omega) \quad \int_\Omega \nabla u \cdot \nabla v dx - \int_\Omega f v dx = 0 \ \ \forall v \in H^1(\Omega) \tag{15.4.55}$$

It still follows that u must be a weak solution of $-\Delta u = f$, and by the same argument as in Example 15.3, $\frac{\partial u}{\partial n} = 0$ on $\partial\Omega$. Thus critical points of J over $H^1(\Omega)$ provide us with solutions of

$$-\Delta u = f \quad x \in \Omega \quad \frac{\partial u}{\partial n} = 0 \quad x \in \partial\Omega \tag{15.4.56}$$

We must first recognize that we can no longer expect a solution to exist for arbitrary choices of $f \in L^2(\Omega)$, since if we choose $v \equiv 1$ we obtain the condition

$$\int_\Omega f dx = 0 \tag{15.4.57}$$

which is thus a necessary condition for solvability. Likewise, if a solution exists it will not be unique, since any constant could be added to it. From another point of view, if we examine the proof of Theorem 15.7, we see that the infimum of J is clearly equal to $-\infty$, unless $\int_\Omega f dx = 0$, since we can choose u to be an arbitrary constant function. Thus the minimum of J cannot be achieved by any function $u \in H^1(\Omega)$.

To work around this difficulty, we make use of the closed subspace of zero mean functions in $H^1(\Omega)$, which we denote by

$$H^1_*(\Omega) = \left\{ u \in H^1(\Omega) \colon \int_\Omega u dx = 0 \right\} \tag{15.4.58}$$

Here the inner product and norm will simply be the restriction of the usual ones in H^1 to H^1_*. Analogous to the Poincaré inequality, Proposition 13.1 we have the following proposition.

Proposition 15.1. *If Ω is a bounded, connected, open set in $\mathbb{R}^N$ with sufficiently smooth boundary then there exists a constant C, depending only on Ω, such that*

$$\|u\|_{L^2(\Omega)} \le C \|\nabla u\|_{L^2(\Omega)} \quad \forall u \in H^1_*(\Omega) \tag{15.4.59}$$

See Exercise 15.6 for the proof. The key point is that H^1_* contains no constant functions other than zero. Now if we regard the functional J in Eq. (15.4.43) as defined only on the Hilbert space $H^1_*(\Omega)$, then the proof of Theorem 15.7 can be modified in an obvious way to obtain that for any $f \in L^2(\Omega)$ there exists

$$u \in H^1_*(\Omega) \quad \int_\Omega \nabla u \cdot \nabla v dx - \int_\Omega f v dx = 0 \;\; \forall v \in H^1_*(\Omega) \tag{15.4.60}$$

For any $v \in H^1(\Omega)$ let $\mu = \frac{1}{m(\Omega)} \int_\Omega v dx$ be the mean value of v, so that $v - \mu \in H^1_*(\Omega)$. If in addition we assume that the necessary condition (15.4.57) holds, it follows that

$$\int_\Omega \nabla u \cdot \nabla v dx = \int_\Omega \nabla u \cdot \nabla (v - \mu) dx = \int_\Omega f(v - \mu) dx = \int_\Omega f v dx \tag{15.4.61}$$

for any $v \in H^1(\Omega)$. Thus u satisfies Eq. (15.4.55) and so is a weak solution of Eq. (15.4.56). It is unique within the subspace $H^1_*(\Omega)$, but by adding any constant we obtain the general solution $u(x) + C$ in $H^1(\Omega)$.

15.5 OTHER PROBLEMS IN THE CALCULUS OF VARIATIONS

We now consider a more general type of optimization problem. Let $\mathcal{L} = \mathcal{L}(x, u, p)$ be a sufficiently smooth function on the domain $\{(x, u, p) \colon x \in \Omega, u \in \mathbb{R}, p \in \mathbb{R}^N\}$ where as usual $\Omega \subset \mathbb{R}^N$, and set

$$J(u) = \int_\Omega \mathcal{L}(x, u(x), \nabla u(x)) dx \tag{15.5.62}$$

The function $\mathcal{L}$ is called the *Lagrangian* in this context. We consider the problem of finding critical points of J, and for the moment proceed formally, without regard to the precise spaces of functions involved. Expanding $J(u + \alpha v)$ in powers of α, we get

$$J(u + \alpha v) = \int_\Omega \mathcal{L}(x, u(x) + \alpha v(x), \nabla u(x) + \alpha \nabla v(x)) dx \tag{15.5.63}$$

$$= \int_\Omega \mathcal{L}(x, u(x), \nabla u(x)) dx + \alpha \int_\Omega \frac{\partial \mathcal{L}}{\partial u}(x, u(x), \nabla u(x)) v(x) dx \tag{15.5.64}$$

$$+ \alpha \int_\Omega \sum_{j=1}^N \frac{\partial \mathcal{L}}{\partial p_j}(x, u(x), \nabla u(x)) \frac{\partial v}{\partial x_j}(x) dx + o(\alpha) \tag{15.5.65}$$

Thus the critical point condition reduces to

$$0 = \int_\Omega \left[\frac{\partial \mathcal{L}}{\partial u}(\cdot, u, \nabla u) v + \sum_{j=1}^N \frac{\partial \mathcal{L}}{\partial p_j}(\cdot, u, \nabla u) \frac{\partial v}{\partial x_j} \right] dx \tag{15.5.66}$$

for all suitable v's. Among the choices of v we can make, we certainly expect to find those which satisfy $v = 0$ on $\partial\Omega$. By an integration by parts we then get

$$0 = \int_{\Omega} \left[\frac{\partial \mathcal{L}}{\partial u}(\cdot, u, \nabla u) - \sum_{j=1}^{N} \frac{\partial}{\partial x_j} \frac{\partial \mathcal{L}}{\partial p_j}(\cdot, u, \nabla u) \right] v dx \tag{15.5.67}$$

Since v is otherwise arbitrary, we conclude that

$$\frac{\partial \mathcal{L}}{\partial u} - \sum_{j=1}^{N} \frac{\partial}{\partial x_j} \frac{\partial \mathcal{L}}{\partial p_j} = 0 \tag{15.5.68}$$

is a necessary condition for a critical point of J, that is to say, Eq. (15.5.68) is the corresponding Euler-Lagrange equation. Typically it amounts to a partial differential equation for u, or an ordinary differential equation if $N = 1$.

The fact that Eq. (15.5.67) leads to Eq. (15.5.68) is often referred to loosely as the *Fundamental lemma of the Calculus of Variations*, resulting formally from the intuition that we may (approximately) choose v to be equal to the bracketed term in Eq. (15.5.67) which it multiplies, so that v must have L^2 norm equal to zero. Despite using the term "lemma," it is not a precise statement of anything unless some specific assumptions are made on $\mathcal{L}$ and the function spaces involved.

Example 15.5. The functional J in Eq. (15.4.43) comes from the Lagrangian

$$\mathcal{L}(x, u, p) = \frac{1}{2}|p|^2 - f(x)u \tag{15.5.69}$$

Thus $\frac{\partial \mathcal{L}}{\partial u} = -f(x)$ and $\frac{\partial \mathcal{L}}{\partial p_j} = p_j$, so Eq. (15.5.68) becomes, upon substituting $p = \nabla u$,

$$-f(x) - \sum_{j=1}^{N} \frac{\partial}{\partial x_j} \frac{\partial u}{\partial x_j} = 0 \tag{15.5.70}$$

which is obviously the same as Eq. (15.4.45). □

Example 15.6. A very classical problem in the calculus of variations is that of finding the shape of a hanging uniform chain, constrained by given fixed locations for its two endpoints. The physical principle which we invoke is that the shape must be such that the potential energy is minimized. To find an expression for the potential energy, let the shape be given by a function $h = u(x), a < x < b$. Observe that the contribution to the total potential energy from a short segment of the chain is $gh\Delta m$ where g is the gravitational constant and Δm is the mass of the segment, and so may be given as $\rho \Delta s$ where ρ is the (constant) density, and Δs is the length of the segment. Since $\Delta s = \sqrt{1 + u'(x)^2}\Delta x$, we are led in the usual way to the potential energy functional

$$J(u) = \int_a^b u(x)\sqrt{1 + u'(x)^2} dx \tag{15.5.71}$$

to minimize. Applying Eq. (15.5.68) with $\mathcal{L}(x,u,p) = u\sqrt{1+p^2}$ gives the Euler-Lagrange equation

$$\frac{\partial \mathcal{L}}{\partial u} - \frac{d}{dx}\frac{\partial \mathcal{L}}{\partial p} = \sqrt{1+u'^2} - \frac{d}{dx}\left(\frac{uu'}{\sqrt{1+u'^2}}\right) = 0 \tag{15.5.72}$$

To solve this nonlinear ODE, we first multiply the equation through by $\frac{uu'}{\sqrt{1+u'^2}}$ to get

$$uu' - \frac{1}{2}\frac{d}{dx}\left(\frac{uu'}{\sqrt{1+u'^2}}\right)^2 = 0 \tag{15.5.73}$$

so

$$u^2 - \left(\frac{uu'}{\sqrt{1+u'^2}}\right)^2 = C \tag{15.5.74}$$

for some constant C. After some obvious algebra we get the separable first-order ODE

$$u' = \pm\sqrt{\left(\frac{u}{C}\right)^2 - 1} \tag{15.5.75}$$

which is readily integrated to obtain the general solution

$$u(x) = C\cosh\left(\frac{x}{C} + D\right) \tag{15.5.76}$$

The two constants C, D are determined by the endpoints $(a, u(a))$ and $(b, u(b))$, but in all cases the hanging chain is seen to assume the "catenary" shape, determined by the hyperbolic cosine function. □

Example 15.7. Another important class of examples comes from the theory of minimal surfaces. A function $u = u(x)$ defined on a domain $\Omega \subset \mathbb{R}^2$ may be regarded as defining a surface in $\mathbb{R}^3$, and the corresponding surface area is

$$J(u) = \int_\Omega \sqrt{1+|\nabla u|^2}dx \tag{15.5.77}$$

Suppose we seek the surface of least possible area, subject to the requirement that $u(x) = g(x)$ on $\partial\Omega$, where g is a prescribed function. Such a surface is said to span the bounding curve $\Gamma = \{(x_1, x_2, g(x_1, x_2))\colon (x_1, x_2) \in \partial\Omega\}$. The problem of finding a minimal surface with a given boundary curve is known as *Plateau's problem*.

For this discussion we assume that g is the restriction to $\partial\Omega$ of some function in $H^1(\Omega)$ and then let $\mathcal{X} = \{u \in H^1(\Omega)\colon u - g \in H_0^1(\Omega)\}$. Thus in looking at $J(u+\alpha v)$ we should always assume that $v \in H_0^1(\Omega)$, as in the discussion leading to Eq. (15.5.67). With $\mathcal{L}(x,u,p) = \sqrt{1+|p|^2}$ we obtain

$$\frac{\partial \mathcal{L}}{\partial p_j} = \frac{p_j}{\sqrt{1+|p|^2}} \tag{15.5.78}$$

The resulting Euler-Lagrange equation is then the *minimal surface equation*

$$\sum_{j=1}^{2}\left(\frac{u_{x_j}}{\sqrt{1+|\nabla u|^2}}\right)_{x_j}=0 \tag{15.5.79}$$

It turns out that the expression on the left-hand side is the so-called *mean curvature*[2] of the surface defined by $u(x,y)$, so a minimal surface always has zero mean curvature. □

Let us finally consider an example in the case of constrained optimization,

$$\min_{H(u)=C} J(u) \tag{15.5.80}$$

where J is defined as in Eq. (15.5.62) and H is another functional of the same sort, say

$$H(u)=\int_\Omega \mathcal{N}(x,u(x),\nabla u(x))dx \tag{15.5.81}$$

As discussed in Section 15.3 we should seek critical points of $J-\lambda H$, which we may regard as coming from the *augmented Lagrangian* $\mathcal{M}:=\mathcal{L}-\lambda\mathcal{N}$. The Euler-Lagrange equation for a solution will then be

$$\frac{\partial \mathcal{M}}{\partial u}-\sum_{j=1}^{N}\frac{\partial}{\partial x_j}\frac{\partial \mathcal{M}}{\partial p_j}=0 \qquad \int_\Omega \mathcal{N}(x,u(x),\nabla u(x))dx=C \tag{15.5.82}$$

Example 15.8. (Dido's problem) Consider the area A in the (x,y) plane between $y=0$ and $y=u(x)$, where $u(x)\ge 0$, $u(0)=u(1)=0$. If the curve $y=u(x)$ is fixed to have length L, how should we choose the shape of the curve to maximize the area A? This is an example of a so-called *isoperimetric problem* because the total perimeter of the boundary of A is fixed to be $1+L$. Clearly the mathematical expression of this problem may be written in the form (15.5.80) with

$$J(u)=\int_0^1 u(x)dx \quad H(u)=\int_0^1\sqrt{1+u'(x)^2}dx \quad C=L \tag{15.5.83}$$

so that

$$\mathcal{M}=u-\lambda\sqrt{1+p^2} \tag{15.5.84}$$

The first equation in Eq. (15.5.82) thus gives

$$\left(\frac{u'}{\sqrt{1+u'^2}}\right)'=\frac{1}{\lambda} \tag{15.5.85}$$

2. It is equal to the average of the principal curvatures.

From straightforward algebra and integration we obtain

$$u' = \pm \frac{x - x_0}{\sqrt{\lambda^2 - (x - x_0)^2}} \tag{15.5.86}$$

for some x_0, which subsequently leads to the result, intuitively expected, that the curve must be an arc of a circle,

$$(u - u_0)^2 + (x - x_0)^2 = \lambda^2 \tag{15.5.87}$$

for some x_0, u_0. From the boundary conditions $u(0) = u(1) = 0$ it is easy to see that $x_0 = 1/2$, and the length constraint implies

$$L = \int_0^1 \sqrt{1 + u'^2}dx = \lambda \int_0^1 \frac{dx}{\sqrt{\lambda^2 - (x - \frac{1}{2})^2}} = \lambda \sin^{-1}\left(\frac{x - \frac{1}{2}}{\lambda}\right)\Bigg|_0^1$$
$$= 2\lambda \sin^{-1}\frac{1}{2\lambda} \tag{15.5.88}$$

By elementary calculus techniques we may verify that a unique $\lambda \geq 1/2$ exists for any $L \in (1, \frac{\pi}{2}]$. The restriction $L > 1$ is of course a necessary one for the curve to connect the two endpoints and enclose a positive area, but $L \leq \frac{\pi}{2}$ is only an artifact due to us requiring that the curve be given in the form $y = u(x)$. If instead we allow more general curves (e.g., given parametrically) then any $L > 1$ is possible, see Exercise 15.17. □

15.6 THE EXISTENCE OF MINIMIZERS

We turn now to some discussion of conditions which guarantee the existence of a solution of a minimization problem. We emphasize that Eq. (15.5.68) is only a necessary condition for a solution, and some different kind of argument is needed to establish that a given minimization problem actually has a solution. Let $\mathbf{H}$ be a Hilbert space, $\mathcal{X} \subset \mathbf{H}$ an admissible subset of $\mathbf{H}$, $J: \mathcal{X} \to \mathbb{R}$ and consider the problem

$$\min_{x \in \mathcal{X}} J(x) \tag{15.6.89}$$

One result which is immediate from applying Theorem 3.4 to $-J$ is that a solution exists provided $\mathcal{X}$ is compact and J is continuous. It is unfortunately the case for many interesting problems that one or both of these conditions fails to be true, thus some other considerations are needed. We will use the following definitions.

Definition 15.1. J is *coercive* if $J(x) \to +\infty$ as $\|x\| \to \infty, x \in \mathcal{X}$.

Definition 15.2. J is *lower semicontinuous* if $J(x) \leq \liminf_{n\to\infty} J(x_n)$ whenever $x_n \in \mathcal{X}$, $x_n \to x$, and *weakly lower semicontinuous* if $J(x) \leq \liminf_{n\to\infty} J(x_n)$ whenever $x_n \in \mathcal{X}$, $x_n \overset{w}{\to} x$.

Definition 15.3. J is *convex* if $J(tx + (1 - t)y) \le tJ(x) + (1 - t)J(y)$ whenever $0 \le t \le 1$ and $x, y \in \mathcal{X}$.

Recall also that $\mathcal{X}$ is weakly closed if $x_n \in \mathcal{X}$, $x_n \stackrel{w}{\rightarrow} x$ implies that $x \in \mathcal{X}$.

Theorem 15.8. *If J: $\mathcal{X} \rightarrow \mathbb{R}$ is coercive and weakly lower semicontinuous, and $\mathcal{X} \subset \mathbf{H}$ is weakly closed, then there exists a solution of Eq. (15.6.89). If J is convex then it is only necessary to assume that J is lower semicontinuous rather than weakly lower semicontinuous.*

Proof. Let $d = \inf_{x \in \mathcal{X}} J(x)$. If $d \neq -\infty$ then there exists $R > 0$ such that $J(x) \ge d + 1$ if $x \in \mathcal{X}$, $\|x\| > R$, while if $d = -\infty$ there exists $R > 0$ such that $J(x) \ge 0$ if $x \in \mathcal{X}$, $\|x\| > R$. Either way, the infimum of J over $\mathcal{X}$ must be the same as the infimum over $\{x \in \mathcal{X}\colon \|x\| \le R\}$. Thus there must exist a sequence $x_n \in \mathcal{X}$, $\|x_n\| \le R$ such that $J(x_n) \rightarrow d$.

By the second part of Theorem 12.1 and the weak closedness of $\mathcal{X}$, it follows that there is a subsequence $\{x_{n_k}\}$ and a point $x \in \mathcal{X}$ such that $x_{n_k} \stackrel{w}{\rightarrow} x$. In particular $J(x) = d$ must hold, since

$$d \le J(x) \le \liminf_{n_k \rightarrow \infty} J(x_{n_k}) = d \tag{15.6.90}$$

Thus d must be finite, and the infimum of J is achieved at x, so x is a solution of Eq. (15.6.89).

The final statement is a consequence of the following lemma, which is of independent interest. □

Lemma 15.1. *If J is convex and lower semicontinuous then it is weakly lower semicontinuous.*

Proof. If

$$E_\alpha = \{x \in \mathbf{H}\colon J(x) \le \alpha\} \tag{15.6.91}$$

then E_α is closed since $x_n \in E_\alpha, x_n \rightarrow x$ implies that $J(x) \le \liminf_{n \rightarrow \infty} J(x_n) \le \alpha$. Also, E_α is convex since if $x, y \in E_\alpha$ and $t \in [0, 1]$, then $J(tx + (1 - t)y) \le tJ(x) + (1 - t)J(y) \le t\alpha + (1 - t)\alpha = \alpha$. Now by part 3 of Theorem 12.1 (Mazur's theorem) we get that E_α is weakly closed.

Now if $x_n \stackrel{w}{\rightarrow} x$ and $\alpha = \liminf_{n \rightarrow \infty} J(x_n)$, we may find $n_k \rightarrow \infty$ such that $J(x_{n_k}) \rightarrow \alpha$. If $\alpha \neq -\infty$ and $\epsilon > 0$ we must have $x_{n_k} \in E_{\alpha+\epsilon}$ for sufficiently large n_k, and so $x \in E_{\alpha+\epsilon}$ by the weak closedness. Since ϵ is arbitrary, we must have $J(x) \le \alpha$, as needed. The proof is similar if $\alpha = -\infty$. □

15.7 CALCULUS IN BANACH SPACES

In this final section we discuss some notions, which are often used in formalizing the general ideas already used in this chapter.

Let $\mathbf{X}, \mathbf{Y}$ be Banach spaces and F: $D(F) \subset \mathbf{X} \rightarrow \mathbf{Y}$ be a mapping, nonlinear in general, and let x_0 be an interior point of $D(F)$.

Definition 15.4. If there exists a linear operator $A \in \mathcal{B}(\mathbf{X}, \mathbf{Y})$ such that

$$\lim_{x \to x_0} \frac{\|F(x) - F(x_0) - A(x - x_0)\|}{\|x - x_0\|} = 0 \tag{15.7.92}$$

then we say F is *Fréchet differentiable* at x_0, and $A =: DF(x_0)$ is the Fréchet derivative of F at x_0.

It is easy to see that there is at most one such operator A, see Exercise 15.20. It is also immediate that if $DF(x_0)$ exists then F must be continuous at x_0.

Note that Eq. (15.7.92) is equivalent to

$$F(x) = F(x_0) + DF(x_0)(x - x_0) + o(\|x - x_0\|) \quad x \in D(F) \tag{15.7.93}$$

This general concept of differentiability of a mapping at a given point amounts to the property that the mapping may be approximated in a precise sense by a linear map[3] in the vicinity of the given point x_0. The difference

$$E(x, x_0) := F(x) - F(x_0) - DF(x_0)(x - x_0) = o(\|x - x_0\|) \tag{15.7.94}$$

will be referred to as the *linearization error*, and approximating $F(x)$ by $F(x_0) + DF(x_0)(x - x_0)$ as *linearization* of F at x_0.

Example 15.9. If $F\colon \mathbf{H} \to \mathbb{R}$ is defined by $F(x) = \|x\|^2$ on a real Hilbert space $\mathbf{H}$ then

$$F(x) - F(x_0) = \|x_0 + (x - x_0)\|^2 - \|x_0\|^2 = 2\langle x_0, x - x_0\rangle + \|x - x_0\|^2 \tag{15.7.95}$$

It follows that Eq. (15.7.93) holds with $DF(x_0) = A \in \mathcal{B}(\mathbf{H}, \mathbb{R}) = \mathbf{H}^*$ given by

$$Az = 2\langle x_0, z\rangle \tag{15.7.96}$$

□

Example 15.10. Let $F\colon \mathbb{R}^N \to \mathbb{R}^M$ be defined as

$$F(x) = F(x_1, \dots, x_N) = \begin{bmatrix} f_1(x_1, \dots, x_N) \\ \vdots \\ f_M(x_1, \dots, x_N) \end{bmatrix} \tag{15.7.97}$$

If the component functions $f_1, \dots, f_M$ are continuously differentiable on some open set containing x_0, then

$$f_k(x) = f_k(x_0) + \sum_{j=1}^{N} \frac{\partial f_k}{\partial x_j}(x_0)(x_j - x_{0j}) + o(\|x - x_0\|) \tag{15.7.98}$$

Therefore

$$F(x) = F(x_0) + A(x_0)(x - x_0) + o(\|x - x_0\|) \tag{15.7.99}$$

---

3. Here we will temporarily use the word linear to refer to what might more properly be called an affine function, $F(x_0) + A(x - x_0)$ which differs from the linear function $x \to Ax$ by the constant $F(x_0) - Ax_0$.

with $A(x_0) \in \mathcal{B}(\mathbb{R}^N, \mathbb{R}^M)$ given by the Jacobian matrix of the transformation F at x_0; that is, the $M \times N$ matrix whose k,j entry is $\frac{\partial f_k}{\partial x_j}(x_0)$. It follows that $DF(x_0)$ is the linear mapping defined by the matrix $A(x_0)$, or more informally $DF(x_0) = A(x_0)$. □

Example 15.11. If $A \in \mathcal{B}(\mathbf{X}, \mathbf{Y})$ and $F(x) = Ax$ then $F(x) = F(x_0) + A(x - x_0)$ so $DF(x_0) = A$; that is, the derivative of a linear map is itself. □

Example 15.12. If $J\colon \mathbf{X} \to \mathbb{R}$ is a functional on $\mathbf{X}$, and if $DJ(x_0)$ exists then

$$DJ(x_0)y = \frac{d}{d\alpha} J(x_0 + \alpha y)\bigg|_{\alpha=0} \tag{15.7.100}$$

since

$$J(x_0 + \alpha y) - J(x_0) = DJ(x_0)(\alpha y) + E(x_0 + \alpha y, x_0) \tag{15.7.101}$$

Dividing both sides by α and letting $\alpha \to 0$, we get Eq. (15.7.100). The right-hand side of Eq. (15.7.100) has the interpretation of being the directional derivative of J at x_0 in the y-direction, and in this context is often referred to as the *Gâteaux* derivative. The observation just made is simply that the Gâteaux derivative coincides with Fréchet derivative if the latter exists. From another point of view, it says that if the Fréchet derivative exists, a formula for it may be found by computing the Gâteaux derivative. It is, however, possible that J has a derivative in the Gâteaux sense, but not in the Fréchet sense, see Exercise 15.21. In any case we see that if J is differentiable in the Fréchet sense, then the Euler-Lagrange equation for a critical point of J amounts to $DJ(x_0) = 0$. □

With a notion of derivative at hand, we can introduce several additional useful concepts. We denote by $C(\mathbf{X}, \mathbf{Y})$ the vector space of continuous mappings from $\mathbf{X}$ to $\mathbf{Y}$. The mapping $DF\colon x_0 \to DF(x_0)$ is evidently itself a mapping between Banach spaces, namely $DF\colon \mathbf{X} \to \mathcal{B}(\mathbf{X}, \mathbf{Y})$, and we say $F \in C^1(\mathbf{X}, \mathbf{Y})$ if this map is continuous with respect to the usual norms. Furthermore, we then denote $D^2F(x_0)$ as the Fréchet derivative of DF at x_0, if it exists, in which case $D^2F(x_0) \in \mathcal{B}(\mathbf{X}, \mathcal{B}(\mathbf{X}, \mathbf{Y}))$.

There is a natural isomorphism between $\mathcal{B}(\mathbf{X}, \mathcal{B}(\mathbf{X}, \mathbf{Y}))$ and $\mathcal{B}(\mathbf{X} \times \mathbf{X}, \mathbf{Y})$, namely if $A \in \mathcal{B}(\mathbf{X}, \mathcal{B}(\mathbf{X}, \mathbf{Y}))$ there is an associated $\tilde{A} \in \mathcal{B}(\mathbf{X} \times \mathbf{X}, \mathbf{Y})$ related by

$$\tilde{A}(x, z) = A(x)z \quad x, z \in \mathbf{X} \tag{15.7.102}$$

Thus it is natural to regard $D^2F(x_0)$ as a continuous bilinear map, and the action of the map will be denoted as $D^2F(x_0)(x, z) \in \mathbf{Y}$. We say $F \in C^2(\mathbf{X}, \mathbf{Y})$ if $x_0 \to D^2F(x_0)$ is continuous. It can be shown that $D^2F(x_0)$ must be symmetric if $F \in C^2(\mathbf{X}, \mathbf{Y})$—this amounts to the equality of mixed partial derivatives in the case $\mathbf{X} = \mathbb{R}^N$.

In general, we may inductively define $D^kF(x_0)$ to be the Fréchet derivative of $D^{k-1}F$ at x_0, if it exists, which will then be a k-linear mapping of $\underbrace{\mathbf{X} \times \cdots \times \mathbf{X}}_{k \text{ times}}$ into $\mathbf{Y}$.

Example 15.13. If $\mathbf{H}$ is a real Hilbert space and $F(x) = \|x\|^2$, recall we have seen that $DF(x_0)z = 2\langle x_0, z\rangle$. Thus

$$DF(x)z - DF(x_0)z = 2\langle x - x_0, z\rangle = D^2F(x_0)(x - x_0, z) + o(\|x - x_0\|) \quad (15.7.103)$$

provided $D^2F(x_0)(x, z) = 2\langle x, z\rangle$, and obviously the error term is exactly zero. □

Example 15.14. If $F: \mathbb{R}^N \to \mathbb{R}$ then by Example 15.10 $DF(x_0)$ is given by the gradient of F, that is

$$DF(x_0) \in \mathcal{B}(\mathbb{R}^N, \mathbb{R}) \quad DF(x_0)z = \sum_{j=1}^{N} \frac{\partial F}{\partial x_j}(x_0)z_j \quad (15.7.104)$$

Therefore we may regard $DF: \mathbb{R}^N \to \mathbb{R}^N$ and so $D^2F(x_0) \in \mathcal{B}(\mathbb{R}^N, \mathbb{R}^N)$, given by (now using Example 15.10 in the case $M = N$) the Jacobian of the gradient of F, that is

$$D^2F(x_0)(z, w) = \sum_{j,k=1}^{N} H_{jk}(x_0)z_j w_k = \sum_{j,k=1}^{N} \frac{\partial^2 F}{\partial x_k \partial x_j}(x_0)z_j w_k \quad (15.7.105)$$

where H is the usual Hessian matrix. □

Certain calculus rules are valid and may be proved in essentially the same way as in the finite dimensional case.

Theorem 15.9. *(Chain rule for Fréchet derivative) Assume that $\mathbf{X}, \mathbf{Y}, \mathbf{Z}$ are Banach spaces and*

$$F: D(F) \subset \mathbf{X} \to \mathbf{Y} \quad G: D(G) \subset \mathbf{Y} \to \mathbf{Z} \quad (15.7.106)$$

Assume that x_0 is an interior point of $D(F)$ and $DF(x_0)$ exists, $y_0 = F(x_0)$ is an interior point of $D(G)$ and $DG(y_0)$ exists. Then $G \circ F: \mathbf{X} \to \mathbf{Z}$ is Fréchet differentiable at x_0 and

$$D(G \circ F)(x_0) = DG(y_0)DF(x_0) \quad (15.7.107)$$

Proof. Let

$$\begin{aligned} E_F(x, x_0) &= F(x) - F(x_0) - DF(x_0)(x - x_0) \\ E_G(y, y_0) &= G(y) - G(y_0) - DG(y_0)(y - y_0) \end{aligned} \quad (15.7.108)$$

so that

$$G(F(x)) - G(F(x_0)) = DG(y_0)DF(x_0)(x - x_0) + DG(y_0)E_F(x, x_0) + E_G(F(x), y_0) \quad (15.7.109)$$

for x sufficiently close to x_0.

By the differentiability of F, G we have

$$\|E_F(x, x_0)\| = o(\|x - x_0\|) \quad \|E_G(F(x), y_0)\| = o(\|F(x) - F(x_0)\|) = o(\|x - x_0\|) \quad (15.7.110)$$

Since also $DG(y_0)$ is bounded, the conclusion follows. □

It is a familiar fact in one space dimension that a bound on the derivative of a function implies Lipschitz continuity. See Exercise 15.27 for an analogous property for mappings between Banach space. We also mention here a formulation of the second derivative test, which is valid in this abstract situation.

Theorem 15.10. *(Second derivative test) Let X be a Banach space and $J \in C^2(X, \mathbb{R})$. If J achieves its minimum at $x_0 \in X$ then $D^2J(x_0)$ must be positive semidefinite, that is, $D^2J(x_0)(z,z) \geq 0$ for all $z \in X$. Conversely if x_0 is a critical point of J at which D^2J is positive definite, $D^2J(x_0)(z,z) > 0$ for $z \neq 0$, then x_0 is a local minimum of J.*

15.8 EXERCISES

15.1. Using the trial function

$$\phi(x) = 1 - \frac{|x|^2}{R^2}$$

compute an upper bound for the first Dirichlet eigenvalue of $-\Delta$ in the ball $B(0, R)$ of $\mathbb{R}^N$. Compare to the exact value of λ_1 in dimensions 2 and 3. (Zeros of Bessel functions can be found, e.g., in tables, or by means of a root finding routine in Matlab.)

15.2. Consider the Sturm-Liouville problem

$$u'' + \lambda u = 0 \quad 0 < x < 1$$
$$u'(0) = u(1) = 0$$

whose eigenvalues are the critical points of

$$J(u) = \frac{\int_0^1 u'(x)^2 dx}{\int_0^1 u(x)^2 dx}$$

on the space $H = \{u \in H^1(0,1): u(1) = 0\}$. Use the Rayleigh-Ritz method to estimate the first two eigenvalues, and compare to the exact values. For best results, choose polynomial trial functions which resemble what the first two eigenfunctions should look like.

15.3. Use the result of Exercise 13.14 to give an alternate derivation of the fact the Dirichlet quotient achieves its minimum at ψ_1. (Hint: For $u \in H_0^1(\Omega)$ compute $\|u\|^2_{H_0^1(\Omega)}$ and $\|u\|^2_{L^2(\Omega)}$ by expanding in the eigenfunction basis.)

15.4. Let T be the integral operator

$$Tu(x) = \int_0^1 |x-y|u(y)dy$$

on $L^2(0,1)$. Show that

$$\frac{1}{3} \leq \|T\| \leq \frac{1}{\sqrt{6}}$$

(Suggestion: The lower bound can be obtained using a simple choice of trial function in the corresponding Rayleigh quotient.)

15.5. Let A be an $M \times N$ real matrix, $b \in \mathbb{R}^M$ and define $J(x) = \|Ax - b\|_2$ for $x \in \mathbb{R}^N$. (Here $\|x\|_2$ denotes the two norm, the usual Euclidean distance on $\mathbb{R}^M$).

(a) What is the Euler-Lagrange equation for the problem of minimizing J?

(b) Under what circumstances does the Euler-Lagrange equation have a unique solution?

(c) Under what circumstances will the solution of the Euler-Lagrange equation also be a solution of $Ax = b$?

15.6. Prove the version of the Poincaré inequality stated in Proposition 15.1. (Suggestions: If no such C exists show that we can find sequence $u_k \in H^1_*(\Omega)$ with $\|u_k\|_{L^2(\Omega)} = 1$ such that $\|\nabla u_k\|_{L^2(\Omega)} \le \frac{1}{k}$. The Rellich-Kondrachov compactness property, Theorem 13.4, remains valid with $H^1_0(\Omega)$ replaced by $H^1(\Omega)$ provided $\partial\Omega$ is C^1, see Theorem 1, Section 5.7 of [11]. Use this to obtain a convergent subsequence whose limit must have contradictory properties.)

15.7. Fill in the details of the following alternate proof that there exists a weak solution of the Neumann problem

$$-\Delta u = f \quad x \in \Omega \quad \frac{\partial u}{\partial n} = 0 \quad x \in \partial\Omega \quad \text{(NP)}$$

(as usual, Ω is a bounded open set in $\mathbb{R}^N$) provided $f \in L^2(\Omega)$, and $\int_\Omega f(x)dx = 0$:

(a) Show that for any $\epsilon > 0$ there exists a (suitably defined) unique weak solution u_ϵ of

$$-\Delta u + \epsilon u = f \quad x \in \Omega \quad \frac{\partial u}{\partial n} = 0 \quad x \in \partial\Omega$$

(b) Show that $\int_\Omega u_\epsilon(x)dx = 0$ for any such ϵ.

(c) Show that there exists $u \in H^1(\Omega)$ such that $u_\epsilon \to u$ weakly in $H^1(\Omega)$ as $\epsilon \to 0$, and u is a weak solution of (NP).

15.8. Consider a Lagrangian of the form $\mathcal{L} = \mathcal{L}(u, p)$ (i.e., it happens not to depend on the space variable x) when $N = 1$. Show that if u is a solution of the Euler-Lagrange equation then

$$\mathcal{L}(u, u') - u' \frac{\partial \mathcal{L}}{\partial p}(u, u') = C$$

for some constant C. In this way we are able to achieve a reduction of order from a second-order ODE to a first-order ODE. Use this observation to redo the derivation of the solution of the hanging chain problem.

15.9. Find the function $u(x)$ which minimizes

$$J(u) = \int_0^1 (u'(x) - u(x))^2 dx$$

among all functions $u \in H^1(0,1)$ satisfying $u(0) = 0, u(1) = 1$.

15.10. The area of a surface obtained by revolving the graph of $y = u(x)$, $0 < x < 1$ about the x-axis, is

$$J(u) = 2\pi \int_0^1 u(x)\sqrt{1 + u'(x)^2} dx$$

Assume that u is required to satisfy $u(0) = a, u(1) = b$, where $0 < a < b$.

(a) Find the Euler-Lagrange equation for the problem of minimizing this surface area.

(b) Show that

$$\frac{u(u')^2}{\sqrt{1 + (u')^2}} - u\sqrt{1 + (u')^2}$$

is a constant function for any such minimal surface. (Hint: Use Exercise 15.8.)

(c) Solve the first-order ODE in part (b) to find the minimal surface. Make sure to compute all constants of integration.

15.11. Find a functional on $H^1(\Omega)$ for which the Euler-Lagrange equation is

$$-\Delta u = f \quad x \in \Omega \qquad -\frac{\partial u}{\partial n} = k(x)u \quad x \in \partial\Omega$$

15.12. Find the Euler-Lagrange equation for minimizing

$$J(u) = \int_\Omega |\nabla u(x)|^q dx$$

subject to the constraint

$$H(u) = \int_\Omega |u(x)|^r dx = 1$$

where $q, r > 1$.

15.13. Let $\Omega \subset \mathbb{R}^N$ be a bounded open set, $\rho \in C(\overline{\Omega})$, $\rho(x) > 0$ in Ω, and

$$J(u) = \frac{\int_\Omega |\nabla u(x)|^2 dx}{\int_\Omega \rho(x) u(x)^2 dx}$$

(a) Show that any nonzero critical point $u \in H_0^1(\Omega)$ of J is a solution of the eigenvalue problem

$$-\Delta u = \lambda \rho(x) u \quad x \in \Omega$$
$$u = 0 \quad x \in \partial\Omega$$

(b) Show that all eigenvalues are positive.

(c) If $\rho(x) \geq 1$ in Ω and λ_1 denotes the smallest eigenvalue, show that $\lambda_1 < \lambda_1^*$, where λ_1^* is the corresponding first eigenvalue of $-\Delta$ in Ω.

15.14. Define

$$J(u) = \frac{1}{2}\int_\Omega (\Delta u)^2 dx + \int_\Omega fu dx$$

What PDE problem is satisfied by a critical point of J over $\mathcal{X} = H^2(\Omega) \cap H_0^1(\Omega)$? Make sure to specify any relevant boundary conditions. What is different if instead we let $\mathcal{X} = H_0^2(\Omega)$?

15.15. Let Ω be a bounded open set in $\mathbb{R}^N$ with sufficiently smooth boundary. If $p < N$, a special case of the Sobolev embedding theorem states that there exists a constant $C = C(\Omega, p, q)$ such that

$$\|u\|_{L^q(\Omega)} \leq C\|u\|_{W^{1,p}(\Omega)} \quad 1 \leq q \leq \frac{Np}{N-p}$$

Use this to show that Eq. (15.4.54) holds for $N \geq 3$, $p = \frac{2N}{N+2}$, and so the problem (15.4.45) has a solution obtainable by the variational method, for all f in this L^p space.

15.16. Formulate and derive a replacement for Eq. (15.5.68) for the case that u is a vector function.

15.17. Redo Dido's problem (Example 15.8) but allowing for an arbitrary curve $(x(t), y(t))$ in the plane connecting the points $(0,0)$ and $(1,0)$. Since there are now two unknown functions, the result of Exercise 15.16 will be relevant.

15.18. Show that if Ω is a bounded domain in $\mathbb{R}^N$ and $f \in L^2(\Omega)$, then the problem of minimizing

$$J(u) = \frac{1}{2}\int_\Omega |\nabla u|^2 dx - \int_\Omega fu dx$$

over $H_0^1(\Omega)$ satisfies all of the conditions of Theorem 15.8. What goes wrong if we replace $H_0^1(\Omega)$ by $H^1(\Omega)$?

15.19. We say that $J: \mathcal{X} \to \mathbb{R}$ is strictly convex if

$$J(tx + (1-t)y) < tJ(x) + (1-t)J(y) \quad x, y \in \mathcal{X} \quad 0 < t < 1$$

If J is strictly convex, show that the minimization problem (15.6.89) has at most one solution.

15.20. Show that the Fréchet derivative, if it exists, must be unique.

15.21. If $F: \mathbb{R}^2 \to \mathbb{R}$ is defined by

$$F(x,y) = \begin{cases} \frac{xy^2}{x^2+y^4} & (x,y) \neq (0,0) \\ 0 & (x,y) = (0,0) \end{cases}$$

show that F is Gâteaux differentiable but not Fréchet differentiable at the origin.

15.22. Let F be a C^1 mapping of a Banach space $\mathbf{X}$ into itself. Give a formal derivation of Newton's method

$$x_{n+1} = x_n - DF(x_n)^{-1}(F(x_n) - y)$$

for solving $F(x) = y$.

15.23. If A is a bounded linear operator on a Banach space $\mathbf{X}$, discuss the differentiability of the map $t \to e^{tA}$, regarded as a mapping from $\mathbb{R}$ into $\mathcal{B}(\mathbf{X})$. (Recall that the exponential of a bounded linear operator was defined in Exercise 4.10 of Chapter 4.)

15.24. Prove the second derivative test Theorem 15.10.

15.25. Verify the critical point condition (15.2.28).

15.26. If $\mathbf{X}$ is a Banach space and f: $[a, b] \to \mathbf{X}$ is continuous, show that $\int_a^b f(t)dt$ can be uniquely defined in $\mathbf{X}$ as a limit of Riemann sums. More precisely, there exists a unique $z \in \mathbf{X}$ such that the following is true: for any $\epsilon > 0$ there exists $\delta > 0$ such that for any partition $a = t_0 < t_1, \dots, < t_n = b$ of $[a, b]$ with $|t_j - t_{j-1}| \le \delta$ for all j, and for any choice of $t_j^* \in [t_{j-1}, t_j]$,

$$\left\| \sum_{j=1}^n f(t_j^*)(t_j - t_{j-1}) - z \right\| < \epsilon$$

Show furthermore that

$$\left\| \int_a^b f(t)dt \right\| \le \int_a^b \|f(t)\|dt$$

and

$$\phi\left(\int_a^b f(t)dt \right) = \int_a^b \phi(f(t))dt \quad \forall \phi \in \mathbf{X}^*$$

(We remark that the integral can also be defined under much weaker assumptions on f, by introducing a suitable notion of measurability.)

15.27. Let $\mathbf{X}, \mathbf{Y}$ be Banach spaces, F: $D(F) \subset \mathbf{X} \to \mathbf{Y}$, and let $x, x_0 \in D(F)$ be such that $tx + (1 - t)x_0 \in D(F)$ for $t \in [0, 1]$. If

$$M := \sup_{0 \le t \le 1} \|DF(tx + (1 - t)x_0)\|$$

show that

$$\|F(x) - F(x_0)\| \le M\|x - x_0\|$$

(Suggestion: Justify and use a suitable version of the fundamental theorem of calculus.)

Chapter 16

Weak Solutions of Partial Differential Equations

For simplicity, and because it is all that is needed in most applications, we will continue to assume that all abstract and function spaces are real; that is, only real-valued functions and scalars are allowed.

16.1 LAX-MILGRAM THEOREM

The main goal of this final chapter is to develop further tools, which will allow us to answer some basic questions about second-order linear PDEs with variable coefficients. Beginning our discussion with the elliptic case, there are actually two natural ways to write such an equation, namely

$$Lu := -\sum_{j,k=1}^{N} a_{jk}(x)\frac{\partial^2 u}{\partial x_j \partial x_k} + \sum_{j=1}^{N} b_j(x)\frac{\partial u}{\partial x_j} + c(x)u = f(x) \quad x \in \Omega \quad (16.1.1)$$

and

$$Lu := -\sum_{j,k=1}^{N} \frac{\partial}{\partial x_j}\left(a_{jk}(x)\frac{\partial u}{\partial x_k}\right) + \sum_{j=1}^{N} b_j(x)\frac{\partial u}{\partial x_j} + c(x)u = f(x) \quad x \in \Omega \quad (16.1.2)$$

A second-order PDE is said to be *elliptic* if it can be written in one of the forms (16.1.1), (16.1.2) and there exists $\theta > 0$ (the *ellipticity constant*) such that

$$\sum_{j,k=1}^{N} a_{jk}(x)\xi_j\xi_k \geq \theta|\xi|^2 \quad \forall x \in \Omega \quad \forall \xi \in \mathbb{R}^N \quad (16.1.3)$$

That is to say, the matrix with entries $a_{jk}(x)$ is uniformly positive definite on Ω. It is easy to verify that this use of the term "elliptic" is consistent with all previous usages. We will in addition always assume that the coefficients a_{jk}, b_j, c belong to $L^\infty(\Omega)$.

Techniques of Functional Analysis for Differential and Integral Equations
http://dx.doi.org/10.1016/B978-0-12-811426-1.00016-7

The structure of these two equations is referred to, respectively, as *non-divergence form* and *divergence form* since in the second case the leading order sum could be written as $\nabla \cdot v$ if v is the vector field with components $v_j = \sum_{k=1}^N a_{jk} u_{x_k}$. The minus sign in the leading order term is included for later convenience, for the same reason that Poisson's equation is typically written as $-\Delta u = f$. Also for notational simplicity we will from here on adopt the *summation convention*; that is, repeated indices are summed. Thus the two forms of the PDE may be written instead as

$$-a_{jk}(x)u_{x_k x_j} + b_j(x)u_{x_j} + c(x)u = f(x) \quad x \in \Omega \tag{16.1.4}$$

$$-\left(a_{jk}(x)u_{x_k}\right)_{x_j} + b_j(x)u_{x_j} + c(x)u = f(x) \quad x \in \Omega \tag{16.1.5}$$

There is obviously an equivalence between the two forms provided the leading coefficients a_{jk} are differentiable in an appropriate sense, so that

$$\left(a_{jk}(x)u_{x_k}\right)_{x_j} = a_{jk}(x)u_{x_k x_j} + (a_{jk})_{x_j} u_{x_j} \tag{16.1.6}$$

is valid, but one of the main reasons to maintain the distinction is that there may be situations where we do not want to make any such differentiability assumption. In such a case we cannot expect classical solutions to exist, and will rely instead on a notion of weak solution, which generalizes Eq. (13.4.80) for the case of the Poisson equation.

A second reason, therefore, for direct consideration of the PDE in divergence form is that a suitable definition of weak solution arises in a very natural way. The formal result of multiplying the equation by a test function v and integrating over Ω is that

$$\int_\Omega [a_{jk}(x)u_{x_k}(x)v_{x_j}(x) + b_j(x)u_{x_j}(x)v(x) + c(x)u(x)v(x)]dx = \int_\Omega f(x)v(x)dx \tag{16.1.7}$$

If we also wish to impose the Dirichlet boundary condition $u = 0$ for $x \in \partial\Omega$ then as in the case of the Laplace equation we interpret this as the requirement that $u \in H_0^1(\Omega)$. Assuming that $f \in L^2(\Omega)$ the integrals in Eq. (16.1.7) are all defined and finite for $v \in H_0^1(\Omega)$ and so we are motivated to make the following definition.

Definition 16.1. If $f \in L^2(\Omega)$ we say that u is a weak solution of the Dirichlet problem

$$-\left(a_{jk}(x)u_{x_k}\right)_{x_j} + b_j(x)u_{x_j} + c(x)u = f(x) \quad x \in \Omega \tag{16.1.8}$$

$$u = 0 \quad x \in \partial\Omega \tag{16.1.9}$$

if $u \in H_0^1(\Omega)$ and Eq. (16.1.7) holds for every $v \in H_0^1(\Omega)$.

In deciding whether a certain definition of weak solution for a PDE is an appropriate one, the following considerations should be borne in mind

- If the definition is too narrow, then a solution need not exist.
- If the definition is too broad, then many solutions will exist.

Thus if both existence and uniqueness can be proved, it is an indication that the balance is just right; that is, the requirements for a weak solution are neither too narrow nor too broad, so that the definition is suitable.

Here is a special case for which uniqueness is simple to prove.

Proposition 16.1. *Let Ω be a bounded domain in $\mathbb{R}^N$. There exists $\epsilon > 0$ depending only on the domain Ω and the ellipticity constant θ in Eq. (16.1.3), such that if*

$$c(x) \geq 0 \;\; x \in \Omega \quad \textit{and} \quad \max_j \|b_j\|_{L^\infty(\Omega)} < \epsilon \tag{16.1.10}$$

then there is at most one weak solution of the Dirichlet problem (16.1.8) and (16.1.9).

Proof. If u_1, u_2 are both weak solutions then $u = u_1 - u_2$ is a weak solution with $f \equiv 0$. We may then choose $v = u$ in Eq. (16.1.7) to get

$$\int_\Omega [a_{jk}(x)u_{x_k}(x)u_{x_j}(x) + b_j(x)u_{x_j}(x)u(x) + c(x)u(x)^2]dx = 0 \tag{16.1.11}$$

By the ellipticity assumption we have $a_{jk}u_{x_k}u_{x_j} \geq \theta|\nabla u|^2$ and recalling that $c \geq 0$ there results

$$\theta\|u\|^2_{H^1_0(\Omega)} \leq \epsilon\sqrt{N}\|u\|_{L^2(\Omega)}\|u\|_{H^1_0(\Omega)} \tag{16.1.12}$$

Now if $C = C(\Omega)$ denotes a constant for which Poincaré's inequality (13.4.82) holds, we obtain either $u \equiv 0$ or $\theta \leq \epsilon\sqrt{N}C$. Thus any $\epsilon < \theta/C\sqrt{N}$ has the required properties. □

The smallness restriction on the b_j's can be weakened considerably, but the nonnegativity assumption on $c(x)$ is more essential. For example in the case of

$$-\Delta u + c(x)u = 0 \quad x \in \Omega \quad u = 0 \quad x \in \partial\Omega \tag{16.1.13}$$

uniqueness fails if $c(x) = -\lambda_n$, if λ_n is any Dirichlet eigenvalue of $-\Delta$, since then any corresponding eigenfunction is a nontrivial solution.

Now turning to the question of the existence of weak solutions, our strategy will be to adapt the argument that occurs in Proposition 13.2 showing that the operator T is onto. Consider first the special case

$$-\left(a_{jk}(x)u_{x_k}\right)_{x_j} = f(x) \quad x \in \Omega \quad u = 0 \quad x \in \partial\Omega \tag{16.1.14}$$

where as before we assume the ellipticity property (16.1.3), $a_{jk} \in L^\infty(\Omega), f \in L^2(\Omega)$ and in addition the symmetry property $a_{jk} = a_{kj}$ for all j, k. Define

$$A[u, v] = \int_\Omega a_{jk}(x)u_{x_k}(x)v_{x_j}(x)dx \tag{16.1.15}$$

We claim that A is a valid inner product on the real Hilbert space $H_0^1(\Omega)$. Note that

$$A[u, v] \le C\|u\|_{H_0^1(\Omega)}\|v\|_{H_0^1(\Omega)} \tag{16.1.16}$$

for some constant C depending on $\max_{j,k} \|a_{j,k}\|_{L^\infty(\Omega)}$, so $A[u, v]$ is defined for all $u, v \in H_0^1(\Omega)$, and

$$A[u, u] \ge \theta\|u\|^2_{H_0^1(\Omega)} \tag{16.1.17}$$

by the ellipticity assumption. Thus the inner product axioms [H1] and [H2] hold. The symmetry axiom [H4] follows from the assumed symmetry of a_{jk}, and the remaining inner product axioms are obvious. If we let $\psi(v) = \int_\Omega fv dx$ then just as in the proof of Proposition 13.2 we have that ψ is a continuous linear functional on $H_0^1(\Omega)$. We conclude that there exists $u \in H_0^1(\Omega)$ such that $A[u, v] = \psi(v)$ for every $v \in H_0^1(\Omega)$, which is precisely the definition of weak solution of Eq. (16.1.14).

The argument just given seems to rely in an essential way on the symmetry assumption, but it turns out that with a somewhat different proof we can eliminate that hypothesis. This result, in its most abstract form, is the so-called *Lax-Milgram theorem.* Note that even if we had no objection to the symmetry assumption on a_{jk}, it would still not be possible to allow for the presence of first-order terms in any obvious way in the above argument.

Definition 16.2. If $\mathbf{H}$ is a Hilbert space and A: $\mathbf{H} \times \mathbf{H} \to \mathbb{R}$, we say A is

- *bilinear* if it is linear in each argument separately;
- *bounded* if there exists a constant M such that $A[u, v] \le M\|u\|\|v\|$ for all $u, v \in \mathbf{H}$; and
- *coercive* if there exists $\gamma > 0$ such that $A[u, u] \ge \gamma\|u\|^2$ for all $u \in \mathbf{H}$.

Theorem 16.1. *(Lax-Milgram) Assume that* A: $\mathbf{H}\times\mathbf{H} \to \mathbb{R}$ *is bilinear, bounded, and coercive on the Hilbert space* $\mathbf{H}$, *and that* ψ *belongs to the dual space* $\mathbf{H}^*$. *Then there exists a unique* $u \in \mathbf{H}$ *such that*

$$A[u, v] = \psi(v) \quad \forall v \in \mathbf{H} \tag{16.1.18}$$

Proof. Let

$$E = \{y \in \mathbf{H}: \exists u \in \mathbf{H} \text{ such that } A[u, v] = \langle y, v\rangle \ \forall v \in \mathbf{H}\} \tag{16.1.19}$$

If u is an element corresponding to some $y \in E$ we then have

$$\gamma\|u\|^2 \le A[u, u] = \langle y, u\rangle \le \|y\|\|u\| \tag{16.1.20}$$

so $\gamma\|u\| \le \|y\|$. In particular u is uniquely determined by y and E is closed. We claim that $E = \mathbf{H}$. If not, then there exists $z \in E^\perp$, $z \ne 0$. If we let $\phi(v) = A[z, v]$ then $\phi \in \mathbf{H}^*$, so by the Riesz Representation Theorem (Theorem 5.6) there exists $w \in \mathbf{H}$ such that $\phi(v) = \langle w, v\rangle$, or $A[z, v] = \langle w, v\rangle$, for all v.

Thus $w \in E$, but since $z \in E^{\perp}$ we find $\gamma \|z\|^2 \le A[z,z] = \langle z, w\rangle = 0$, a contradiction.

Finally if $\psi \in \mathbf{H}^*$, using Theorem 5.6 again, we obtain $y \in \mathbf{H}$ such that $\psi(v) = \langle y, v\rangle$ for every v, and since $y \in E = \mathbf{H}$ there exists $u \in \mathbf{H}$ such that $\psi(v) = A[u,v]$ for all $v \in \mathbf{H}$, as needed.

The element w is unique, since if $A[u_1, v] = A[u_2, v]$ for all $v \in \mathbf{H}$ then choosing $v = u_1 - u_2$ we get $A[v,v] = 0$ and consequently $v = u_1 - u_2 = 0$. □

Since there is no need for any assumption of symmetry, we can use the Lax-Milgram theorem to prove a more general result about the existence of weak solutions, under the same assumptions we used to prove uniqueness above.

Theorem 16.2. *Let Ω be a bounded domain in $\mathbb{R}^N$. There exists $\epsilon > 0$ depending only on Ω and the ellipticity constant θ such that if $c(x) \ge 0$ in Ω and* $\max_j \|b_j\|_{L^\infty(\Omega)} < \epsilon$ *then there exists a unique weak solution of the Dirichlet problem (Eqs. 16.1.8, 16.1.9) for any $f \in L^2(\Omega)$.*

Proof. In the real Hilbert space $\mathbf{H} = H_0^1(\Omega)$ let

$$A[u,v] = \int_\Omega [a_{jk}(x)u_{x_k}(x)v_{x_j}(x) + b_j(x)u_{x_j}(x)v(x) + c(x)u(x)v(x)]dx \tag{16.1.21}$$

for $u, v \in H_0^1(\Omega)$. It is immediate that A is bilinear and bounded. By the ellipticity and other assumptions made on the coefficients we get

$$A[u,u] = \int_\Omega [a_{jk}(x)u_{x_k}(x)u_{x_j}(x) + b_j(x)u_{x_j}(x)u(x) + c(x)u(x)^2]dx \tag{16.1.22}$$

$$\ge \theta \|u\|^2_{H_0^1(\Omega)} - \epsilon \|u\|_{L^2(\Omega)} \|u\|_{H_0^1(\Omega)} \tag{16.1.23}$$

$$\ge \gamma \|u\|^2_{H_0^1(\Omega)} \tag{16.1.24}$$

if $\gamma = \theta/2$ and $\epsilon = \gamma/C$, where $C = C(\Omega)$ is a constant for which the Poincaré inequality (13.4.82) is valid. Finally since $\psi(v) = \int_\Omega f v dx$ defines an element of $\mathbf{H}^*$, the conclusion follows from the Lax-Milgram theorem. □

As another application of the Lax-Milgram theorem, we can establish the existence of eigenvalues and eigenfunctions of more general elliptic operators. Let

$$Lu = -(a_{jk}u_{x_k})_{x_j} \tag{16.1.25}$$

Here we will assume the ellipticity condition (16.1.3), $a_{jk} \in L^\infty(\Omega)$ and the symmetry property $a_{jk} = a_{kj}$.

For $f \in L^2(\Omega)$ let $w = Sf$ be the unique weak solution $w \in H_0^1(\Omega)$ of

$$Lw = f \quad x \in \Omega \quad w = 0 \quad x \in \partial\Omega \tag{16.1.26}$$

whose existence is guaranteed by Theorem 16.2, that is, $w \in H_0^1(\Omega)$ and $A[w,v] = \int_\Omega f v dx$ for all $v \in H_0^1(\Omega)$, where

$$A[w,v] = \int_\Omega a_{jk} w_{x_k} v_{x_j} dx \tag{16.1.27}$$

Choosing $v = w$, using the ellipticity and the Poincaré inequality gives

$$\theta \|w\|^2_{H_0^1(\Omega)} \leq C\|f\|_{L^2(\Omega)} \|w\|_{H_0^1(\Omega)} \tag{16.1.28}$$

Thus S: $L^2(\Omega) \to H_0^1(\Omega)$ is bounded and consequently compact as a linear operator on $L^2(\Omega)$ by Rellich's theorem. We claim next that S is self-adjoint on $L^2(\Omega)$. To see this, suppose $f, g \in L^2(\Omega)$, $v = Sf$, and $w = Sg$. Then

$$\langle Sf, g\rangle = \langle v, g\rangle = \langle g, v\rangle = A[w,v] \tag{16.1.29}$$

$$\langle f, Sg\rangle = \langle f, w\rangle = A[v,w] \tag{16.1.30}$$

But $A[w,v] = A[v,w]$ by our symmetry assumption, so it follows that S is self-adjoint. It now follows from Theorem 12.10 that there exists an orthonormal basis $\{\psi_n\}_{n=1}^\infty$ of $L^2(\Omega)$ consisting of eigenfunctions of S, corresponding to real eigenvalues $\{\mu_n\}_{n=1}^\infty$, $\mu_n \to 0$. The eigenvalues of S are all strictly positive, since $Su = \mu u$ implies that $A[\mu u, \mu u] = \int_\Omega \mu u^2 dx$. If $\lambda_n = \mu_n^{-1}$ then ψ_n is evidently a weak solution of

$$L\psi_n = \lambda_n \psi_n \quad x \in \Omega \quad \psi_n = 0 \quad x \in \partial\Omega \tag{16.1.31}$$

and we may assume the ordering

$$0 < \lambda_1 \leq \lambda_2 \leq \cdots \leq \lambda_n \to +\infty \tag{16.1.32}$$

To summarize, we have obtained the following generalization of Theorem 13.5.

Theorem 16.3. *Assume that the ellipticity condition* (16.1.3) *holds,* $a_{jk} = a_{kj}$, *and* $a_{jk} \in L^\infty(\Omega)$ *for all* j,k. *Then the operator*

$$Tu = -(a_{jk}(x)u_{x_k})_{x_j} \quad D(T) = \{u \in H_0^1(\Omega): (a_{jk}(x)u_{x_k})_{x_j} \in L^2(\Omega)\} \tag{16.1.33}$$

has an infinite sequence of real eigenvalues of finite multiplicity,

$$0 < \lambda_1 \leq \lambda_2 \leq \lambda_3 \leq \cdots \leq \lambda_n \to +\infty \tag{16.1.34}$$

and corresponding eigenfunctions $\{\psi_n\}_{n=1}^\infty$, *which may be chosen as an orthonormal basis of* $L^2(\Omega)$.

As an immediate application, we can derive a formal series solution for the parabolic problem with time-independent coefficients

$$u_t - (a_{jk}(x)u_{x_k})_{x_j} = 0 \qquad x \in \Omega \quad t > 0 \tag{16.1.35}$$

$$u(x,t) = 0 \qquad x \in \partial\Omega \quad t > 0 \tag{16.1.36}$$

$$u(x,0) = f(x) \quad x \in \Omega \tag{16.1.37}$$

Making the same assumptions on a_{jk} as in the theorem, so that an orthonormal basis $\{\psi_n\}_{n=1}^\infty$ of eigenfunctions exists in $L^2(\Omega)$, we can obtain the solution in the form

$$u(x,t) = \sum_{n=1}^\infty \langle f, \psi_n \rangle e^{-\lambda_n t}\psi_n(x) \tag{16.1.38}$$

in precisely the same way as was done to derive Eq. (13.4.107) for the heat equation. The smallest eigenvalue λ_1 again plays a distinguished role in determining the overall decay rate for typical solutions.

16.2 MORE FUNCTION SPACES

In this section we will introduce some more useful function spaces. Recall that the Sobolev space $W_0^{k,p}(\Omega)$ is the closure of $C_0^\infty(\Omega)$ in the norm of $W^{k,p}(\Omega)$.

Definition 16.3. We define the negative-order Sobolev space $W^{-k,p'}(\Omega)$ to be the dual space of $W_0^{k,p}(\Omega)$. That is to say,

$$W^{-k,p'}(\Omega) = \left\{T \in \mathcal{D}'(\Omega) \colon \exists C \text{ such that } |Tv| \le C\|v\|_{W_0^{k,p}(\Omega)} \quad \forall v \in C_0^\infty(\Omega)\right\} \tag{16.2.39}$$

We emphasize that we are defining the dual of $W_0^{k,p}(\Omega)$, not $W^{k,p}(\Omega)$. The notation suggests that T may be regarded as the "$-k$th derivative" (i.e., a k-fold integral) of a function in $L^{p'}(\Omega)$, where p' is the usual Hölder conjugate exponent, and we will make some more precise statement along these lines in later discussion. When $p = 2$ the alternative notation $H^{-k}(\Omega)$ is commonly used. The same notation was also used in the case $\Omega = \mathbb{R}^N$ in which case a different definition using the Fourier transform was given in Exercise 8.1. One can check that the definitions are equivalent. The norm of an element in $W^{-k,p'}(\Omega)$ is defined in the usual way for dual spaces, namely

$$||T||_{W^{-k,p'}(\Omega)} = \sup_{\phi \ne 0} \frac{|T\phi|}{||\phi||_{W_0^{k,p}(\Omega)}} \tag{16.2.40}$$

If $\phi \in W_0^{k,p}(\Omega)$ and $T \in W^{-k,p'}(\Omega)$ then it is common to use the "inner product-like" notation $\langle T, \phi \rangle$ in place of $T\phi$, and we may refer to this value as the *duality pairing* of T and ϕ.

Example 16.1. If $x_0 \in (a,b)$ and $T\phi = \phi(x_0)$ (i.e., $T = \delta_{x_0}$), then $T \in H^{-1}(a,b)$. To see this, observe that for $\phi \in C_0^\infty(a,b)$ we have obviously

$$|T\phi| = |\phi(x_0)| = \left|\int_a^{x_0} \phi'(x)dx\right| \le \sqrt{|b-a|}\|\phi'\|_{L^2(a,b)} = \sqrt{|b-a|}\|\phi\|_{H_0^1(a,b)} \tag{16.2.41}$$

It is essential that $\Omega = (a,b)$ is one dimensional here. If $\Omega \subset \mathbb{R}^N$ and $x_0 \in \Omega$ it can be shown that $\delta_{x_0} \in W^{-k,p'}(\Omega)$ if and only if $k > N/p$. □

Let us next observe that in the proof of Theorem 16.2, the only property of f which we actually used was that $\psi(v) = \int_\Omega f v dx$ defines an element in the dual space of $H_0^1(\Omega)$. Thus it should be possible to obtain similar conclusions if we replace the assumption $f \in L^2(\Omega)$ by $f \in H^{-1}(\Omega)$. To make this precise, we will first make the obvious definition that for $T \in H^{-1}(\Omega)$ and L a divergence form operator as in Eq. (16.1.2) with associated bilinear form (16.1.21), u is a weak solution of

$$Lu = T \quad x \in \Omega \quad u = 0 \quad x \in \partial\Omega \tag{16.2.42}$$

provided

$$u \in H_0^1(\Omega) \quad A[u,v] = Tv \quad \forall v \in H_0^1(\Omega) \tag{16.2.43}$$

We then have, by essentially the same proof as for Theorem 16.2, the following theorem.

Theorem 16.4. *Let Ω be a bounded domain in $\mathbb{R}^N$. There exists $\epsilon > 0$ depending only on Ω and the ellipticity constant θ such that if $c(x) \geq 0$ in Ω and* $\max_j \|b_j\|_{L^\infty(\Omega)} < \epsilon$ *then there exists a unique weak solution of the Dirichlet problem (Eq. 16.2.42) for any $T \in H^{-1}(\Omega)$.*

Another related point of interest concerns the case when $L = \Delta$.

Proposition 16.2. *If $T \in H^{-1}(\Omega)$ and $u \in H_0^1(\Omega)$ is the corresponding weak solution of*

$$-\Delta u = T \quad x \in \Omega \quad u = 0 \quad x \in \partial\Omega \tag{16.2.44}$$

then

$$\|u\|_{H_0^1(\Omega)} = \|T\|_{H^{-1}(\Omega)} \tag{16.2.45}$$

Proof. The definition of weak solution here is

$$\int_\Omega \nabla u \cdot \nabla v = Tv \quad \forall v \in H_0^1(\Omega) \tag{16.2.46}$$

so it follows that if u is the weak solution whose existence is assured by Theorem 16.4,

$$|Tv| \leq \|u\|_{H_0^1(\Omega)} \|v\|_{H_0^1(\Omega)} \tag{16.2.47}$$

and therefore $\|T\|_{H^{-1}(\Omega)} \leq \|u\|_{H_0^1(\Omega)}$. But choosing $v = u$ in the same identity gives

$$\|u\|_{H_0^1(\Omega)}^2 = Tu \leq \|T\|_{H^{-1}(\Omega)} \|u\|_{H_0^1(\Omega)} \tag{16.2.48}$$

and the conclusion follows. □

In particular we see that the map $T \to u$, which is commonly denoted by $(-\Delta)^{-1}$, is an isometric isomorphism of $H^{-1}(\Omega)$ onto $H_0^1(\Omega)$, thus is a specific example of the correspondence between a Hilbert space and its dual space, as is guaranteed by Theorem 5.6. Using this map we can also give a convenient characterization of $H^{-1}(\Omega)$.

Corollary 16.1. *$T \in H^{-1}(\Omega)$ if and only if there exists $f_1 \ldots f_N \in L^2(\Omega)$ such that*

$$T = \sum_{j=1}^{N} \frac{\partial f_j}{\partial x_j} \tag{16.2.49}$$

in the sense of distributions on Ω.

Proof. Given $T \in H^{-1}(\Omega)$ we let $u = (-\Delta)^{-1}T \in H_0^1(\Omega)$ in which case $f_j := u_{x_j}$ has the required properties. Conversely, if $f_1, \ldots, f_N \in L^2(\Omega)$ are given and T is defined as a distribution by Eq. (16.2.49) it follows that

$$Tv = \sum_{j=1}^{N} \int_\Omega f_j v_{x_j} dx \tag{16.2.50}$$

for any test function v. Therefore

$$|Tv| \le \sum_{j=1}^{N} \|f_j\|_{L^2(\Omega)} \|v_{x_j}\|_{L^2(\Omega)} \le C\|v\|_{H_0^1(\Omega)} \tag{16.2.51}$$

which implies that $T \in H^{-1}(\Omega)$. □

The spaces $W^{-k,p'}$ for finite $p \neq 2$ can be characterized in a similar way, see Theorem 3.10 of [1].

A second kind of space we introduce arises very naturally in cases when there is a distinguished variable, such as time t in the heat equation or wave equation. If $\mathbf{X}$ is any Banach space and $[a,b] \subset \mathbb{R}$, we denote

$$C([a,b]: \mathbf{X}) = \{f: [a,b] \to X: f \text{ is continuous on } [a,b]\} \tag{16.2.52}$$

Continuity here is with respect to the obvious topologies, that is, for any $\epsilon > 0$ there exists $\delta > 0$ such that $\|f(t) - f(t')\|_{\mathbf{X}} \le \epsilon$ if $|t - t'| < \delta$, $t, t' \in [a,b]$. One can readily verify that

$$\|f\|_{C([a,b]:\mathbf{X})} = \max_{a \le t \le b} \|f(t)\|_{\mathbf{X}} \tag{16.2.53}$$

defines a norm with respect to which $C([a,b] : \mathbf{X})$ is a Banach space. The definition may be modified in the usual way for the case that $[a,b]$ is replaced by an open, semiopen, or infinite interval, although of course it need not then be a Banach space.

A related collection of spaces is defined by means of the norm defined as

$$\|f\|_{L^p([a,b]:\mathbf{X})} := \left(\int_a^b \|f(t)\|_{\mathbf{X}}^p dt \right)^{1/p} \tag{16.2.54}$$

for $1 \le p < \infty$. To avoid questions of measurability we will simply define $L^p([a,b] : \mathbf{X})$ to be the closure of $C([a,b] : \mathbf{X})$ with respect to this norm. See, for example, Section 5.9.2 of [11] or Section 39 of [38] for more details, and for the case $p = \infty$.

If $\mathbf{X}$ is a space of functions and $u = u(x,t)$ is a function for which $u(\cdot,t) \in \mathbf{X}$ for every (or almost every) $t \in [a,b]$, then we will often regard u as being the map u: $[a,b] \to \mathbf{X}$ defined by $u(t)(x) = u(x,t)$. Thus u is viewed as a "curve" in the space $\mathbf{X}$.

The following example illustrates a typical use of such spaces in a PDE problem. According to the discussion of Example 13.4, if Ω is a bounded open set in $\mathbb{R}^N$ and $f \in L^2(\Omega)$ then the unique solution $u = u(x,t)$ of

$$u_t - \Delta u = 0 \qquad x \in \Omega \;\; t > 0 \tag{16.2.55}$$

$$u(x,t) = 0 \qquad x \in \partial\Omega \;\; t > 0 \tag{16.2.56}$$

$$u(x,0) = f(x) \qquad x \in \Omega \tag{16.2.57}$$

is given by

$$u(x,t) = \sum_{n=1}^{\infty} c_n e^{-\lambda_n t} \psi_n(x) \tag{16.2.58}$$

Here $\lambda_n > 0$ is the nth Dirichlet eigenvalue of $-\Delta$ in Ω, $\{\psi_n\}_{n=1}^{\infty}$ is a corresponding orthonormal eigenfunction basis of $L^2(\Omega)$, and $c_n = \langle f, \psi_n \rangle$.

Theorem 16.5. *The solution u satisfies*

$$u(\cdot,t) \in H_0^1(\Omega) \quad \forall t > 0 \tag{16.2.59}$$

and

$$u \in C([0,T] : L^2(\Omega)) \cap L^2([0,T] : H_0^1(\Omega)) \tag{16.2.60}$$

for any $T > 0$.

Proof. Pick $0 \le t < t' \le T$ and observe by Bessel's equality that

$$\|u(\cdot,t) - u(\cdot,t')\|_{L^2(\Omega)}^2 = \sum_{n=1}^{\infty} |c_n|^2 (e^{-\lambda_n t} - e^{-\lambda_n t'})^2 \le \sum_{n=1}^{\infty} |c_n|^2 (1 - e^{-\lambda_n (t'-t)})^2 \tag{16.2.61}$$

Since $f \in L^2(\Omega)$ we know that $\{c_n\} \in \ell^2$, so for given $\epsilon > 0$ we may pick an integer N such that

$$\sum_{n=N+1}^{\infty} |c_n|^2 < \frac{\epsilon}{2} \tag{16.2.62}$$

Next, pick $M > 0$ such that $|c_n|^2 \le M$ for all n and then $\delta > 0$ such that

$$|e^{-\lambda_n \delta} - 1|^2 \le \frac{\epsilon}{2NM} \tag{16.2.63}$$

for $n = 1, \ldots, N$. If $0 \le t < t' \le t + \delta$ we then have

$$\|u(\cdot,t) - u(\cdot,t')\|^2_{L^2(\Omega)} \le \sum_{n=1}^{\infty} |c_n|^2 (1 - e^{-\lambda_n(t'-t)})^2 \tag{16.2.64}$$

$$\le \sum_{n=1}^{N} |c_n|^2 (1 - e^{-\lambda_n \delta})^2 + \sum_{n=N+1}^{\infty} |c_n|^2 (1 - e^{-\lambda_n \delta})^2 \tag{16.2.65}$$

$$\le \sum_{n=1}^{N} M \frac{\epsilon}{2NM} + \sum_{n=N+1}^{\infty} |c_n|^2 < \epsilon \tag{16.2.66}$$

This completes the proof that $u \in C([0,T] : L^2(\Omega))$.

To verify Eq. (16.2.59) we use the fact that

$$\|v\|^2_{H^1_0(\Omega)} = \sum_{n=1}^{\infty} \lambda_n |\langle v, \psi_n \rangle|^2 \tag{16.2.67}$$

for $v \in H^1_0(\Omega)$, see Exercise 13.14. Thus it is enough to show that

$$\sum_{n=1}^{\infty} \lambda_n |\langle u(\cdot,t), \psi_n \rangle|^2 = \sum_{n=1}^{\infty} \lambda_n |\langle f, \psi_n \rangle|^2 e^{-2\lambda_n t} < \infty \tag{16.2.68}$$

for fixed $t > 0$. By means of elementary calculus it is easy to check that $se^{-s} \le e^{-1}$ for $s \ge 0$, hence

$$\lambda_n e^{-2\lambda_n t} \le \frac{1}{2et} \quad n = 1, 2, \ldots \tag{16.2.69}$$

Thus

$$\sum_{n=1}^{\infty} \lambda_n |\langle f, \psi_n \rangle|^2 e^{-2\lambda_n t} \le \frac{\sum_{n=1}^{\infty} |\langle f, \psi_n \rangle|^2}{2et} = \frac{\|f\|^2_{L^2(\Omega)}}{2et} < \infty \tag{16.2.70}$$

as needed, as long as $t > 0$.

Finally,

$$\|u\|^2_{L^2([0,T]:H_0^1(\Omega))} = \int_0^T \|u(\cdot,t)\|^2_{H_0^1(\Omega)} = \int_0^T \sum_{n=1}^{\infty} \lambda_n e^{-2\lambda_n t} |\langle f, \psi_n\rangle|^2 dt \tag{16.2.71}$$

$$= \sum_{n=1}^{\infty} \lambda_n \int_0^T e^{-2\lambda_n t} dt |\langle f, \psi_n\rangle|^2 \tag{16.2.72}$$

$$= \frac{1}{2}\sum_{n=1}^{\infty} (1 - e^{-2\lambda_n T}) |\langle f, \psi_n\rangle|^2 \le \frac{1}{2}\|f\|^2_{L^2(\Omega)} \tag{16.2.73}$$

This completes the proof. □

Note that the proof actually establishes the quantitative estimates

$$\|u(\cdot,t)\|_{H_0^1(\Omega)} \le \frac{\|f\|_{L^2(\Omega)}}{\sqrt{2et}} \quad \forall t > 0 \tag{16.2.74}$$

$$\|u\|_{L^2([0,T]:H_0^1(\Omega))} \le \frac{\|f\|_{L^2(\Omega)}}{\sqrt{2}} \quad \forall T > 0 \tag{16.2.75}$$

The fact that $u(\cdot,t) \in H_0^1(\Omega)$ for $t > 0$ even though f is only assumed to belong to $L^2(\Omega)$ is sometimes referred to as a *regularizing effect*—the solution becomes instantaneously smoother than it starts out being. With more advanced methods one can actually show that u is infinitely differentiable, with respect to both x and t for $t > 0$. The conclusion $u(\cdot,t) \in H_0^1(\Omega)$ for $t > 0$ also gives a precise meaning for the boundary condition (16.2.56), and similarly $u \in C([0,T] : L^2(\Omega))$ provides a specific sense in which the initial condition (16.2.57) holds, namely $u(\cdot,t) \to f$ in $L^2(\Omega)$ as $t \to 0+$.

The preceding discussion is very specific to the heat equation—on physical grounds alone one may expect rather different behavior for solutions of the wave equation, see Exercise 16.9.

16.3 GALERKIN'S METHOD

For PDE problems of the form $Lu = f$, $u_t = Lu$, or $u_{tt} = Lu$, we can obtain very explicit solution formulas involving the eigenvalues and eigenfunctions of a suitable operator T corresponding to L, provided there exist such eigenvalues and eigenfunctions. But there are situations of interest when this is not necessarily the case, for example if T is not symmetric. Another case which may arise for time-dependent problems is when the expression for L, and hence the corresponding T, is itself t dependent. Even if the symmetry property were assumed to hold for each fixed t, it would then still not be possible to obtain solution formulas by means of a suitable eigenvalue/eigenfunction series.

An alternative but closely related method, which will allow for such generalizations, is *Galerkin's method*, which we will now discuss in the context of the abstract problem

$$u \in \mathbf{H} \quad A[u,v] = \psi(v) \quad \forall v \in \mathbf{H} \tag{16.3.76}$$

under the same assumptions as in the Lax-Milgram theorem (Theorem 16.1). Recall this means we assume that A is bilinear, bounded, and coercive on the real Hilbert space $\mathbf{H}$ and $\psi \in \mathbf{H}^*$.

We start by choosing an arbitrary basis $\{v_m\}_{m=1}^{\infty}$ of $\mathbf{H}$, that is, a basis which has no specific connection to the bilinear form A, and look for an approximate solution (the *Galerkin approximation*) in the form

$$u_n = \sum_{m=1}^{n} c_m v_m \tag{16.3.77}$$

If u_n happened to be the exact solution we would have $A[u_n, v] = \psi(v)$ for any $v \in \mathbf{H}$ and in particular

$$A[u_n, v_\ell] = \sum_{m=1}^{n} c_m A[v_m, v_\ell] = \psi(v_\ell) \quad \forall \ell \tag{16.3.78}$$

However, this amounts to infinitely many equations for $c_1, \dots, c_n$, so cannot be satisfied in general. Instead, we require it only for $\ell = 1, \dots, n$, and so obtain an $n \times n$ linear system for these unknowns. The resulting system

$$\sum_{m=1}^{n} c_m A[v_m, v_\ell] = \psi(v_\ell) \quad \ell = 1, \dots, n \tag{16.3.79}$$

is guaranteed nonsingular under our assumptions. Indeed, if

$$\sum_{m=1}^{n} d_m A[v_m, v_\ell] = 0 \quad \ell = 1, \dots, n \tag{16.3.80}$$

and $w = \sum_{m=1}^{n} d_m v_m$ then

$$A[w, v_\ell] = 0 \quad \ell = 1, \dots, n \tag{16.3.81}$$

and so multiplying the ℓth equation by d_ℓ and summing we get $A[w,w] = 0$. By the coercivity assumption it follows that $w = 0$ and so $d_1, \dots, d_n = 0$ by the linear independence of the v_m's.

If we set $E_n = \text{span}\{v_1, \dots, v_n\}$ then the previous discussion amounts to defining u_n to be the unique solution of

$$u_n \in E_n \quad A[u_n, v] = \psi(v) \quad \forall v \in E_n \tag{16.3.82}$$

which may be obtained by solving the finite system (16.3.79). It now remains to study the behavior of u_n as $n \to \infty$, with the hope that this sequence is convergent to $u \in \mathbf{H}$, which is the solution of Eq. (16.3.76).

The identity $A[u_n, u_n] = \psi(u_n)$, obtained by choosing $v = u_n$ in Eq. (16.3.82), together with the coercivity assumption, gives

$$\gamma \|u_n\|^2 \le \|\psi\| \|u_n\| \tag{16.3.83}$$

Thus the sequence u_n is bounded in $\mathbf{H}$ and so has a weakly convergent subsequence $u_{n_k} \stackrel{w}{\to} u$ in $\mathbf{H}$. We may now pass to the limit as $n_k \to \infty$, taking into account the meaning of weak convergence, in the relation

$$A[u_{n_k}, v_\ell] = \psi(v_\ell) \tag{16.3.84}$$

for any fixed ℓ, obtaining $A[u, v_\ell] = \psi(v_\ell)$ for every ℓ. It then follows that Eq. (16.3.76) holds, because finite linear combinations of the v_m's are dense in $\mathbf{H}$. Also, since u is the unique solution of Eq. (16.3.76) the entire sequence u_n must be weakly convergent to u.

We remark that in a situation like Eq. (16.1.14) in which, at least formally, $A[u, v] = \langle Lu, v\rangle_{\mathbf{H}_1}$ and $\psi(v) = \langle f, v\rangle_{\mathbf{H}_1}$ for some second Hilbert space $\mathbf{H}_1 \supset \mathbf{H}$, then the system (16.3.79) amounts to the requirement that $Lu_n - f = L(u_n - u) \in E_n^\perp$, where the orthogonality is with respect to the $\mathbf{H}_1$ inner product. If also the embedding of $\mathbf{H}$ into $\mathbf{H}_1$ is compact (think of $\mathbf{H} = H_0^1(\Omega)$ and $\mathbf{H}_1 = L^2(\Omega)$) then we also obtain immediately that $u_n \to u$ strongly in $\mathbf{H}_1$.

The Galerkin approximation technique can become a very powerful and effective computational technique if the basis $\{v_m\}$ is chosen in a good way, and in particular much more specific and refined convergence results can be proved for special choices of the basis. For example in the *finite element method*, approximations to solutions of PDE problems are obtained in the form (16.3.77) by solving Eq. (16.3.79), where the v_n's are chosen to be certain piecewise polynomial functions.

The Galerkin approach can also be adapted to the case of time-dependent problems. We sketch the approach by consideration of the parabolic problem

$$u_t = (a_{jk}(x,t)u_{x_k})_{x_j} \quad x \in \Omega \quad 0 < t < T \tag{16.3.85}$$

$$u(x,t) = 0 \quad x \in \partial\Omega \quad 0 < t < T \tag{16.3.86}$$

$$u(x,0) = f(x) \quad x \in \Omega \tag{16.3.87}$$

Here as usual we assume that Ω is a bounded open set in $\mathbb{R}^N$, $a_{jk} \in L^\infty(\Omega \times (0,T))$ for all j, k and there exists a constant $\theta > 0$ such that $a_{jk}(x,t)\xi_j\xi_k \ge \theta|\xi|^2$ for all $\xi \in \mathbb{R}^N$, $(x,t) \in \Omega \times (0,T)$.

Define the time-dependent bilinear form

$$\mathcal{A}[u, v; t] = \int_\Omega a_{jk}(x,t)u_{x_k}(x)v_{x_j}(x)dx \tag{16.3.88}$$

for $u, v \in H_0^1(\Omega)$. Choose $\{v_m\}_{m=1}^\infty$ to be a basis of $L^2(\Omega)$ which is also a basis of $H_0^1(\Omega)$ (e.g., eigenfunctions of the Dirichlet Laplacian), and seek an approximate solution in the form

$$u_n(x,t) = \sum_{m=1}^{n} c_m(t) v_m(x) \tag{16.3.89}$$

By reasoning analogous to that leading to Eq. (16.3.82), we require that

$$\frac{d}{dt}\langle u_n, v\rangle_{L^2} + \mathcal{A}[u_n, v; t] = 0 \tag{16.3.90}$$

for any $v \in \text{span}\{v_1, \dots, v_n\}$ and $t \geq 0$, or equivalently

$$\sum_{m=1}^{n} c_m'(t)\langle v_m, v_\ell\rangle_{L^2} + \sum_{m=1}^{n} c_m(t)\mathcal{A}[v_m, v_\ell; t] = 0 \quad t \geq 0 \tag{16.3.91}$$

This is a linear ODE system for $c_1(t), \dots, c_n(t)$, and note that the Gram matrix $\langle v_m, v_\ell\rangle_{L^2}$ is nonsingular by our assumptions. With initial values $c_m(0) = c_{m0}$ obtained from

$$\sum_{m=1}^{\infty} c_{m0} v_m(x) = f(x)$$

there will exist a unique solution $c_1(t), \dots, c_n(t)$ on any interval $[0, T]$, and so the approximate solution u_n is well defined. Bounds independent of n and convergence of the approximate solutions, in an appropriate sense, can then be established, but is beyond the scope of this book, see for example Theorem 3, Section 7.1 of [11].

16.4 INTRODUCTION TO LINEAR SEMIGROUP THEORY

The material in this final section concerns another widely used approach to time-dependent problems, so-called *semigroup* methods. To motivate the discussion, let us first consider the initial value problem for a constant coefficient linear ODE system

$$\mathbf{x}' = A\mathbf{x} \quad \mathbf{x}(0) = \mathbf{x}_0 \tag{16.4.92}$$

for some $n \times n$ matrix A and $\mathbf{x}_0 \in \mathbb{R}^n$. The solution to this problem may be written as

$$\mathbf{x}(t) = e^{tA}\mathbf{x}(0) \tag{16.4.93}$$

where e^{tA} is the matrix exponential

$$e^{tA} = \sum_{k=0}^{\infty} \frac{(tA)^k}{k!} \tag{16.4.94}$$

(see Exercise 4.10 in Chapter 4). If we define $S(t) = e^{tA}$ then $\mathcal{S}_A := \{S(t)\}_{t\in\mathbb{R}}$ is a family of matrices with the properties

$$S(t_1 + t_2) = S(t_1)S(t_2) \quad t_1, t_2 \in \mathbb{R} \tag{16.4.95}$$

$$S(0) = I \tag{16.4.96}$$

In particular $\mathcal{S}_A$ may be regarded as a group of matrices, with usual matrix multiplication as the group operation. Alternatively, and more appropriately from our point of view, we may regard $\mathcal{S}_A$ as a group of bounded linear operators on $\mathbb{C}^n$ with operator composition as the group operation. The solution of the initial value problem (16.4.92) may then be viewed as being defined by the group of operators $\mathcal{S}_A$, with the solution $\mathbf{x}(t)$ at time t corresponding to initial state $\mathbf{x}_0$ being given by

$$\mathbf{x}(t) = S(t)\mathbf{x}_0 \tag{16.4.97}$$

Next let us look at the PDE problem in Example 13.4 with unique solution $u(x,t)$ given by the infinite series (13.4.105). For a given initial state $f \in L^2(\Omega)$ and time $t \geq 0$ define the mapping $S(t)$ by

$$(S(t)f)(x) = u(x,t) \tag{16.4.98}$$

If we think of $u(\cdot,t)$ as the "solution state" for fixed t, then $S(t)$ is again the mapping from the initial state of the solution, to the solution state $u(\cdot,t) \in L^2(\Omega)$ at a later time t. Correspondingly we adopt the viewpoint that the solution u becomes a "curve" $\{u = u(t), 0 \leq t < \infty\}$ in the "state space" $L^2(\Omega)$, defined by $u(t) = u(\cdot,t)$, that is to say, $u(t)(x) = u(x,t)$.

It is readily verified (see the discussion of Example 13.4) that $S(t)$: $L^2(\Omega) \to L^2(\Omega)$ and is linear for every $t \geq 0$. Furthermore the properties (16.4.95), (16.4.96) hold, but only for $t_1, t_2 \geq 0$ in the case of Eq. (16.4.95). Thus instead of being a group of linear operators, $\{S(t)\}_{t\geq 0}$ is a *semigroup*, that is to say, inverse elements need not exist, as would be the case for a group of operators. This is a typical situation for a time-dependent PDE which is not time reversible, such as the heat equation. It is natural and common to use the notation

$$S(t) = e^{t\Delta} \tag{16.4.99}$$

here, but it is necessary to always keep in mind that the exponential now is *not* defined by means of an infinite series as in Eq. (16.4.94), instead, at least for now, it is just a convenient symbolic notation for Eq. (13.4.105).

As an abstraction of this situation we may suppose that an ODE or PDE problem is given which can be formally expressed as an *abstract Cauchy problem*

$$\frac{du}{dt} = Au \quad t \geq 0 \quad u(0) = u_0 \tag{16.4.100}$$

for some linear operator A acting on a normed linear space $\mathbf{X}$ and $u_0 \in \mathbf{X}$. The meaning of u_t here is the natural one,

$$\frac{du}{dt}(t) = \lim_{h\to 0} \frac{u(t+h) - u(t)}{h} \tag{16.4.101}$$

whenever the limit exists in $\mathbf{X}$.

Definition 16.4. We say u is a solution of Eq. (16.4.100) if

- $u \in C([0, \infty) : X)$
- $u(0) = u_0$
- $u(t) \in D(A)$, $u_t(t)$ exists and $u_t(t) = Au(t)$ for all $t > 0$

We then wish to ask questions such as whether there exists a corresponding semigroup of linear operators $\mathcal{S}_A = \{S(t)\}_{t \geq 0}$ for which the solution of Eq. (16.4.100) is given by $u(t) = S(t)u_0$, and what properties this solution has. One may anticipate that a central concern is the interplay between the properties of A and those of $S(t)$.

The case of a bounded linear operator on a Banach space is relatively easy to handle, almost the same as the finite dimensional case, and we leave the details as an exercise.

Theorem 16.6. *If $A \in \mathcal{B}(\mathbf{X})$ define e^{tA} by the infinite series* (16.4.94). *Then*

(a) *$t \to e^{tA}$ is a continuous mapping from $\mathbb{R}$ into $\mathcal{B}(\mathbf{X})$.*
(b) *$\|e^{tA}\| \leq e^{|t|\|A\|}$ for all $t \in \mathbb{R}$.*
(c) *The group properties* (16.4.95), (16.4.96) *are satisfied with $S(t) = e^{tA}$.*
(d) *$u(t) = e^{tA}u_0$ is the unique solution of Eq. (16.4.100) for any $u_0 \in \mathbf{X}$.*
(e) *$Au_0 = \frac{d}{dt}(e^{tA}u_0)\Big|_{t=0}$ for all $u_0 \in \mathbf{X}$.*

For the case of an unbounded operator A, we can no longer use the infinite series definition, and as indicated above should no longer expect that e^{tA} exists for all $t \in \mathbb{R}$.

Definition 16.5. If $\mathbf{X}$ is a Banach space and $\mathcal{S} = \{S(t)\colon t \geq 0\}$ is a family of bounded linear operators on $\mathbf{X}$ then we say $\mathcal{S}$ is a C_0 *semigroup* if

$$S(t_1 + t_2) = S(t_1)S(t_2) \quad t_1, t_2 \in [0, \infty) \tag{16.4.102}$$

$$\lim_{t \to 0+} S(t)x = x \quad \forall x \in \mathbf{X} \tag{16.4.103}$$

If in addition $\|S(t)\| \leq 1$ for all $t \geq 0$ then we say $\mathcal{S}$ is a C_0 contraction[1] semigroup.

Definition 16.6. Let $\mathcal{S}$ be a C_0 semigroup, and define the linear operator

$$Ax = \lim_{t \to 0+} \frac{S(t)x - x}{t} \tag{16.4.104}$$

on the domain $D(A)$ consisting of all $x \in \mathbf{X}$ for which the indicated limit exists. We say that A is the *infinitesimal generator* of $\mathcal{S}$.

---

1. Ordinarily one would expect $\|S(t)\| < 1$ in referring to $S(t)$ as a contraction, but this is nevertheless the standard terminology.

Obviously $0 \in D(A)$ but more than that we cannot yet say. Note that the last part of Theorem 16.6 amounts to the statement that A is the infinitesimal generator of e^{tA} in the case that A is bounded. Thus we may anticipate that to solve the abstract Cauchy problem (16.4.100) we will be seeking a C_0 semigroup whose infinitesimal generator is the given operator A. In most interesting applications A will be unbounded.

The following property will turn out to be a useful one it what follows.

Definition 16.7. Let $\mathbf{X}$ be a Banach space. A linear operator A: $D(A) \subset \mathbf{X} \to \mathbf{X}$ is said to be *dissipative* in $\mathbf{X}$ if

$$\|\lambda x - Ax\| \geq \lambda \|x\| \quad \forall x \in D(A) \ \ \forall \lambda > 0 \tag{16.4.105}$$

If also there exists $\lambda_0 > 0$ such that

$$R(\lambda_0 I - A) = \mathbf{X} \tag{16.4.106}$$

we say A is *m-dissipative*.[2]

Note that when A is m-dissipative it follows immediately that $\lambda_0 \in \rho(A)$, the resolvent set of A, and in particular A is closed. In the special case of a Hilbert space, Eq. (16.4.105) can be replaced by the following simpler condition, whose proof is left for the exercises.

Proposition 16.3. *If* $\mathbf{H}$ *is a real Hilbert space and* A: $D(A) \subset \mathbf{H} \to \mathbf{H}$ *is a linear operator, then* A *is dissipative if and only if*

$$\langle Ax, x \rangle \leq 0 \quad \forall x \in D(A) \tag{16.4.107}$$

Example 16.2. Let $\Omega \subset \mathbb{R}^N$ be a bounded open set, $\mathbf{H} = L^2(\Omega)$ and

$$Au = \Delta u \tag{16.4.108}$$

on the domain

$$D(A) = \{u \in H_0^1(\Omega) : \Delta u \in L^2(\Omega)\} \tag{16.4.109}$$

Then A is m-dissipative in $\mathbf{H}$. To see this, first note that the existence of a unique solution u of $\lambda u - Au = f$ for any $f \in L^2(\Omega)$ and any $\lambda > 0$ is a special case of Theorem 16.2. Since

$$\langle Au, u \rangle = \int_\Omega u \Delta u dx = -\int_\Omega |\nabla u|^2 dx \leq 0 \tag{16.4.110}$$

it follows from Proposition 16.3 that A is m-dissipative. □

We can now state one of the central theorems of linear semigroup theory. This theorem is capable of a great many elaborations, variations, and

2. An alternative terminology which is also widely used is that A is *accretive* (respectively, *m-accretive*) if $-A$ is dissipative (respectively, m-dissipative).

generalizations, see for example [27] for a thorough treatment with many applications. As will be discussed more below, the main implication of this theorem is the existence of a suitable weak solution of the abstract Cauchy problem (16.4.100) under some hypotheses.

Theorem 16.7. *If A is a densely defined, m-dissipative linear operator on a Banach space $\mathbf{X}$, then there exists a C_0 contraction semigroup $\mathcal{S} = \{S_A(t)\}_{t\geq 0}$ whose infinitesimal generator is A.*

This theorem is a part of what is usually called the Lumer-Phillips theorem, which is itself mainly a consequence of the more well-known Hille-Yosida theorem.

The proof of this theorem is constructive, involving a special kind of approximation of the operator A by bounded linear operators, which we discuss next.

Proposition 16.4. *Let A be m-dissipative on a Banach space $\mathbf{X}$, $\lambda > 0$ and set*

$$E_\lambda = \lambda(\lambda I - A)^{-1} \quad A_\lambda = AE_\lambda = \lambda(E_\lambda - I) \tag{16.4.111}$$

Then $E_\lambda, A_\lambda \in \mathcal{B}(\mathbf{X})$ and

1. $\|E_\lambda\| \leq 1$.
2. $\|A_\lambda\| \leq 2\lambda$ *and* $\|e^{tA_\lambda}\| \leq 1$ *for any* $t \geq 0$.
3. $\lim_{\lambda\to\infty} E_\lambda x = x$ *for all* $x \in \mathbf{X}$.
4. $\lim_{\lambda\to\infty} A_\lambda x = Ax$ *for all* $x \in D(A)$.

The approximation A_λ of A is usually known as the *Yosida approximation*.

Proof. We begin by showing that $(0,\infty) \subset \rho(A)$. As observed above, $\rho(A)$ contains a point $\lambda_0 \in (0,\infty)$, and since by Theorem 11.1 $\rho(A)$ is open, we need only show that $\rho(A)$ is relatively closed in $(0,\infty)$. So suppose that $\lambda_n \in \rho(A) \cap (0,\infty)$ and $\lambda_n \to \lambda > 0$. For $f \in \mathbf{X}$ there must exist $x_n \in D(A)$ such that $\lambda_n x_n - Ax_n = f$, and so by Eq. (16.4.105)

$$\|x_n\| \leq \frac{\|f\|}{\lambda_n} \tag{16.4.112}$$

In particular the sequence $\{x_n\}$ is bounded in $\mathbf{X}$. For any m, n we then have, using Eq. (16.4.105) again, that

$$\lambda_m\|x_n - x_m\| \leq \|\lambda_m(x_n - x_m) - A(x_n - x_m)\| = |\lambda_n - \lambda_m|\|x_n\| \tag{16.4.113}$$

Thus $\{x_n\}$ is a Cauchy sequence in $\mathbf{X}$. If $x_n \to x$ then $Ax_n \to \lambda x - f$, and since A is closed, $\lambda x - Ax = f$. Since $\lambda I - A$ is one-to-one and onto, $\lambda \in \rho(A)$.

The statements that $E_\lambda, A_\lambda \in \mathcal{B}(\mathbf{X})$, as well as the upper bounds for their norms now follow immediately. To obtain the bound for the norm of e^{tA_λ} we use

$$\|e^{tA_\lambda}\| = \|e^{-\lambda t}e^{t\lambda E_\lambda}\| \leq e^{-\lambda t}e^{t\lambda\|E_\lambda\|} \leq 1 \tag{16.4.114}$$

Next, if $x \in D(A)$,

$$\|E_\lambda x - x\| = \frac{\|A_\lambda x\|}{\lambda} = \frac{\|E_\lambda Ax\|}{\lambda} \le \frac{\|Ax\|}{\lambda} \tag{16.4.115}$$

so $E_\lambda x \to x$ as $\lambda \to \infty$. Here we have used the fact that A and E_λ commute, see Exercise 11.3 of Chapter 11. The same conclusion now follows for any $x \in \overline{D(A)} = \mathbf{X}$, since $\|E_\lambda\| \le 1$. Finally, if $x \in D(A)$ then $A_\lambda x = E_\lambda Ax \to Ax$. □

Corollary 16.2. *A linear operator A on a Banach space* **X** *is m-dissipative if and only if* $(0,\infty) \subset \rho(A)$ *and for* $\lambda > 0$ *the resolvent operator* $R_\lambda = (\lambda I - A)^{-1}$ *satisfies*

$$\|R_\lambda\| \le \frac{1}{\lambda} \tag{16.4.116}$$

Proposition 16.5. *If A is a densely defined, m-dissipative linear operator on a Banach space* **X**, *then*

$$\lim_{\lambda\to\infty} e^{tA_\lambda}x := S(t)x \tag{16.4.117}$$

exists for every $x \in \mathbf{X}$ *and fixed* $t \ge 0$.

Proof. First observe[3] that if $\lambda, \mu > 0$ and $t \ge 0$ then

$$\left\| e^{tA_\lambda}x - e^{tA_\mu}x \right\| = \left\| \int_0^1 \frac{d}{ds}(e^{tsA_\lambda}e^{t(1-s)A_\mu}x)ds \right\| \tag{16.4.118}$$

$$= \left\| \int_0^1 te^{tsA_\lambda}A_\lambda e^{t(1-s)A_\mu}x - te^{tsA_\lambda}e^{t(1-s)A_\mu}A_\mu x\, ds \right\| \tag{16.4.119}$$

$$\le t\int_0^1 \left\| e^{tsA_\lambda}e^{t(1-s)A_\mu}(A_\lambda x - A_\mu x) \right\| ds \tag{16.4.120}$$

$$\le t\|A_\lambda x - A_\mu x\| \tag{16.4.121}$$

For $x \in D(A)$ the last term tends to zero as $\lambda, \mu \to \infty$, and therefore the limit in Eq. (16.4.117) exists for all such x. Since $\|e^{tA_\lambda}\|$ is bounded independently of λ, the limit also exists for all $x \in \overline{D(A)} = \mathbf{X}$. □

From the inequalities stated in the above proof, it also follows by letting $\mu \to \infty$ that

$$\|e^{tA_\lambda}x - S(t)x\| \le T\|A_\lambda x - Ax\| \tag{16.4.122}$$

if $x \in D(A)$ and $t \le T$.

We can now complete the proof of Theorem 16.7.

3. See for example Exercise 15.26 in Chapter 15 for properties of integrals of Banach space valued functions used here.

Proof. (Proof of Theorem 16.7) It is immediate from Propositions 16.4 and 16.5 that for every $t \geq 0$, $S(t)$ defined by Eq. (16.4.117) is a linear operator on $\mathbf{X}$ with $\|S(t)\| \leq 1$ and Eq. (16.4.102) holds.

If $\epsilon > 0$ and $x \in \mathbf{X}$, we can pick $z \in D(A)$ such that $\|x - z\| < \epsilon$ and then $\lambda > 0$ such that $\|A_\lambda z - Az\| < \epsilon$. It follows, taking into account Eq. (16.4.122), that

$$\|S(t)x - x\| \tag{16.4.123}$$

$$\leq \|S(t)x - S(t)z\| + \|S(t)z - e^{tA_\lambda}z\| \tag{16.4.124}$$

$$+ \|e^{tA_\lambda}z - z\| + \|z - x\| \tag{16.4.125}$$

$$\leq 2\|z - x\| + \|A_\lambda z - Az\| + \|e^{tA_\lambda}z - z\| \tag{16.4.126}$$

$$\leq 3\epsilon + \|e^{tA_\lambda}z - z\| \tag{16.4.127}$$

if $t \leq 1$. Finally choosing t sufficiently small, recalling that $t \to e^{tA_\lambda}z$ is continuous on $\mathbb{R}$ for any fixed λ, we obtain that $\|S(t)x - x\| < 4\epsilon$ for small enough $t > 0$, that is, Eq. (16.4.103) holds.

Thus $\{S(t)\}_{t\geq 0}$ is a C_0 contraction semigroup, and it remains to show that A is its infinitesimal generator.

Let B be the infinitesimal generator of $\{S(t)\}_{t\geq 0}$ as in Definition 16.6. For $x \in D(A)$, we have

$$S(t)x - x = \lim_{\lambda\to\infty} (e^{tA_\lambda}x - x) \tag{16.4.128}$$

$$= \lim_{\lambda\to\infty} \int_0^t \frac{d}{ds}(e^{sA_\lambda}x)ds \tag{16.4.129}$$

$$= \lim_{\lambda\to\infty} \int_0^t e^{sA_\lambda}A_\lambda x ds \tag{16.4.130}$$

$$= \int_0^t S(s)Ax ds \tag{16.4.131}$$

Dividing by t and letting $t \to 0+$, and taking account of the continuity of $t \to S(t)Ax$, it follows that $Bx = Ax$, so B is an extension of A. On the other hand, the resolvent set of B must contain the positive real axis (see Exercise 16.13), as does the resolvent set of A, by Corollary 16.2. In particular $1 \in \rho(A) \cap \rho(B)$, so $(I - B)^{-1} \subset (I - A)^{-1}$. Since both operators are everywhere defined, it follows that they are equal, hence also $A = B$. □

Under the assumptions of Theorem 16.7 we interpret $u(t) = S(t)x$ as being a suitable weak solution of the abstract Cauchy problem (16.4.100) for any $x \in \mathbf{X}$, although without further assumptions it need not be a classical solution of Eq. (16.4.100). This will, however, be the case if $x \in D(A)$.

Theorem 16.8. *If A is a densely defined m-dissipative linear operator on a Banach space $\mathbf{X}$, $\{S(t)\}_{t\geq 0}$ is the corresponding C_0 contraction semigroup, and $x \in D(A)$, then $u(t) = S(t)x$ is a classical solution of Eq. (16.4.100).*

Proof. We must verify that $u(t) \in D(A)$, $u'(t)$ exists and $u'(t) = Au(t)$ for $t > 0$. First note that for $t, h > 0$,

$$\frac{u(t+h) - u(t)}{h} = \frac{S(h) - I}{h} S(t)x \tag{16.4.132}$$

$$= S(t)\left(\frac{S(h) - I}{h}\right) x \to S(t)Ax \tag{16.4.133}$$

as $h \to 0$. This already shows that $u(t) \in D(A)$ for $t > 0$, $S(t)Ax = AS(t)x$, and that the right-hand derivative of u exists at t and is equal to $Au(t)$. We leave the proof for the left-hand derivative as an exercise. □

We have already observed that the Dirichlet Laplacian (Example 16.2 or Eq. 13.4.75) is densely defined and m-dissipative, thus we obtain a weak solution of the corresponding abstract Cauchy problem, which is the formally the same as the initial and boundary value problem for the heat equation in Example 13.4. We conclude with one final example.

Example 16.3. Consider the initial and boundary value problem for the wave equation

$$u_{tt} - \Delta u = 0 \qquad x \in \Omega \quad t > 0 \tag{16.4.134}$$

$$u(x,t) = 0 \qquad x \in \partial\Omega \quad t > 0 \tag{16.4.135}$$

$$u(x,0) = f(x) \qquad u_t(x,0) = g(x) \qquad x \in \Omega \tag{16.4.136}$$

Observe that if we let $v_1 = u$ and $v_2 = u_t$ then this second-order PDE is the same as the first-order system

$$v_{1t} = v_2 \quad v_{2t} = \Delta v_1 \tag{16.4.137}$$

With this in mind, let $\mathbf{H}$ be the Hilbert space $H_0^1(\Omega) \times L^2(\Omega)$ and let A be the densely defined linear operator on $\mathbf{H}$ defined by

$$A(v_1, v_2) = (v_2, \Delta v_1) \tag{16.4.138}$$

with domain

$$D(A) = \{(v_1, v_2) \in \mathbf{H} : v_2 \in H_0^1(\Omega), \Delta v_1 \in L^2(\Omega)\} \tag{16.4.139}$$

Our problem is then equivalent to

$$v' = Av \quad t \geq 0 \quad v(0) = (f, g) \tag{16.4.140}$$

if $f \in H_0^1(\Omega), g \in L^2(\Omega)$.

To verify that A is m-dissipative, first observe that

$$\langle A(v_1, v_2), (v_1, v_2)\rangle_{\mathbf{H}} = \langle (v_2, \Delta v_1), (v_1, v_2)\rangle_{\mathbf{H}} \tag{16.4.141}$$

$$= \langle v_2, v_1\rangle_{H_0^1(\Omega)} + \langle \Delta v_1, v_2\rangle_{L^2(\Omega)} \tag{16.4.142}$$

$$= \int_{\Omega} \nabla v_2 \cdot \nabla v_1 dx - \int_{\Omega} \nabla v_1 \cdot \nabla v_2 dx \tag{16.4.143}$$

$$= 0 \tag{16.4.144}$$

Thus A is dissipative on $\mathbf{H}$ by Proposition 16.3. The operator equation $\lambda v - Av = f = (f_1, f_2)$, after an obvious elimination of v_2, is seen to have a solution, provided there is a solution $v_1 \in H_0^1(\Omega)$ of

$$\lambda^2 v_1 - \Delta v_1 = f_2 + \lambda f_1 \tag{16.4.145}$$

Since $f_2 + \lambda f_1 \in L^2(\Omega)$, this is true for the same reason as in Example 16.2, and so we conclude that A is m-dissipative. By Theorem 16.7, if $f \in H_0^1(\Omega)$ and $g \in L^2(\Omega)$ we therefore obtain the existence of a solution $u \in C([0,\infty) : H_0^1(\Omega))$ such that $u_t \in C([0,\infty) : L^2(\Omega))$. □

16.5 EXERCISES

16.1. Verify that the definition of ellipticity (16.1.3) is consistent with the one given in Definition 8.4, and with the one given for the special case (1.3.74). That is to say, for any PDE for which more than one of these definitions is applicable, the definitions are equivalent.

16.2. Let λ_1 be the smallest Dirichlet eigenvalue for $-\Delta$ in Ω, assume that $c \in C(\overline{\Omega})$ and $c(x) > -\lambda_1$ in $\overline{\Omega}$. If $f \in L^2(\Omega)$ prove the existence of a solution of

$$-\Delta u + c(x)u = f \quad x \in \Omega \quad u = 0 \quad x \in \partial\Omega \tag{16.5.146}$$

16.3. Let $\lambda > 0$ and define

$$A[u, v] = \int_{\Omega} a_{jk}(x) u_{x_k}(x) v_{x_j}(x) dx + \lambda \int_{\Omega} uv dx \tag{16.5.147}$$

for $u, v \in H^1(\Omega)$. Assume the ellipticity property (16.1.3) and that $a_{jk} \in L^\infty(\Omega)$. If $f \in L^2(\Omega)$ show that there exists a unique solution of

$$u \in H^1(\Omega) \quad A[u, v] = \int_{\Omega} fv dx \quad \forall v \in H^1(\Omega) \tag{16.5.148}$$

Justify that u may be regarded as the weak solution of

$$-(a_{jk} u_{x_k})_{x_j} + \lambda u = f(x) \quad x \in \Omega \quad a_{jk} u_{x_k} n_j = 0 \quad x \in \partial\Omega \tag{16.5.149}$$

The above boundary condition is said to be of *conormal* type.

16.4. If $f \in L^2(0,1)$ we say that u is a weak solution of the fourth-order problem

$$u'''' + u = f \quad 0 < x < 1$$

$$u''(0) = u'''(0) = u''(1) = u'''(1) = 0$$

if $u \in H^2(0,1)$ and

$$\int_0^1 (u''(x)\zeta''(x) + u(x)\zeta(x))dx = \int_0^1 f(x)\zeta(x)dx \quad \text{for all } \zeta \in H^2(0,1)$$

Discuss why this is a reasonable definition and use the Lax-Milgram theorem to prove that there exists a weak solution. The following fact may be useful here: there exists a finite constant C such that

$$\|\phi'\|^2_{L^2(0,1)} \le C\left(\|\phi\|^2_{L^2(0,1)} + \|\phi''\|^2_{L^2(0,1)}\right) \quad \forall \phi \in H^2(0,1)$$

see for example Lemma 4.10 of [1].

16.5. Let $\Omega \subset \mathbb{R}^N$ be a bounded open set containing the origin. Show that $\delta \in H^{-1}(\Omega)$ if and only if $N = 1$.

16.6. Let f and g be in $L^2(0,1)$. Use the Lax-Milgram theorem to prove there is a unique weak solution $\{u, v\} \in H_0^1(0,1) \times H_0^1(0,1)$ to

$$-u'' + u + v' = f$$
$$-v'' + v + u' = g,$$

where $u(0) = v(0) = 0$, $u(1) = v(1) = 0$. (Hint: Start by defining the bilinear form

$$A[(u,v),(\phi,\psi)] = \int_0^1 (u'\phi' + u\phi + v'\phi + v'\psi' + v\psi + u'\psi)dx$$

on $H_0^1(0,1) \times H_0^1(0,1)$.)

16.7. If $\mathbf{X}$ is a Banach space prove that $C([a,b] : \mathbf{X})$ is also a Banach space with norm defined in Eq. (16.2.53).

16.8. Let L be the divergence form elliptic operator $Lv = -\left(a_{jk}(x)v_{x_k}\right)_{x_j}$ in a bounded open set $\Omega \subset \mathbb{R}^N$, $u_0 \in L^2(\Omega)$, and let u be a solution of the parabolic problem

$$\begin{aligned} u_t + Lu &= 0 & x \in \Omega, \quad & t > 0 \\ u(x,t) &= 0 & x \in \partial\Omega, \quad & t > 0 \\ u(x,0) &= u_0(x) & x \in \Omega & \end{aligned}$$

Let ϕ be a C^2 convex function on $\mathbb{R}$ with $\phi'(0) = 0$.

(a) Show that

$$\int_\Omega \phi(u(x,t))dx \le \int_\Omega \phi(u_0(x))dx$$

for any $t > 0$.

(b) By choosing $\phi(s) = |s|^p$ and letting $p \to \infty$, show that

$$\|u(\cdot,t)\|_{L^\infty} \le \|u_0\|_{L^\infty}$$

16.9. Let $f \in H_0^1(\Omega)$ and $g \in L^2(\Omega)$, where Ω is a bounded open set in $\mathbb{R}^N$. Find a solution of

$$u_{tt} = \Delta u \quad x \in \Omega \quad 0 < t < T$$
$$u(x,t) = 0 \quad x \in \partial\Omega \quad 0 < t < T$$
$$u(x,0) = f(x) \quad u_t(x,0) = g(x) \quad x \in \Omega$$

in the form of an eigenfunction expansion, and identify spaces of the form $L^p(0,T : \mathbf{X})$ containing u or u_t.

16.10. What is the dual space of $L^p((a,b) : L^q(\Omega))$ for $p,q \in (1,\infty)$?

16.11. If $\{S(t) : t \geq 0\}$ is a C_0 semigroup on a Banach space $\mathbf{X}$, and $u(t) = S(t)x$, show that $u \in C([0,T] : \mathbf{X})$ for any $T > 0$ and any $x \in \mathbf{X}$.

16.12. Let A be a densely defined linear operator on a Hilbert space $\mathbf{H}$. If both A and A^* are dissipative, show that A is m-dissipative.

16.13. If A is the infinitesimal generator of a C_0 contraction semigroup $\{S(t)\}_{t\geq 0}$, show that the resolvent set of A must contain $(0,\infty)$. (Suggestion: Define the "Laplace transform" of the semigroup,

$$R_\lambda x = \int_0^\infty e^{-\lambda t} S(t)x dt$$

and show that R_λ has all the necessary properties to be the resolvent operator of A.)

16.14. Let $\mathbf{X} = L^p(\mathbb{R}^N)$ for some $p \in [1,\infty)$, $h \in \mathbb{R}^N$, and define

$$(S(t)f)(x) = f(x + th) \quad f \in \mathbf{X}$$

Show that $\{S(t) : t \geq 0\}$ is a C_0 contraction semigroup on $\mathbf{X}$ and find its infinitesimal generator.

16.15. Prove Proposition 16.3.

16.16. Verify the claim made in the proof of Theorem 16.8 about the existence of the left-hand derivative. (Suggestion: Start with the identity

$$\frac{u(t) - u(t-h)}{h} - S(t)Ax =$$
$$S(t-h)\left(\frac{S(h)x - x}{h} - Ax\right) + (S(t-h)Ax - S(t)Ax)$$

for sufficiently small $h > 0$.)

Appendix

A.1 LEBESGUE MEASURE AND THE LEBESGUE INTEGRAL

In the Riemann theory of integration, as is typically taught in a calculus class, the value of an integral $\int_a^b f(x)\,dx$ is obtained as a limit of so-called Riemann sums. Although quite sufficient from the point of view of being able to compute the value of integrals which commonly arise, it is inadequate as a general definition of integral for several reasons. For example, a, b must be finite numbers, f must be a bounded function, the set of f's for which integral is defined turns out to be more limited than one would like, and certain limit procedures are more awkward than necessary. The Lebesgue theory of integration largely dispenses with all of these problems by defining the integral in a somewhat different way. A careful development of these ideas requires a whole book (see e.g., [30, 32, 40]) or course, but it will be enough for the purpose of this book for the reader to be familiar with certain key definitions, concepts, and theorems.

Definition A.1. A set $E \subset \mathbb{R}^N$ is said to have Lebesgue measure zero if for any $\epsilon > 0$ there exist points $x_k \in \mathbb{R}^N$, $r_k > 0$, $k = 1, 2, \ldots$ such that

$$E \subset \bigcup_{k=1}^{\infty} B(x_k, r_k) \qquad \sum_{k=1}^{\infty} r_k^N < \epsilon \tag{A.1.1}$$

This property amounts to requiring that the set E can be enclosed in a countable union of balls in $\mathbb{R}^N$ whose total volume is an arbitrarily small positive number. Any countable set $E = \{y_k\}_{k=1}^{\infty}$ in $\mathbb{R}^N$ is of measure zero since we could take $x_k = y_k, r_k = \frac{\epsilon}{2^k}$. As another example, any line in $\mathbb{R}^2$, or more generally any $N-1$ dimensional surface in $\mathbb{R}^N$, is of measure zero.

A property which holds except on a set of measure zero is said to hold *almost everywhere* (a.e.).

Example A.1. Let

$$f(x) = \begin{cases} 0 & x \in \mathbb{Q} \\ 1 & x \notin \mathbb{Q} \end{cases} \tag{A.1.2}$$

Since $\mathbb{Q}$ is countable it is a set of measure zero, hence $f(x) = 1$ a.e. Note also that f is discontinuous at every point but is a.e. equal to a function (namely $g(x) \equiv 1$) which is continuous at every point.

The concept of a set of measure zero arises in a key place even in Riemann integration theory. □

Theorem A.1. *(Theorem 5.54 in [40]) If f is a bounded function on $[a, b] \subset \mathbb{R}$, then f is Riemann integrable if and only if f is continuous a.e.*

Next we introduce the concepts of measurable set and measurable function. The definition we are about to state is usually given as a theorem, based on a different definition, but it is known to be equivalent to any standard definition, see Theorem (3.28) of [40]. We use it to minimize the need for additional technical concepts, as it will not be important to have the most common definition available to us.

Definition A.2. A set $E \subset \mathbb{R}^N$ is measurable if there exist open sets O_n, $n = 1, 2, \ldots$ and a set Z of measure zero, such that

$$E = \cap_{n=1}^{\infty} O_n \backslash Z \tag{A.1.3}$$

In particular, any open set is measurable and any set of measure zero is measurable. A countable intersection of open sets is sometimes called a G_δ set, so the measurability condition is that E can be written as a G_δ set with a set of measure zero excised. There exist nonmeasurable sets but they are somewhat pathological.

For any measurable set E, the measure[1] of E, which we will denote by $m(E)$, may now be defined as a nonnegative real number or $+\infty$, as follows.

Definition A.3. If $E \subset \mathbb{R}^N$ is a measurable set then

$$m(E) = \inf_{E \subset \cup_{n=1}^{\infty} I_n} \sum_{n=1}^{\infty} vol(I_n) \tag{A.1.4}$$

where here each I_n is an open "cube" of the form $(a_1, b_1) \times \cdots \times (a_N, b_N)$ and $vol(I_n)$ is the ordinary volume, $vol(I_n) = \prod_{k=1}^{N} (b_k - a_k)$.

The right-hand side of Eq. (A.1.4) is always defined, and known as the outer measure of E, whether or not E is measurable, but is only called the measure of E in the case that E is measurable. Measure is a way to assign a "size" to a set, and has "size-like" properties, in particular:

1. $m(E) = vol(E)$, the usual volume of E, if E is a ball or cube in $\mathbb{R}^N$.
2. If E_1, E_2 are measurable sets and $E_1 \subset E_2$ then $m(E_1) \le m(E_2)$.
3. If E is measurable so is E^c. In particular, any countable union of closed sets is measurable.
4. If $E_1, E_2, \ldots$ are measurable sets then $\left(\bigcup_{n=1}^{\infty} E_n\right)$ and $\left(\bigcap_{n=1}^{\infty} E_n\right)$ are also measurable.
5. $m(E_1 \cup E_2) = m(E_1) + m(E_2) - m(E_1 \cap E_2)$ whenever E_1, E_2 are measurable sets of finite measure.
6. If $E_1, E_2, \ldots$ are disjoint measurable sets then

$$m\left(\bigcup_{n=1}^{\infty} E_n\right) = \sum_{n=1}^{\infty} m(E_n) \tag{A.1.5}$$

1. Or *Lebesgue measure* of E if it is necessary to distinguish it from other measures.

Now we define what is meant by a measurable function.

Definition A.4. If $E \subset \mathbb{R}^N$ is a measurable set and f: $E \to [-\infty, \infty]$, we say f is a measurable function on E if for any open set $\mathcal{O} \subset \mathbb{R}$ the inverse image $f^{-1}(\mathcal{O})$ is measurable in $\mathbb{R}^N$.

Next, we say s: $E \to \mathbb{R}$ is a simple function if there exist disjoint measurable sets $E_1, \dots, E_n \subset E$ and finite constants $a_1, \dots, a_n$ such that

$$E = \bigcup_{n=1}^{\infty} E_n \quad s(x) = \sum_{k=1}^{n} a_k \chi_{E_k}(x) \tag{A.1.6}$$

(Recall the χ_A is the indicator function of the set A.) For a simple function s, the integral of s over E is defined in the natural way as

$$\int_E s\,dx = \sum_{k=1}^{n} a_k\, m(E_k) \tag{A.1.7}$$

provided at most one of the sets E_k has measure $+\infty$.

If $E \subset \mathbb{R}^N$ is a measurable set and f: $E \to [0, \infty]$ is a measurable function, then the Lebesgue integral $\int_E f\,dx$ is defined as either a finite nonnegative number or $+\infty$ by the formula

$$\int_E f\,dx = \sup_{s \in S(f)} \int_E s\,dx \tag{A.1.8}$$

where $S(f)$ denotes the class of simple functions for which $0 \le s(x) \le f(x)$ and $\int_E s\,dx < \infty$.

If f: $E \to [-\infty, \infty]$ is measurable and at least one of the numbers $\int_E f_+\,dx$, $\int_E f_-\,dx$ is finite, then $\int_E f\,dx$ is defined by

$$\int_E f\,dx = \int_E f_+\,dx - \int_E f_-\,dx \tag{A.1.9}$$

If $\int_E f\,dx$ is finite then so is $\int_E |f|\,dx = \int_E f_+\,dx + \int_E f_-\,dx$ and in this case we say f is integrable or $f \in L^1(E)$. Integrals of complex valued functions may also be defined in the natural way,

$$\int_E f\,dx = \int_E u\,dx + i\int_E v\,dx \tag{A.1.10}$$

if $f = u + iv$ and u, v are real-valued functions in $L^1(E)$.

When $\Omega = [a, b] \subset \mathbb{R}$ this definition is consistent with the Riemann definition of integral in the following sense (see Theorem 5.52 of [40]).

Theorem A.2. *If f is a bounded function on $[a, b] \subset \mathbb{R}$ which is Riemann integrable, then $f \in L^1([a, b])$ and the two definitions of integral coincide.*

We conclude this section by stating a number of useful and important properties in the form of a theorem. All of the stated results may be found, for example, in Rudin [32], Royden and Fitzpatrick [30], or Wheeden and Zygmund [40].

Theorem A.3. *Let $E \subset \mathbb{R}^N$ be a measurable set.*

(1) *If $f, g \in L^1(E)$ and α, β are constants then $\alpha f + \beta g \in L^1(E)$ and*

$$\int_E (\alpha f + \beta g)\, dx = \alpha \int_E f\, dx + \beta \int_E g\, dx \tag{A.1.11}$$

In particular $L^1(E)$ is a vector space.

(2) *If f is measurable on E and ϕ is a continuous function on the range of f then $\phi \circ f$ is measurable on E. In particular $|f|^p$ is measurable for all $p > 0$.*

(3) *If $f \in L^1(E)$ then $|\int_E f\, dx| \le \int_E |f|\, dx$.*

(4) *If f is measurable on E and $f = g$ a.e. then g is also measurable and $\int_E f\, dx = \int_E g\, dx$.*

(5) *If f_n is measurable on E for all n, then so are $\sup_n f_n$, $\inf_n f_n$, $\limsup_{n\to\infty} f_n$, and $\liminf_{n\to\infty} f_n$. In particular if $f_n \to f$ a.e. then f is measurable.*

(6) *(Fatou's lemma) If $f_n \ge 0$ is measurable on E for all n then*

$$\int_E \liminf_{n\to\infty} f_n\, dx \le \liminf_{n\to\infty} \int_E f_n\, dx \tag{A.1.12}$$

(7) *(Lebesgue's dominated convergence theorem) Suppose $f_n \in L^1(E)$ for all n and $f_n \to f$ a.e. Suppose also that there exists $F \in L^1(E)$ such that $|f_n| \le F$ a.e. for every n. Then*

$$\lim_{n\to\infty} \int_E f_n\, dx = \int_E f\, dx \tag{A.1.13}$$

A.2 INEQUALITIES

In this section we state and prove a number of useful inequalities for numbers and functions.

A function ϕ on an interval $(a,b) \subset \mathbb{R}$ is convex if

$$\phi(\lambda x_1 + (1-\lambda)x_2) \le \lambda\phi(x_1) + (1-\lambda)\phi(x_2) \tag{A.2.14}$$

for all $x_1, x_2 \in (a,b)$ and $\lambda \in [0,1]$. A convex function is necessarily continuous (see Theorem 3.2 of [32]). If ϕ is such a function and $c \in (a,b)$ then there always exists a *supporting line* for ϕ at c, more precisely, there exists $m \in \mathbb{R}$ such that if we let $\psi(x) = m(x-c) + \phi(c)$, then $\psi(x) \le \phi(x)$ for all $x \in (a,b)$. If ϕ is differentiable at $x = c$ then $m = \phi'(c)$, otherwise it may be defined in terms of a certain supremum (or infimum) of slopes. If in addition ϕ is twice differentiable then ϕ is convex if and only if $\phi'' \ge 0$.

Proposition A.1. *(Young's Inequality) If $a, b \ge 0$, $1 < p, q < \infty$, and $\frac{1}{p} + \frac{1}{q} = 1$ then*

$$ab \le \frac{a^p}{p} + \frac{b^q}{q} \tag{A.2.15}$$

Proof. If a or b is zero the conclusion is obvious, otherwise, since the exponential function is convex and $1/p + 1/q = 1$ we get

$$ab = e^{(\log a + \log b)} = e^{\left(\frac{\log a^p}{p} + \frac{\log b^q}{q}\right)} \le \frac{e^{(\log a^p)}}{p} + \frac{e^{(\log b^q)}}{q} = \frac{a^p}{p} + \frac{b^q}{q} \quad \text{(A.2.16)}$$

□

In the special case that $p = q = 2$, Eq. (A.2.15) can be proved in an even more elementary way, just by rearranging the obvious inequality $a^2 - 2ab + b^2 = (a - b)^2 \ge 0$.

Corollary A.1. *If* $a, b \ge 0$, $1 < p, q < \infty$, $\frac{1}{p} + \frac{1}{q} = 1$, *and* $\epsilon > 0$ *there holds*

$$ab \le \frac{\epsilon a^p}{p} + \frac{b^q}{q\epsilon^{q/p}} \quad \text{(A.2.17)}$$

Proof. We can write

$$ab = \left(\epsilon^{1/p} a\right)\left(\frac{b}{\epsilon^{1/p}}\right) \quad \text{(A.2.18)}$$

and then apply Proposition A.1. □

Proposition A.2. *(Hölder's Inequality) If* u, v *are measurable functions on* $\Omega \subset \mathbb{R}^N$, $1 \le p, q \le \infty$, *and* $\frac{1}{p} + \frac{1}{q} = 1$ *then*

$$\|uv\|_{L^1(\Omega)} \le \|u\|_{L^p(\Omega)} \|v\|_{L^q(\Omega)} \quad \text{(A.2.19)}$$

Proof. We may assume that $\|u\|_{L^p(\Omega)}, \|v\|_{L^q(\Omega)}$ are finite and nonzero, since otherwise Eq. (A.2.19) is obvious. When $p, q = 1$ or ∞, proof of the inequality is elementary, so assume that $1 < p, q < \infty$. Using Eq. (A.2.17) with $a = |u(x)|$ and $b = |v(x)|$, and integrating with respect to x over Ω gives

$$\int_\Omega |u(x)v(x)|\, dx \le \frac{\epsilon}{p} \int_\Omega |u(x)|^p\, dx + \frac{1}{q\epsilon^{q/p}} \int_\Omega |v(x)|^q\, dx \quad \text{(A.2.20)}$$

By choosing

$$\epsilon = \left(\frac{\int_\Omega |v(x)|^q\, dx}{\int_\Omega |u(x)|^p\, dx}\right)^{1/q} \quad \text{(A.2.21)}$$

the right-hand side of this inequality simplifies to

$$\left(\int_\Omega |u(x)|^p\, dx\right)^{1/p} \left(\int_\Omega |v(x)|^q\, dx\right)^{1/q} \left(\frac{1}{p} + \frac{1}{q}\right) = \|u\|_{L^p(\Omega)} \|v\|_{L^q(\Omega)} \quad \text{(A.2.22)}$$

as needed. □

The special case of Hölder's inequality when $p = q = 2$ is commonly called the Schwarz, or Cauchy-Schwarz inequality. Whenever p, q are related, as in Young's or Hölder's inequality, via $1/p + 1/q = 1$ it is common to refer to $q = p/(p - 1) =: p'$, as the Hölder conjugate exponent of p.

Proposition A.3. *(Minkowski Inequality) If u, v are measurable functions on $\Omega \subset \mathbb{R}^N$ and $1 \le p \le \infty$, then*

$$\|u + v\|_{L^p(\Omega)} \le \|u\|_{L^p(\Omega)} + \|v\|_{L^p(\Omega)} \tag{A.2.23}$$

Proof. We may assume that $\|u\|_{L^p(\Omega)}, \|v\|_{L^p(\Omega)}$ are finite and that $\|u+v\|_{L^p(\Omega)} \neq 0$, since otherwise there is nothing to prove. We have earlier noted in Section 2.1 that $L^p(\Omega)$ is a vector space, so $u + v \in L^p(\Omega)$ also. In the case $1 < p < \infty$ we write

$$\int_\Omega |u(x)+v(x)|^p\,dx \le \int_\Omega |u(x)|\,|u(x)+v(x)|^{p-1}\,dx + \int_\Omega |v(x)|\,|u(x)+v(x)|^{p-1}\,dx \tag{A.2.24}$$

By Hölder's inequality

$$\int_\Omega |u(x)|\,|u(x)+v(x)|^{p-1}\,dx \le \left(\int_\Omega |u(x)|^p\,dx\right)^{1/p} \left(\int_\Omega |u(x)+v(x)|^{(p-1)q}\,dx\right)^{1/q} \tag{A.2.25}$$

where $1/q + 1/p = 1$. Estimating the second term on the right-hand side of Eq. (A.2.24) in the same way, we get

$$\int_\Omega |u(x) + v(x)|^p\,dx \le \left(\int_\Omega |u(x) + v(x)|^p\,dx\right)^{1/q} \left(\|u\|_{L^p(\Omega)} + \|v\|_{L^p(\Omega)}\right) \tag{A.2.26}$$

from which the conclusion (A.2.23) follows by obvious algebra. The two limiting cases $p = 1, \infty$ may be handled in a more elementary manner, and we leave these cases to the reader. □

Both the Hölder and Minkowski inequalities have counterparts

$$\sum_k |a_k b_k| \le \left(\sum_k |a_k|^p\right)^{1/p} \left(\sum_k |b_k|^q\right)^{1/q} \qquad 1 < p, q < \infty \quad \frac{1}{p} + \frac{1}{q} = 1 \tag{A.2.27}$$

$$\left(\sum_k |a_k + b_k|^p\right)^{1/p} \le \left(\sum_k |a_k|^p\right)^{1/p} + \left(\sum_k |b_k|^p\right)^{1/p} \qquad 1 \le p < \infty \tag{A.2.28}$$

(with suitable modification for the case of p or q being ∞) in which the integrals are replaced by finite or infinite sums of real or complex constants—the proofs are otherwise identical.[2]

2. Or from the point of view of abstract measure theory, the proofs are identical because a sum is just a certain kind of integral.

A.3 INTEGRATION BY PARTS

In the elementary integration by parts formula from calculus

$$\int_a^b u(x)v'(x)\,dx = -\int_a^b u'(x)v(x)\,dx + u(x)v(x)|_a^b \tag{A.3.29}$$

one integral is shown to be equal to another integral plus a “boundary term,” where in this case the boundary consists of the two points a, b, namely the boundary of the interval $[a, b]$ over which the integration takes place. In higher dimensional situations we refer to any identity of this general character as being an integration by parts formula. There are a number of such formulas, all more or less equivalent to each other, which are frequently used in applied mathematics, and which we review here.

We will take as a known basic integration by parts formula the divergence theorem

$$\int_\Omega \nabla\cdot\mathbf{F}(x)\,dx = \int_{\partial\Omega} \mathbf{F}\cdot\mathbf{n}(x)\,dS(x) \tag{A.3.30}$$

valid for a C^1 vector field $\mathbf{F}$ and bounded open set $\Omega\subset\mathbb{R}^N$, $N\ge 2$, with C^1 boundary $\partial\Omega$, see for example Theorem 10.51 of [31]. Here $\mathbf{n}(\mathbf{x})$ is the unit outward normal to $\partial\Omega$ at $x\in\partial\Omega$. If we now choose the vector field $\mathbf{F}$ to be zero except for the jth component $F_j(x) = u(x)v(x)$, there results

$$\int_\Omega u(x)\frac{\partial v}{\partial x_j}(x)\,dx = -\int_\Omega \frac{\partial u}{\partial x_j}(x)v(x)\,dx + \int_{\partial\Omega} u(x)v(x)n_j(x)\,dS(x) \tag{A.3.31}$$

Replacing v by v_j, the jth component of a vector function $\mathbf{v}$, and summing on j we next obtain

$$\int_\Omega u(x)(\nabla\cdot\mathbf{v})(x)\,dx = -\int_\Omega \nabla u(x)\cdot\mathbf{v}(x)\,dx + \int_{\partial\Omega} u(x)(\mathbf{v}\cdot\mathbf{n})(x)\,dS(x) \tag{A.3.32}$$

Now choosing $\mathbf{v} = \nabla w$, the gradient of some scalar function w, and noting that $\nabla\cdot(\nabla w) = \Delta w$ we find

$$\int_\Omega u(x)\Delta w(x)\,dx = -\int_\Omega (\nabla u\cdot\nabla w)(x)\,dx + \int_{\partial\Omega} u(x)\frac{\partial w}{\partial n}(x)\,dS(x) \tag{A.3.33}$$

where as usual $\frac{\partial w}{\partial n} = \nabla w\cdot\mathbf{n}$ is the outer normal derivative of w on $\partial\Omega$. Reversing the roles of u and w, and subtracting the resulting expressions, we may then obtain *Green's identity*

$$\int_\Omega (u(x)\Delta w(x) - w(x)\Delta u(x))\,dx = \int_{\partial\Omega}\left(u(x)\frac{\partial w}{\partial n}(x) - w(x)\frac{\partial u}{\partial n}(x)\right)dS(x) \tag{A.3.34}$$

The special case of Eq. (A.3.34) when $u(x) \equiv 1$, namely

$$\int_\Omega \Delta w(x)\, dx = \int_{\partial\Omega} \frac{\partial w}{\partial n}(x)\, dS(x) \tag{A.3.35}$$

is also of interest.

Finally we mention that the classical Green's theorem in the plane,

$$\oint_{\partial A} P\, dx + Q\, dy = \iint_A \left(\frac{\partial Q}{\partial x} - \frac{\partial P}{\partial y} \right) dxdy \tag{A.3.36}$$

is also a special case of Eq. (A.3.30), obtained by choosing the special vector field in $\mathbb{R}^2$, $\mathbf{F} = \langle Q, -P \rangle$.

A.4 SPHERICAL COORDINATES IN $\mathbb{R}^N$

As in the case of $\mathbb{R}^2$ or $\mathbb{R}^3$, it is often convenient to work with spherical coordinates in $\mathbb{R}^N$. Here is how it works:

We denote

$$S_{N-1} = \{x \in \mathbb{R}^N : |x| = 1\}$$

the unit sphere[3] in $\mathbb{R}^N$. Every point $x \in \mathbb{R}^N$ may be expressed as $x = r\omega$ where $r = |x| \geq 0$ and $\omega \in S_{N-1}$, and the representation is unique except for $x = 0$. We then may parameterize S_{N-1} by $N-1$ angle variables $\theta_1, \theta_2, \dots, \theta_{N-1}$, where

$$\begin{cases} x_1 = r \sin\theta_1 \sin\theta_2 \dots \sin\theta_{N-2} \sin\theta_{N-1} \\ x_2 = r \sin\theta_1 \sin\theta_2 \dots \sin\theta_{N-2} \cos\theta_{N-1} \\ \vdots \\ x_{N-1} = r \sin\theta_1 \cos\theta_2 \\ x_N = r \cos\theta_1 \end{cases}$$

Here $0 \leq \theta_j \leq \pi$ for $j = 1, \dots, N-2$ and $0 \leq \theta_{N-1} \leq 2\pi$.

Thus $(r, \theta_1, \theta_2, \dots, \theta_{N-1})$ are spherical coordinates on $\mathbb{R}^N$. The Jacobian determinant of the transformation $(x_1, \dots, x_N) \to (r, \theta_1, \theta_2, \dots, \theta_{N-1})$, needed for integration in spherical coordinates is

$$r^{N-1} \sin^{N-2}\theta_1 \sin^{N-3}\theta_2 \dots \sin\theta_{N-2}$$

Integration of a function f over S_{N-1} may expressed by

$$\int_{S_{N-1}} f(\omega)\, d\omega = \int_0^\pi \dots \int_0^\pi \int_0^{2\pi} f(\theta_1, \dots, \theta_{N-1}) d\sigma$$

---

3. We try to use the terminology "unit ball" for $\{x \in \mathbb{R}^N : |x| < 1\}$, but sometimes "sphere" and "ball" are used interchangeably. Also, S_N is sometimes used as notation for the unit sphere, but S_{N-1} is more common since it is a surface of dimension $N-1$.

where

$$d\sigma = \sin^{N-2}\theta_1 \sin^{N-3}\theta_2 \ldots \sin\theta_{N-2}\, d\theta_{N-1} \ldots d\theta_1$$

Likewise integration of a function f over $\mathbb{R}^N$ is

$$\int_{\mathbb{R}^N} f(x)\,dx = \int_0^\infty \int_{S_{N-1}} f(r\omega)\,d\omega dr$$
$$= \int_0^\infty \int_0^\pi \cdots \int_0^\pi \int_0^{2\pi} f(r,\theta_1,\ldots,\theta_{N-1}) r^{N-1}\,d\sigma\,dr$$

In particular if f is radially symmetric, $f(x) = f(|x|)$, we get

$$\int_{\mathbb{R}^N} f(x)\,dx = \Omega_{N-1} \int_0^\infty f(r) r^{N-1}\,dr \qquad \text{(A.4.37)}$$

where

$$\Omega_{N-1} = \int_{S_{N-1}} d\sigma$$

is the surface area of S_{N-1}.

Bibliography

[1] R.A. Adams, *Sobolev Spaces*, Pure and Applied Mathematics, vol. 65, Academic Press, 1975.

[2] N.I. Akhiezer, I.M. Glazman, *Theory of Linear Operators in Hilbert Space*, Dover Publications Inc., 1993.

[3] N. Bleistein, *Mathematical Methods for Wave Phenomena*, Computer Science and Applied Mathematics, Academic Press Inc., 1984.

[4] F. Brauer, J.A. Nohel, *The Qualitative Theory of Ordinary Differential Equations: An Introduction*, W.A. Benjamin Inc., 1969.

[5] H. Brezis, *Functional Analysis, Sobolev Spaces and Partial Differential Equations*, Springer, 2011.

[6] L. Carleson, On convergence and growth of partial sums of Fourier series, *Acta Math.* 116 (1966) 135–157.

[7] E.A. Coddington, N. Levinson, *Theory of Ordinary Differential Equations*, McGraw-Hill Book Company Inc., 1955.

[8] D. Colton, R. Kress, *Integral Equation Methods in Scattering Theory*, Classics in Applied Mathematics, vol. 72, Society for Industrial and Applied Mathematics (SIAM), 2013.

[9] R. Courant, D. Hilbert, *Methods of Mathematical Physics*, vol. I, Interscience Publishers Inc., 1953.

[10] H. Dym, H.P. McKean, *Fourier Series and Integrals*, Probability and Mathematical Statistics, No. 14, Academic Press, 1972.

[11] L.C. Evans, *Partial Differential Equations*, Graduate Studies in Mathematics, vol. 19, 2nd ed., American Mathematical Society, 2010.

[12] G.B. Folland, *Introduction to Partial Differential Equations*, 2nd ed., Princeton University Press, 1995.

[13] K.O. Friedrichs, The identity of weak and strong extensions of differential operators, *Trans. Am. Math. Soc.* 55 (1944) 132–151.

[14] P.R. Garabedian, *Partial Differential Equations*, John Wiley & Sons Inc., 1964.

[15] G.H. Golub, C.F. Van Loan, *Matrix Computations*, Johns Hopkins Studies in the Mathematical Sciences, 3rd ed., Johns Hopkins University Press, 1996.

[16] H. Hochstadt, *Integral Equations*, Pure and Applied Mathematics, John Wiley & Sons, 1973.

[17] L. Hörmander, *The Analysis of Linear Partial Differential Operators. II*, Grundlehren der Mathematischen Wissenschaften, vol. 256, Springer-Verlag, 1983.

[18] J.K. Hunter, B. Nachtergaele, *Applied Analysis*, World Scientific Publishing Co., Inc., 2001.

[19] F. John, *Partial Differential Equations*, Applied Mathematical Sciences, vol. 1, 4th ed., Springer-Verlag, 1982.

[20] R.K. Juberg, Finite Hilbert transforms in L^p, *Bull. Am. Math. Soc.* 78 (1972) 435–438.

[21] O.D. Kellogg, *Foundations of Potential Theory*, Reprint from the first edition of 1929, Die Grundlehren der Mathematischen Wissenschaften, Band 31, Springer-Verlag, 1967.

[22] R. Kress, *Linear Integral Equations*, Applied Mathematical Sciences, vol. 82, Springer-Verlag, 1989.

[23] P.D. Lax, *Linear Algebra*, Pure and Applied Mathematics, John Wiley & Sons Inc., 1997.

[24] R.C. McOwen, *Partial Differential Equations: Methods and Applications*, 2nd ed., Prentice-Hall, 2003.

[25] N.G. Meyers, J. Serrin, $H = W$, *Proc. Natl Acad. Sci. USA* 51 (1964) 1055–1056.

[26] L.E. Payne, *Improperly Posed Problems in Partial Differential Equations*, Regional Conference Series in Applied Mathematics, No. 22, Society for Industrial and Applied Mathematics, 1975.

[27] A. Pazy, *Semigroups of Linear Operators and Applications to Partial Differential Equations*, Applied Mathematical Sciences, vol. 44, Springer-Verlag, 1983.

[28] M.A. Pinsky, *Introduction to Fourier Analysis and Wavelets*, Brooks/Cole Series in Advanced Mathematics, Brooks/Cole, 2002.

[29] J. Rauch, *Partial Differential Equations*, Graduate Texts in Mathematics, vol. 128, Springer-Verlag, 1991.

[30] H.L. Royden, P.M. Fitzpatrick, *Real Analysis*, 4th ed., Prentice Hall, 2010.

[31] W. Rudin, *Principles of Mathematical Analysis*, International Series in Pure and Applied Mathematics, 3rd ed., McGraw-Hill Book Co., 1976.

[32] W. Rudin, *Real and Complex Analysis*, 3rd ed., McGraw-Hill Book Co., 1987.

[33] W. Rudin, *Functional Analysis*, International Series in Pure and Applied Mathematics, 2nd ed., McGraw-Hill Inc., 1991.

[34] L. Schwartz, *Mathematics for the Physical Sciences*, Hermann/Addison-Wesley Publishing Co., 1966.

[35] I. Stakgold, M. Holst, *Green's Functions and Boundary Value Problems*, Pure and Applied Mathematics, 3rd ed., John Wiley & Sons Inc., 2011.

[36] E.M. Stein, *Singular Integrals and Differentiability Properties of Functions*, Princeton Mathematical Series, No. 30, Princeton University Press, 1970.

[37] E.M. Stein, G. Weiss, *Introduction to Fourier Analysis on Euclidean Spaces*, Princeton Mathematical Series, No. 32, Princeton University Press, 1971.

[38] F. Trèves, *Basic Linear Partial Differential Equations*, Pure and Applied Mathematics, vol. 62, Academic Press, 1975.

[39] H.F. Weinberger, *Variational Methods for Eigenvalue Approximation*, Society for Industrial and Applied Mathematics, 1974.

[40] R.L. Wheeden, A. Zygmund, *Measure and Integral: An Introduction to Real Analysis*, Pure and Applied Mathematics, vol. 43, Marcel Dekker Inc., 1977.

[41] R.M. Young, *An Introduction to Nonharmonic Fourier Series*, 1st ed., Academic Press Inc., 2001.

Index

Symbols

A

B

C

I

K

L

M

N

O

P

Printed in the United States
By Bookmasters

Printed in the United States
By Bookmasters